A Practical Guide to Ore Microscopy—Volume 2

Ricardo Castroviejo

A Practical Guide to Ore Microscopy—Volume 2

Ore Textures and Automated Ore Analysis

Springer

Ricardo Castroviejo
E.T.S. de Ingenieros de Minas y Energía
Universidad Politécnica de Madrid
Madrid, Spain

ISBN 978-3-031-18956-2 ISBN 978-3-031-18954-8 (eBook)
https://doi.org/10.1007/978-3-031-18954-8

This Springer imprint is published by the registered company Springer Nature Switzerland AG
The registered company address is: Gewerbestrasse 11, 6330 Cham, Switzerland

Dedico este libro, fruto de un esfuerzo compartido, a Antonia, mi amiga, esposa y apoyo abnegado en momentos difíciles.

(This book, the fruit of a shared effort, is dedicated to Antonia, my friend, wife and selfless support in difficult times.)

Foreword

Congratulations on acquiring the most useful book on ore microscopy! You'll find the book packed with information relevant to geologists, metallurgists, and others interested in not only ore minerals but all common minerals that are opaque in transmitted light.

Following the adage "when all else fails, look at the rock," this book is a first-rate text on how to maximize the use of a polarizing reflected-light microscope (covered in Part I), identify more than 150 ore minerals (Part II) and more than 50 gangue minerals with which they are commonly associated (Part III), interpret intergrowths of minerals for geological interpretations of ore-forming processes and for applications in geometallurgy and environmental considerations (Part IV), and use the light microscope for image-analysis automation in a much less expensive but complementary approach to scanning electron microscope (SEM)-based techniques (Part V).

The photographs of ore and gangue minerals in all chapters are as accurate in color as possible and provide excellent examples culled from worldwide collections. Each ore mineral in Part II is photographed under the four types of illumination commonly used in reflected-light petrography (plane polarized light in air then with oil plus crossed polarized light in both air and oil). The photographs in Part IV are illustrative of the intergrowths that are essential to accurate characterization of mineral resources and reserves, including potential difficulties in ore processing and waste disposal.

The book comes in two volumes, with Parts I–III in Volume 1 and Parts IV and V plus references and annexes (appendices) in Volume 2. The annexes are exceptionally useful. They address common associations of ore and gangue minerals (#2), techniques for the preparation of polished sections (#3), tables of mineral properties (#4), and well-proven determinative tables (#5).

Professor Castroviejo has spent his career using ore microscopy in research and applications to industry. He demonstrates that reflected light microscopy remains the most cost-effective approach to characterize and interpret ores in terms of genesis, mineral processing, and environmental concerns during and after production. Although SEM, electron microprobe, and other chemical analytical techniques are essential, the ore microscope remains the first and best approach to look at the rocks.

Jonathan G. Price Ph.D.
State Geologist Emeritus
Nevada Bureau of Mines and Geology
University of Nevada, Reno
Nevada, NV, USA

Acknowledgements

This work has benefited from the generous support of numerous individuals and institutions. Being impossible to name them all, I would like to express my gratitude to all of them. The description of minerals, main body of the work, is based on direct observations, mainly on own ore collections and on those of the School of Mines. The former, gathered throughout the author's professional life, have been supplemented with replicas from the *Rehwald-Ramdohr Collection* in Heidelberg, kindly loaned at the time by Prof. G. C. Amstutz. The latter comprise both those of the Department of Geological and Mining Engineering (collected successively by Profs. I. Roso de Luna, T. Febrel, F. Vázquez and J. Samper), and those loaned for study by the Museum of Mineralogy (Dir. Prof. B. Calvo), at the Escuela Técnica Superior de Ingenieros de Minas y Energía, **ETSIME**, of the Universidad Politécnica de Madrid, **UPM**.

I would like to acknowledge the generous contribution of Dr. Heinz-Jürgen Bernhardt, *Institut für Geologie, Mineralogie und Geophysik der Ruhr-Universität Bochum*, **RUB**. He has realized the EMP analyses, shared his own ore collections and, above all, has continuously enriched the author's approach with his advice and discussions. If there is any merit in this work, it is shared with him. Dr. Hans-Peter Schertl, *curator of* the Museum of Mineralogy of the RUB, has provided additional samples. And Ms. Sabine Weisel, Ms. Ellen Kessler and Ms. Tanja Westphal, specialists of the *Sample Preparation* Lab, RUB, have produced polished sections of excellent quality.

Dr. Jonathan G. Price, *Nevada State Geologist* and Director (*em.*) of the **NBNG**, *Nevada Bureau of Mines and Geology* and *McKay School of Mines* (Reno, NV, USA), generously helped me to complete the collection of studied ores by accessing the *Stanford Collection-Mckay School of Mines* and several active gold and silver mines in Nevada. To Prof. Hartmut Beurlen, *Universidade Federal de Pernambuco*, **UFPE**, Recife (Brazil), I owe the Nb, Ta and REE samples, documented with his precise *EMPA* analyses; to Prof. Eric Pirard (*Université de Liège*, Belgium), some CopperBelt ores (DR Congo); and to Dr. María José Correa, *Universidad Nacional de La Plata* (Argentina), the romanechite samples.

The text layout and execution profited from the selfless support of Dr. J. C. Catalina (*author of the* Chap. 2, on automated microscopy) and Ing. U. Grunwald-Romera (*author of Annex 3* on polished sections preparation), who have contributed, together with Ings. P. Romero and D. Alarcón, to data compilation for DB and to digital design. Dr César Cánepa, Universidad Nacional Mayor de San Marcos, **UNMSM**, Lima (Peru), a master for many generations of mineralogists, has given generous and constant help throughout the process. Profs. Soledad Fernández-Santín and Alfredo Hernández-Pacheco, Universidad Complutense de Madrid, **UCM**, have enriched my work with their samples, discussions, and critical reviews of the manuscript. Encouraging support has been provided by the opinions of many colleagues consulted, who have taken the trouble to read and critically comment on my early drafts: Profs. Emil Mackowycki (University of Copenhagen), José Antonio Espí (UPM), Colombo G Tassinari (University of Sao Paulo, **USP**), Eulogio Pardo and Juan Locutura (IGME/UPM), Joaquín Proenza (Univ. Barcelona) and Fernando Gervilla (University of Granada), Drs. Rolando Lastra (CanMet, Ottawa), Laura Pérez-Barnuevo (*COREM*, Quebec) and Thomas Aiglsperger (Luleå), Ing. Jorge Acosta, MSc (*INGEMMET*, Lima), among others.

The inspiration of Prof. Ana García Moreno, UCM, Creator and Director of the digital magazine **REDUCA**—*Recursos Educativos* (*http://www.revistareduca.es*), and the early guidance of Dr. Joan Roca, mathematician and photography researcher, have been very helpful for the digital image processing (to *sRGB color space*, *IEC standard*) and for designing the virtual text.

Dr. Alexis Vizcaíno, *Springer*'s editor, has provided encouraging and patient support in the laborious editing process. The *Hudson Institute of Mineralogy* is acknowledged for the open access mineral database *Mindat.org*, source of credited data referred to for some ores. **EIT** (*European Institute for Innovation and Technology*) funded, through ***EIT-RawMaterials*** Project **no. 15039** (2016–2018), the innovative **AMCO** System (*Automated Microscopic Characterisation of Ores*). The support of the project's colleagues (***University of Liège*** , Belgium; ***ThinSectionLab***, France; ***KGHM***, Poland; ***Cobre Las Cruces***, Spain, **UPM-***Consortium leader*) is gratefully acknowledged.

Last but not least, I wish to thank Prof. T. Febrel (†) for his motivating teaching of Ore Microscopy, at the ETSI Minas-Madrid, as well as the *Deutscher Akademischer Austauschdienst* (**DAAD**, German Academic Exchange Service) for the fellowship making possible my early stay at the *Mineralogisch-Petrographisches Institut der Universität Heidelberg* (Germany), my research training with professors like P. Ramdohr (†) and G. C. Amstutz (†), and the enriching contact with postgraduate colleagues from all over the world. All this is at the root of this work. I am also grateful to UPM, RUB, USP, and NBMG-McKay *School of Mines* for their support of my *sabbatical stay* (2010/2011), which was decisive for this project.

Contents

Part I

Intergrowths. Textural Analysis

1 Intergrowths and Textures

In nature, minerals are usually found in association, forming rocks. Ores are no exception: they are almost always composed of different species, forming intergrowths. The study, description and interpretation of these intergrowths is the objective of textural analysis and is very important for mineralogy, both from a scientific point of view (e.g. mineral genesis) and for a practical approach (e.g. mining exploration and minerallurgy).

Mineral processing or minerallurgy (*ore processing*) involves processing the mineral in industrial plants for its benefit, aiming at the most efficient procedure to sort the useful components (ore) from the tailings to be discarded (gangue). Efficiency also means environmental responsibility or sustainability, implying the capacity to predict the impact, and then to tightly control operations and waste. For this, the mineralogical and textural characterization of the ore provides indispensable information, not only to optimise plant processes (grinding, gravimetry, flotation, etc.) but in earlier phases of the deposit's exploration and evaluation to provide an orientation for the investor. This is due to its potential to predict the behaviour of the mineral in the plant and its appropriate treatment (*predictive mineralogy*). This information is of transcendental importance to defining the viability of the project. These data are also linked to good practices required for quality standards (*cf.* § 1.1, vol. 1).

This information will be briefly discussed below (*cf.* § 1.2.3) in relation to *geometallurgical applications*, but first it is necessary to ensure the correct description and understanding of the textures in their fundamental features (§ 1.1). For this purpose, we first try to find the most suitable criteria to establish a clear and useful classification of the most common ore textures. This should lead to a logical identification scheme, avoiding oversimplification or ambiguity, as well as superfluous intricacies. Next, in applying these criteria, these textures are defined and analysed from the points of view of the mineral deposit researcher, the exploration geologist, the plant engineer or scholar of metallic ores. For this purpose, the scheme of Box 1 will be followed, to summarize in an accessible way the complex universe of the textures of metallic ores, grouping them by three criteria: I Descriptive, either referring to individual grains (Table 1.1) or to aggregates (Table 1.2); II Genetic (Table 1.3); III Geometallurgical (Table 1.4). These tables will be completed when discussed.

R. Castroviejo, *A Practical Guide to Ore Microscopy—Volume 2*,
https://doi.org/10.1007/978-3-031-18954-8_1

Box 1 Summary classification of textures applying descriptive, genetic and geometallurgical criteria (Tables 1.1, 1.2, 1.3 and 1.4).

I Descriptive Classification

Table 1.1 Description of an individual grain	
By its internal structure	Zoning
	Twinning
	Cleavage
	Other forms (emul., spher., etc.)
By its external shape	Idiomorphic/specify *habit*
	Alotriomorphic
	Hypidiomorphic/specify *habit*
	Other forms (pseudom, skel., etc.)
By its size (diam. Ø)	Very coarse (Ø ≥10 mm)
	Coarse (3 mm ≤ Ø <10 mm)
	Medium (0.3 mm ≤ Ø <3 mm)
	Fine (0.03 mm ≤ Ø <0.3 mm)
	Very fine (1 μm ≤ Ø <30 μm)
	Submicroscopic (Ø <1 μm)

Table 1.2 Description of mineral aggregates — Intergrowths		
Crystallinity and grain size	Phaneritic Aphanitic/Microxx Cryptoxx, Framb, etc. Amorphous	
	Equigranular Inequigranular	
Compactness (degree of filling)	Compact	
	Porous	
	Vuggy	
Orientation *vs.* isotropy	Massive	
	Banded/layered	
	Oriented	
	Fibrous	
	Radial	
	Spherulitic	
	Spheroidal	
Type of grain-to-grain contact	Simple (smooth)	
	Complex	Jagged
		Concentric
		Emulsion
		Other
		Irregular
Spatial arrangement of the components	Sandwich	
	Stockwork	
	Encrustation	
	Dissemination	
	Interstitial	
	Intergranular film	

II Genetic Classification

Table 1.3 Genetic classification of textures			
Class	*Processes and resulting textures*		
Primary textures	Cumulate (cumulus, intercumulus, etc)		
	Vein and cavity filling		
	Sequential		
	Colloidal		
	Detrital (various, including oolitic)		
Modified textures	Reequilibration	Annealing, etc.	
		Gel crystallization	
	Metamorphism	Metamorphic (var)	
	Remobilization	Various (infill, etc.)	
	Replacement	Jagged, etc.	
		Relicts	
		Ghosts	
		Supergene	Sulf
			Oxid
	Infiltration	Infiltrative (varied)	
	Deformation	Brittle: breccias, etc.	
		Ductile: mylonit.	
	Fluid-ass. strain	Pressure solution Solution transfer Pressure shadows	

III Geometallurgical Classification

Table 1.4 Basic geometallurgical classification (mill particles)		
Class	*Type*	*Characterisation*
Liberated	Monomineralic	After shape and grain size (Table 1.1)
Middling	Simple contact	Straight and smooth or gently curved
	Irregular contact	Jagged, lobed, jig-saw, interpenetrated
	Stockwork	Network of cross-cutting veins & dissem.
	Corona	Thin layer of ore-2 rimming core of ore-1
	Emulsion	Minor ore in scattered tiny inclusions

1.1 Textural Analysis

Textural analysis of metallic ores is an important and often difficult task. It is important because it can provide valuable clues to the correct exploration and interpretation of an ore body and the design of the industrial beneficiation process, even if it means abandoning it at an early stage if the textural analysis reveals insurmountable difficulties. It is difficult because ores can show varied, complex textures, frequently of an ambiguous or at least debatable interpretation.

In principle, the textural criteria and terminology used in petrography—that is, those used for the description and classification of common rocks—can also be applied to metallic ores since, after all, an ore is a rock of economic interest. However, two important particularities must be taken into account.

- The first is the ease with which metallic ores recrystallize to adapt to change in the physico-chemical conditions of their environment. This means that, unlike common rocks, the textures of ores often evolve significantly after deposition, so that the primary textures may be reduced to a few traces (relicts, ghosts) or be completely unrecognizable: in short, ores are more difficult to interpret because their textures are easily modified.
- The second is that, although ores are rocks, their formation obeys highly specific processes related to the typology of each deposit. Hence, the keys to the interpretation of the textures in mineral deposits often refer to processes that usually receive little attention in petrology, such as hydrothermal transport and filling, weathering or metasomatism. The singularity of these processes forces the introduction of specific criteria from which a rich yet not always rational terminology has been derived.

To be useful, textural analysis should aim at a humble sobriety, applying simple and clear criteria to reduce superfluous terms and intricacies as far as possible even if, since some are widespread, they are not to be ignored.

Such is the aim of this text, and a few examples from the literature may be the best explanation to understand the reasons:

- In his seminal work, after more than 200 pages on intergrowths and textures, Ramdohr (1980) states that the «rather voluminous section on ore intergrowths is not anywhere near complete, and could not be even with very substantial expansion» (p. 2), and concludes that «the more one studies ore textures, the more sceptical one becomes of statements concerning them in the literature, and the more one tends to revise one's earlier views and conclusions» (p. 82). Almost half a century later, Craig (2001) concludes that «ore-mineral textures are of nearly endless varieties; each one presents a tale, but only some are decipherable, at least for the present» (p. 938). This conclusion invites humble realism.
- In the field of petrology, Best (1982) provides an original scheme to characterize the immense variety of igneous, sedimentary and metamorphic textures. It is an extremely simple scheme, comprising only five fundamental types: *sequential, glassy, clastic, deformed and crystalloblastic* textures (note the descriptive character of this classification: for example, sequential texture is possible in igneous and sedimentary rocks).

The option followed here is inspired by the simplicity of Best's sketch without renouncing the singular analysis and interpretation of some relevant types, bearing in mind Ramdohr's warning about premature assertions.

As a realistic starting point for textural analysis, consider the **scale**. The polished section studied under the microscope constitutes only a small part—and not always the most representative part—of the ore deposit. For a correct interpretation, the microscopic texture must be related to the information observed at other scales: mesoscopic (hand sample, outcrop) and megascopic (deposit). Should the reader not be familiar with this approach (from outcrop through to orebody and district scales, to support textural interpretation), the outstanding discussion and examples by Taylor (2009) is an excellent introduction. The whole provides the appropriate framework for description and interpretation, yet microscopic observations can supply decisive clues.

From the conceptual point of view, it is convenient to specify some terms whose indiscriminate use generates no little confusion. Ores, as rocks, occur in mineral aggregates, whose morphology and geometrical relationships are usually described in the general framework of the texture. However, the constituent grains of such aggregates also possess, individually, morphological properties and internal structure. A first important precision allows us to distinguish between **texture** (referred to aggregates or assemblages) and **structure** (referred to single or individual grains).

Structure, in the sense of this text, is the *arrangement in space of the constituent elements of a mineral grain*, comprising both its *external morphology* (crystalline form) and its *internal structuring* (e.g. growth zoning, twinning, cleavage planes, etc.). These are all spatial properties that can be defined independently of the surrounding grains. Also to be considered is the possible absence of any structure observable under the microscope; that is, the indefinable or random spatial arrangement of the constituent elements of the material. This liquid-like arrangement, lacking any defined crystalline structure, defines the *amorphous state*. This state, strictly speaking, is exceptional among known mineral species (e.g. *lechatelierite* or glassy silica, *metastibnite*, dimorph of stibnite). However, to a greater or lesser degree and at various scales of observation it occurs under certain conditions (volcanic glasses, massive colloidal structures, subcrystalline or cryptocrystalline masses), even if the classic *amorphous* character of some minerals is contested (e.g. *opal*,[1] *jordisite*,[2] a dimorph of molybdenite; or the so-called *sulfide glasses* or amorphous hardened sulfide colloids, Ramdohr 1980). In crystalline materials the grains may be homogeneous at microscopic scale (without differentiable internal structure), lacking visible crystal contours: in this case the structure is *unobservable*, even if the material is crystalline, and therefore structured. The precise observation of the structure provides sometimes vital criteria for mineral identification.

Texture, on the other hand, is defined by the *arrangement and relative geometrical properties of some mineral grains with respect to others*; that is, it cannot be defined for each grain independently but only in reference to its environment. Therefore, it is a property that applies to the rock as a whole or, at least, to the mineral aggregates (including the *intergrowths* so common in metallic ores), not to single, isolated grains. By *arrangement* is meant the spatial arrangement of some grains with respect to others. This way, oriented *vs* isotropic (massive), banded *vs* homogeneous, fibrous, graphic, etc. textures are distinguished. The *geometrical properties* include criteria such as the relative size of the grains, the extent of development of crystallization and/or crystalline contours in the contacts between grains, and any morphological modifications due to the geological history of the rock (fragmentation, recrystallization, alterations, etc.). Thus we speak of textures that are granular, porphyritic, microcrystalline, clastic, residual, and so on.

The term texture, as used here, is equivalent to the more general meaning of *fabric*, but strictly speaking (Fettes and Desmons 2007) in metamorphic petrology and structural geology *fabric* refers to the *relative orientation* of the components of a rock. To avoid confusion, only the term texture will be used in this text. It will be used in the sense just given and in no other sense, although there are other possible meanings, such as the very broad one of Barton (1991), defining texture as "the spatial *relations within and among minerals and fluids, regardless of scale or origin*".

Figures 1.1 and 1.2 illustrate, by direct comparison of two images of the same field with and without analyzer, the difference between the texture of an aggregate of silicate and magnetite (Fig. 1.1) and the internal structure of the magnetite grains (Fig. 1.2). They serve to illustrate the sometimes unavoidable ambiguity when applying these criteria to real cases, in which several intergrown phases may coexist in the same grain (Fig. 1.2), in which case it would be possible, *in extremis*, to describe its internal structure as that of an aggregate. Thus, according to Schwartz (1951), Fig. 1.1 would correspond to a *triangular, lamellar or Widmanstätten texture*. It is more coherent and in accordance with the most accepted usage (Schneiderhöhn 1952; Ramdohr 1980) to keep the criterion defined above: in this text, structure refers to the internal geometry or morphology of a grain (even if it is polymineralic); and texture refers to the geometrical properties of an aggregate of various mineral grains.

From the point of view of the technical or geometallurgical classification of textures, note that intergrowths may exist not only in aggregates (the most typical) but also inside a grain (bi- or polymineralic grain, i.e. a *middling*). These deserve specific attention, since the most frequent case of study for the control of plant processes is that of milling products, in which textures have to be defined from observation of (milled) grains.

[1] Cf. https://www.mindat.org/min-3004.html (visited 2021/11/19).
[2] Cf. https://www.mindat.org/min-2114.html (visited 2021/11/19).

Fig. 1.1 Intergrowth of titanomagnetite (light pinkish-brown grey) with silicates (various shades of darker grey) showing micrographic **texture** in basaltic pegmatoid. 070,513, Lanzarote, Canary Islands (*dsp*)

Fig. 1.2 Same image observed with the analyzer (*d* + *p*): the internal lamellar-octahedral **structure** (known as *Widmanstätten's*) is revealed by the appearance of thin lamellae of ilmenite and martite arranged as planes (111) of titanomagnetite

1.2 Classification Criteria for Textural Analysis

The precise definition and correct interpretation of textures provide basic criteria for the investigation of mineral deposits (Metallogeny) and their exploitation; that is, for understanding the behaviour of the ore in the concentration process (Ore Processing or Minerallurgy, cf. § 1.3.4) and for the prevention of environmental pollution by the residues of this process. For this purpose, first and foremost, an unequivocal characterization of each texture must be achieved, based not on debatable interpretations but on objective data and valid for any observer: this is what is known as descriptive or formal classification, essentially geometrical.

In line with these objectives, three fundamental textural classification criteria are applied:

- *Descriptive*, providing an objective, formal definition of single grains, intergrowths and textures (Tables 1.1 and 1.2)
- *Genetic*, seeking to interpret the genesis from textural analysis of the ores (Table 1.3)
- *Technical*, aimed at a predictive explanation of the ore's behaviour in the processing plant in order to contribute, through geometallurgical characterization, to optimization of the process (Table 1.4).

These three classes are also found in the basic schemes of the generally accepted classifications by Schneiderhöhn (1952) and Ramdohr (1980). Nevertheless, this text prefers a substantial simplification, achieved by renouncing many obsolete terms or terms that, for the beginner, represent an unnecessary burden, instead insisting on descriptive aspects. Simplicity should result in greater clarity and a more comprehensible systematization; in any case, the omitted terms can be documented, as in Grigoriev (1928), Schwartz (1951) or the basic texts (e.g. Craig and Vaughan 1994; Park and McDiarmid 1981; and standard works on petrography and mineralogy).

1.2.1 Descriptive Classification

Descriptive classification is based on purely morphological, geometrical criteria, which by definition are objective and reproducible, regardless of which observer or object (individual grains or aggregates).

In the case of **individual grains**, the criteria may refer to the grain's internal structure, external shape or size (Table 1.1). The internal structure may be totally homogeneous, with no observable differentiation features within the grain, or characterized by features such as zoning, twinning and cleavage.

Table 1.1 Description of an individual grain of metallic ores

By its internal structure	Figure 1.3 Zoning
	Figure 1.4 Twinning
	Figure 1.5 Cleavage
	Figure 1.6 Other (spherulitic, *emulsion*, etc.)
By its external shape	Figure 1.7 Idiomorphic/specify habit
	Figure 1.8 Allotriomorphic
	Figure 1.9 Hypidiomorphic/specify habit
	Figure 1.10 Other (pseudomorphic, skeletal, etc.)
By its size (measured diameter = Ø)	Very coarse (Ø $\geq$ 10 mm) Figure 1.11 Coarse (3 mm $\leq$ Ø < 10 mm) Medium (0.3 mm $\leq$ Ø < 3 mm) Fine (0.03 mm $\leq$ Ø < 0.3 mm) Figure 1.12 Very fine (0.001 mm $\leq$ Ø < 0.03 mm) Submicroscopic (Ø < 0.001 mm, 1 μm)

Zoning consists of systematic variation of properties in some zones of the crystal, for example concentric bands parallel to the crystalline faces, which follow one another like envelopes around the crystal core (*growth zoning*, due to the development of the crystal in successive episodes: Fig. 1.3a). Often the zones show compositional variations (*compositional zoning*), sometimes repeated rhythmically from the core to the periphery (*oscillatory zoning*).

Compositional zoning translates into differences in color or reflectance, sometimes subtle (Fig. 1.3b); it is not uncommon, however, for compositional changes to be imperceptible to the optical microscope and visible only by microanalysis (electron microprobe) through variations in the contents of major elements or subtle changes in the distribution of trace elements. At other times, there is no evidence of compositional differences and the zoning is evidenced by some physical property (color, hardness, porosity, etc.), which varies by the sector (concentric or not) of the crystal. Generally the zoning may be observed in plane polarized light (*sp* or single polarizer setting), and in some ores, like skutterudite, the growth zoning is a highly characteristic feature (Fig. 112.3, vol. 1). In some minerals, like löllingite, it is convenient to use the analyzer (*+p* setting, Fig. 1.3c and d); in others any observation of the zoning requires chemical attack (*structural etching*) or simply *air etching* or *tarnishing*; that is, the spontaneous development of a surface oxidation patina on the polished section of the grain, caused only by the action of atmospheric oxygen over time (e.g. pyrite, Fig. 1.3e, f).

(a)

(b)

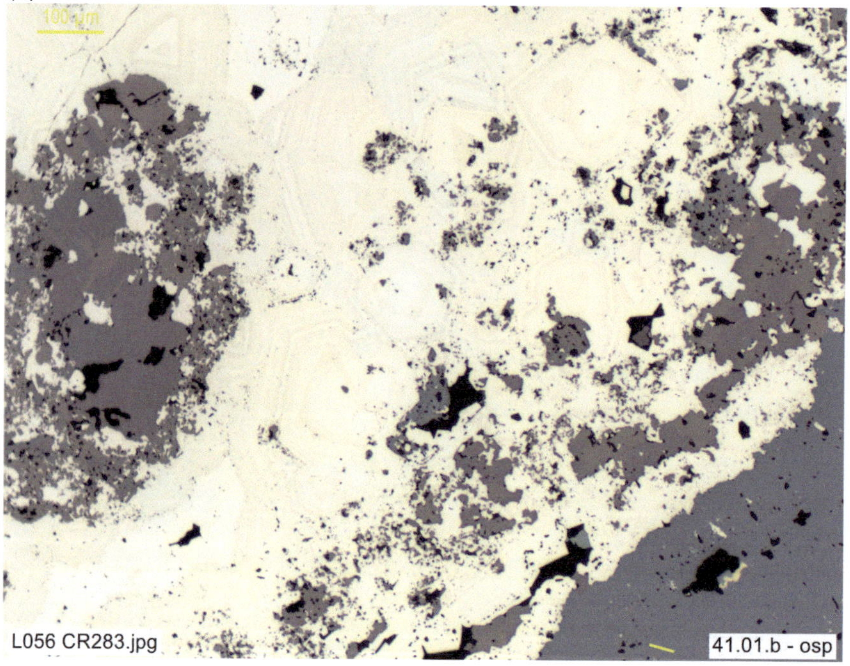

Fig. 1.3 **a** Sulfide concentrate with nickeliferous pyrite (bravoite, yellow to brownish), marcasite (white) and sphalerite (gray): in the center, zoned bravoite crystal. Rub.sn, Maubach, Germany (**osp**). **b** Aggregate of subtly zoned pyrite (with inclusions of famatinite) and enargite (with inclusions of tetrahedrite, both gray). CR283, Peru 251102_1 (**osp**). **c**, **d** (**dsp** and **d + p**, resp.). Zoned crystals of safflorite (*sf*, core) and löllingite (*lo*, periphery), included in groundmass of rammelsbergite, *rmb*, and skutterudite, *sk* (both white). Although conspicuous in *d + p setting*, zoning is invisible in plain polarized light (*dsp*) due to the similarity in color of sf and lo; anisotropism and twinning also identify rmb, in contrast to the extinction of the isotropic sk (cubic). Rub404, Bou Azzer, Morocco. **e**, **f** (**d + p/ci** and **dsp**, resp.). Zoned pyrite porphyroclast. Zoning interrupted by fragmentation, and visible in interference contrast (*d + p/ci, Wollaston prism*), but not in *dsp* setting; interstitial chalcopyrite. Py02, Moeche, Coruña, Spain

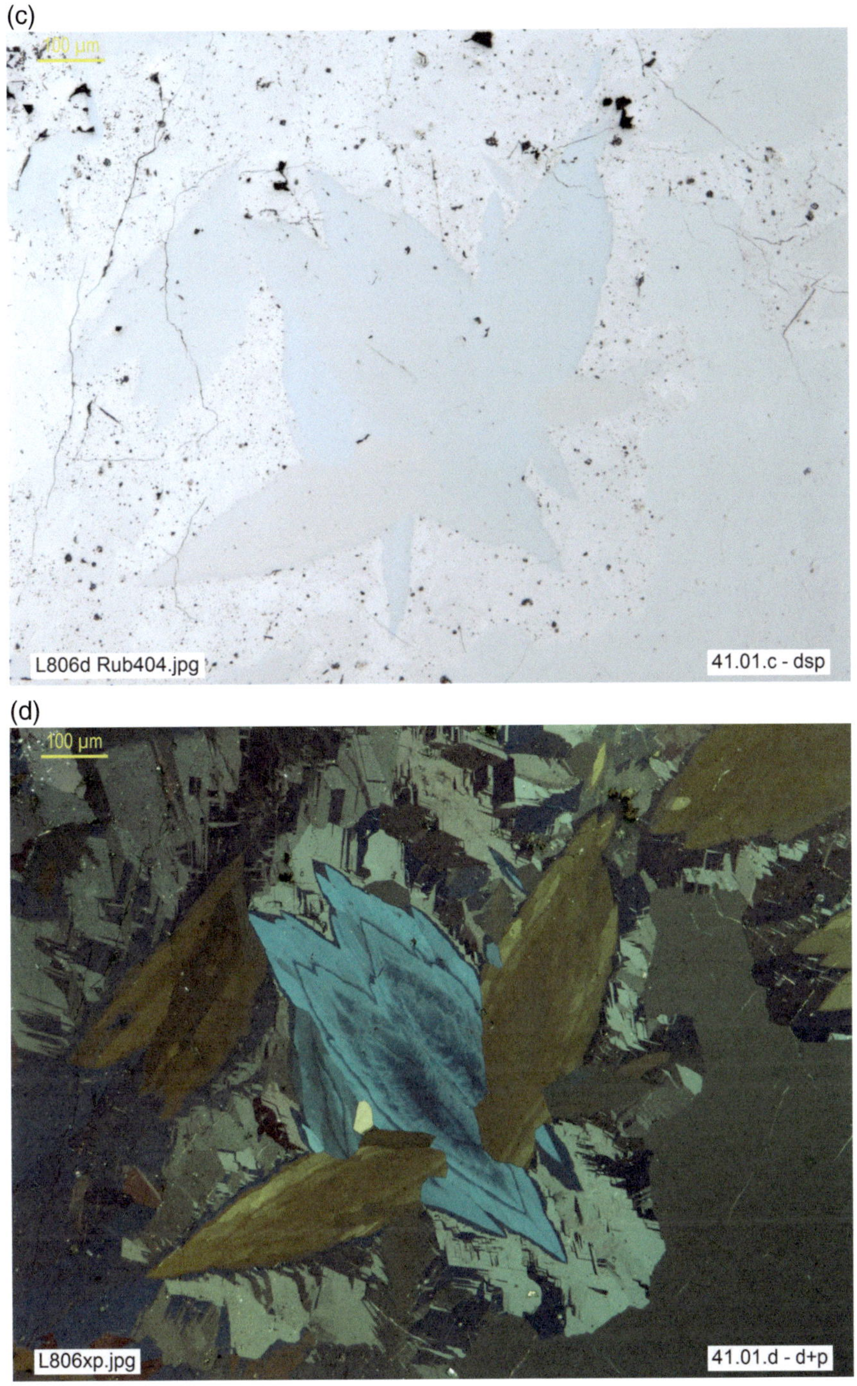

Fig. 1.3 (continued)

Fig. 1.3 (continued)

Twinning (Fig. 1.4a, b) and **cleavage** (Fig. 1.5)—*cf* explanations in vol. 1 § 1.4.2 and 1.4.1—can provide important clues to identification and are specifically discussed when describing each mineral, so the reader is referred to the corresponding section (§ Descriptions, in vol. 1, Part II). There are, in addition, **other** significant peculiar internal structures (cf. mineral descriptions), which often involve intimate intergrowths of two or more phases, and are discussed as complex contacts (*cf.* Table 1.2, and Figs. 1.30 and 1.31). In some cases they may be of diagnostic value: e.g., the *Widmanstätten* structure (magnetite-hematite, *cf.* Fig. 1.2) or the layered cylindrical structure of *cylindrite* (Fig. 1.6a, b). In others (*vbgr* emulsion, Fig. 1.30, or spherulitic, Fig. 1.25), they provide complementary, even genetic, information, and this is discussed in due course.

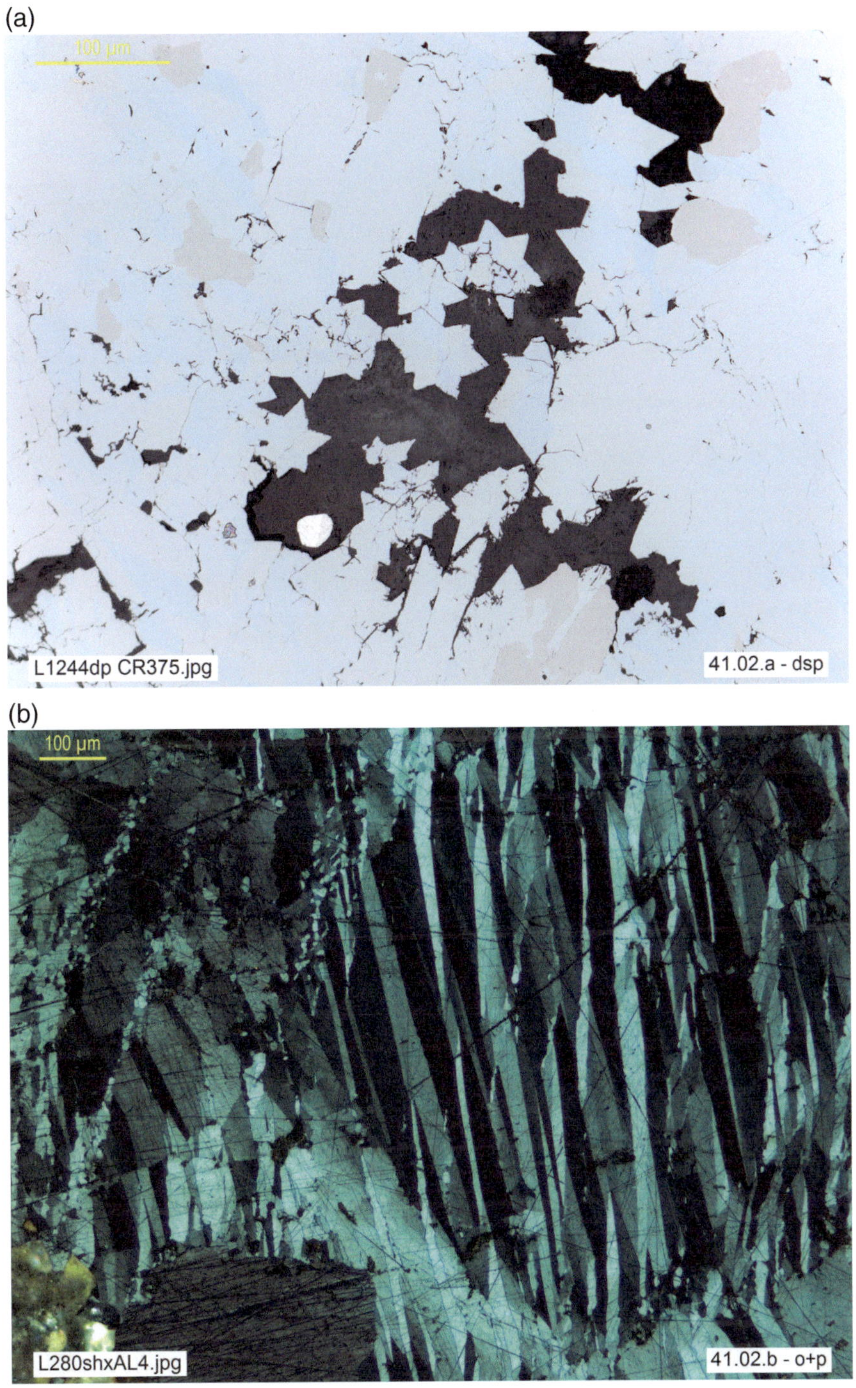

Fig. 1.4 **a** Star-shaped twins of safflorite; the pink mineral is cobaltite. CR 375, Cobalt, Ontario, Canada (**dsp**). **b** Polysynthetic antimonite twins, with evidence of deformation. AL-4, Almuradiel (Spain) (**d + p**)

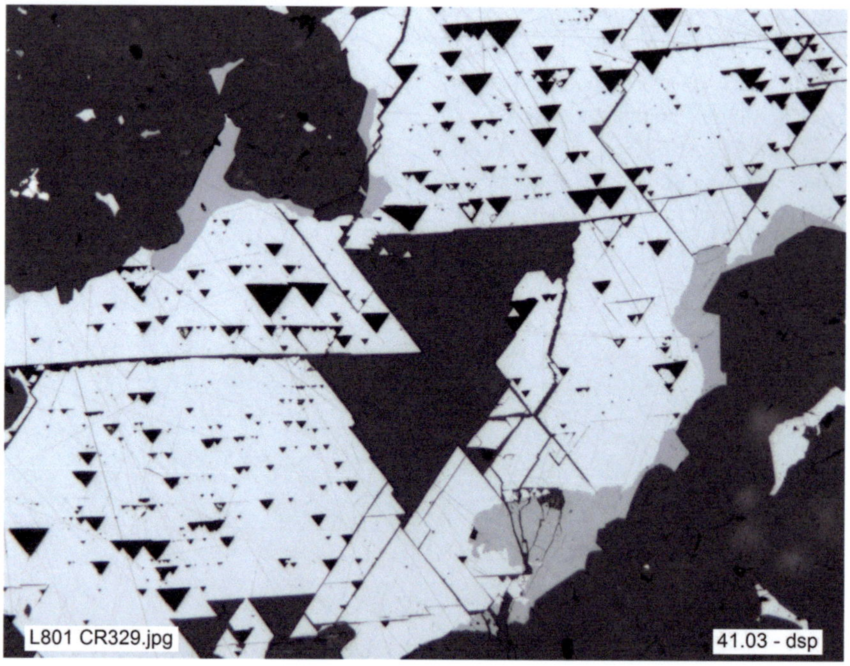

Fig. 1.5 Cubic cleavage (//100) of galena, seen as *triangular pits* due to polishing; the gray mineral is proustite. CR329, Imiter, Morocco (***dsp***)

Fig. 1.6 **a**, **b** Layered helical structure of cylindrite, Rub75hjb, Oruro, Bolivia (**dsp** and **d + p**, *resp.*)

As regards the *external morphology* of the grain, it is characterized by the development of its crystalline faces[3]:

- When they are well developed and are the dominant forms in the whole contour, we speak of an **idiomorphic** grain (or *automorphic* or *euhedral,* Fig. 1.7a–d).
- If they are absent; that is, if the grain acquires the shape imposed by its immediate surroundings, it is said to be **allotriomorphic** (or *xenomorphic* or *anhedral*, Figs. 1.8 and 1.9c).
- If they are only partially developed, the outline of the grain corresponds only in part to its crystalline faces and it is designated as **hypidiomorphic** (or *hypautomorphic* or *subhedral*, Fig. 1.9a–d).

[3] Etymology is recommended to understand and retain scientific terminology, and this is particularly so for textures, so it is pertinent to recall the meanings of the basic Greek terms used: *idios*-properly, *morphé*-form, *autós*-proper, *eu*-well, *hedros*-face, *alotrios*-stranger, *xenos*-foreign, *an*-no, *pan*-all, *hypo*-under, *pseudós*-deception, misunderstanding.

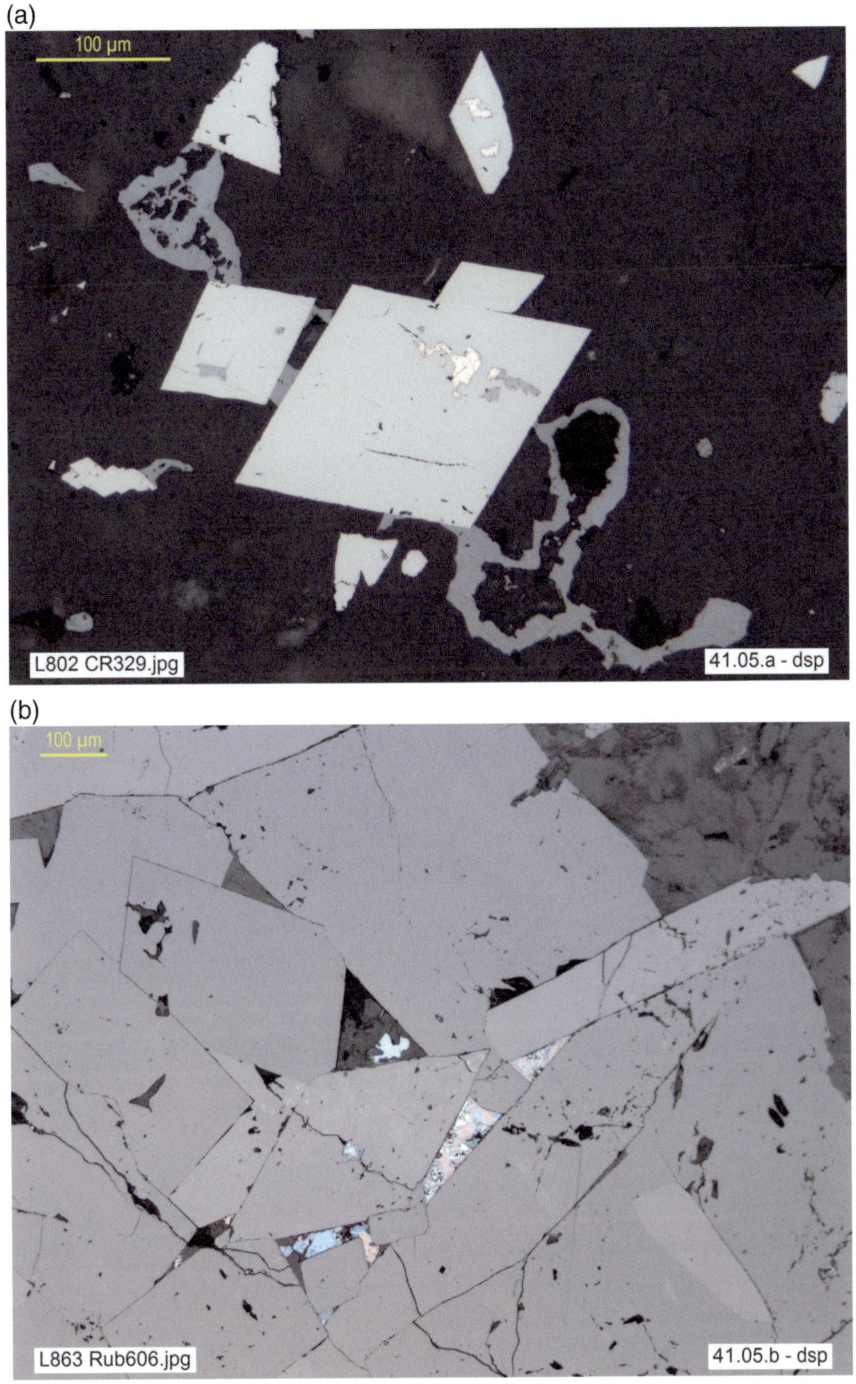

Fig. 1.7 **a** Arsenopyrite idiomorphic crystals, with xenomorphic inclusions of native silver (white) and acanthite (gray); the latter, also disseminated in gangue and coating micro-geodes. CR329, Imiter, Morocco (**dsp**). **b** Holocrystalline vein infill: panidiomorphic cassiterite, with interstitial quartz (black) and minor copper sulfides (colored). Rub606, Geevor Mine, Cornwall, UK (**dsp**). **c** Dissemination of fine-grained idiomorphic pyrite in metapelite. VA14-79.75, Masa Valverde, Huelva, Spain (**dsp**). **d** Fibrous texture, with acicular hematite crystals. Rub226, St. Martin am Silberberg, Huettenberg, Kaernten, Austria (***osp***)

(c)

(d)

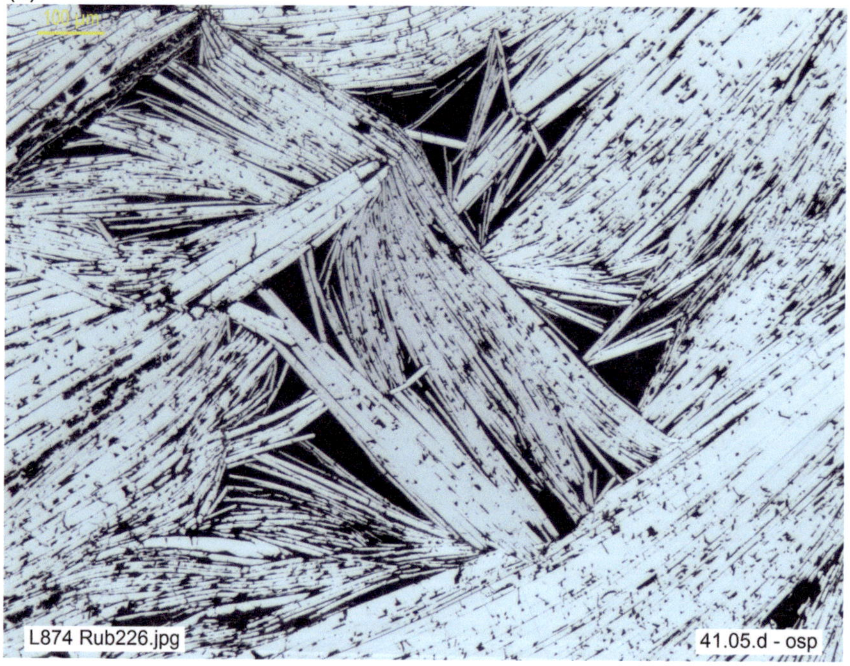

Fig. 1.7 (continued)

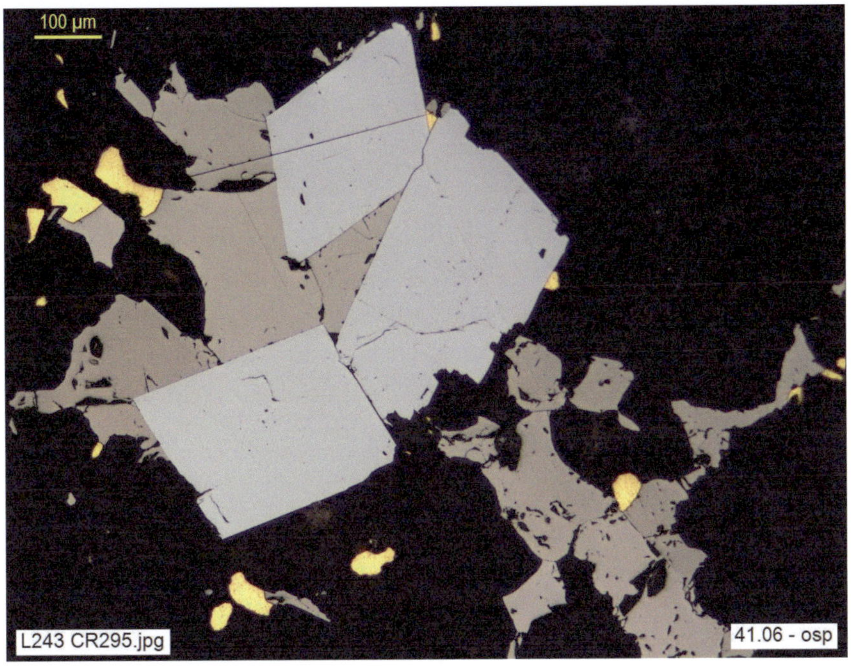

Fig. 1.8 Allotriomorphic gold and pyrrhotite, interstitial in vein infill of idiomorphic arsenopyrite (white) and gangue (black). CR-295, La Rinconada, Puno, Peru (*osp*)

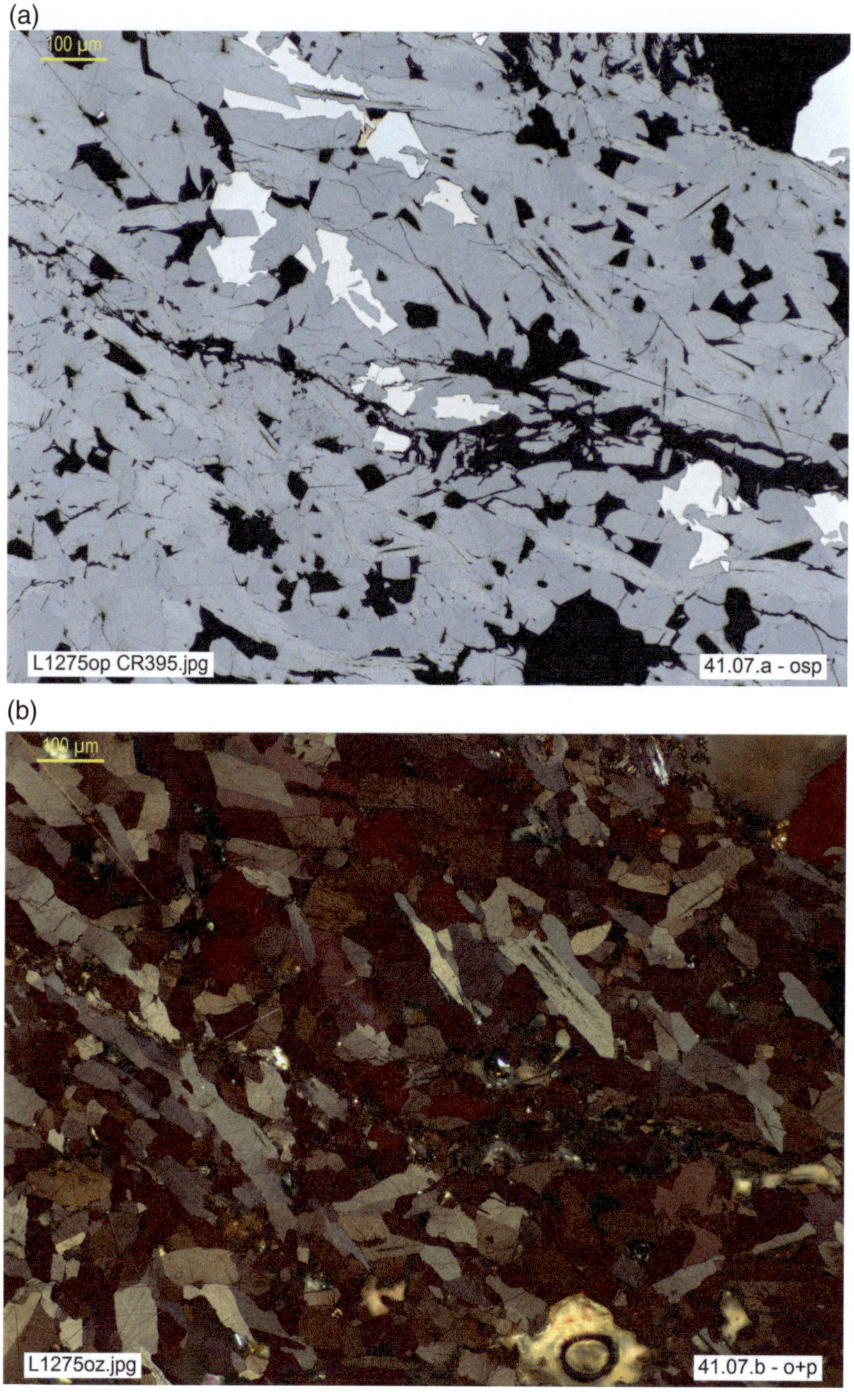

Fig. 1.9 **a**, **b** Hypidiomorphic zinkenite aggregate, with interstitial gudmundite (white), chalcopyrite (yellow) and quartz (black). CR395, Robson Group, Bridge River, British Columbia, Canada (***osp*** and ***o + p***, *resp.*). **c** Microgeode filled with xenomorphic galena (light gray) and chalcopyrite (yellow), in vein infill of pearceite (gray, hypidiomorphic with prismatic habit; intergrown with gn), with quartz and minor, fine-grained inclusions of safflorite (white). CR328, Freiberg, Saxony, Germany (**osp**). **d** Hypidiomorphic inclusion of tetrahedrite (light olive gray, center: six-sided cubic section, rounded) in massive bornite (pink). Both are partially replaced by chalcocite (light blue-gray), which forms a very thin reaction rim (*corona*) along the borders of td. Rub26, Tsumeb, Namibia (**dsp**)

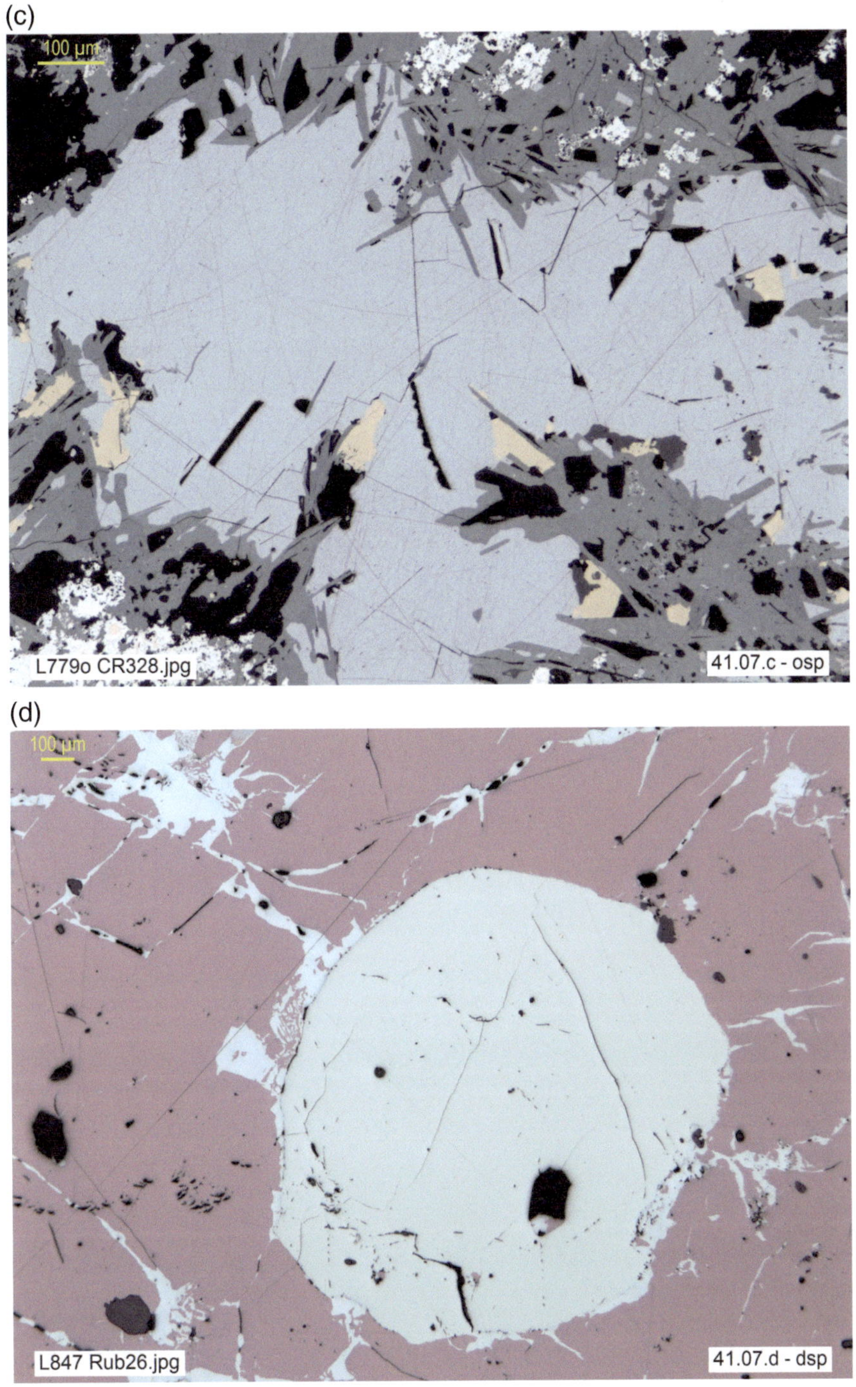

Fig. 1.9 (continued)

In idiomorphic or hypidiomorphic minerals the **habit** (or external form defined by the dominant faces) is sometimes a crucial criterion, and it is convenient to specify it: *prismatic, tabular, leafy, fibrous, acicular*, and so on. It may be that, by alteration or replacement processes, one mineral replaces another (or mineralizes organic matter) and inherits its form, in which case the new mineral is said to be **pseudomorphic** (*false form*, Fig. 1.10a–d). Some authors, such as Ramdohr (1980), distinguish the result of the transformation from an unstable to a more stable form of a polymorph with the term **paramorphic**, for example, marcasite to pyrite (Fig. 1.10e). Other forms are given a particular name: *skeletal*, *dendritic*, and so on (Fig. 1.10f).

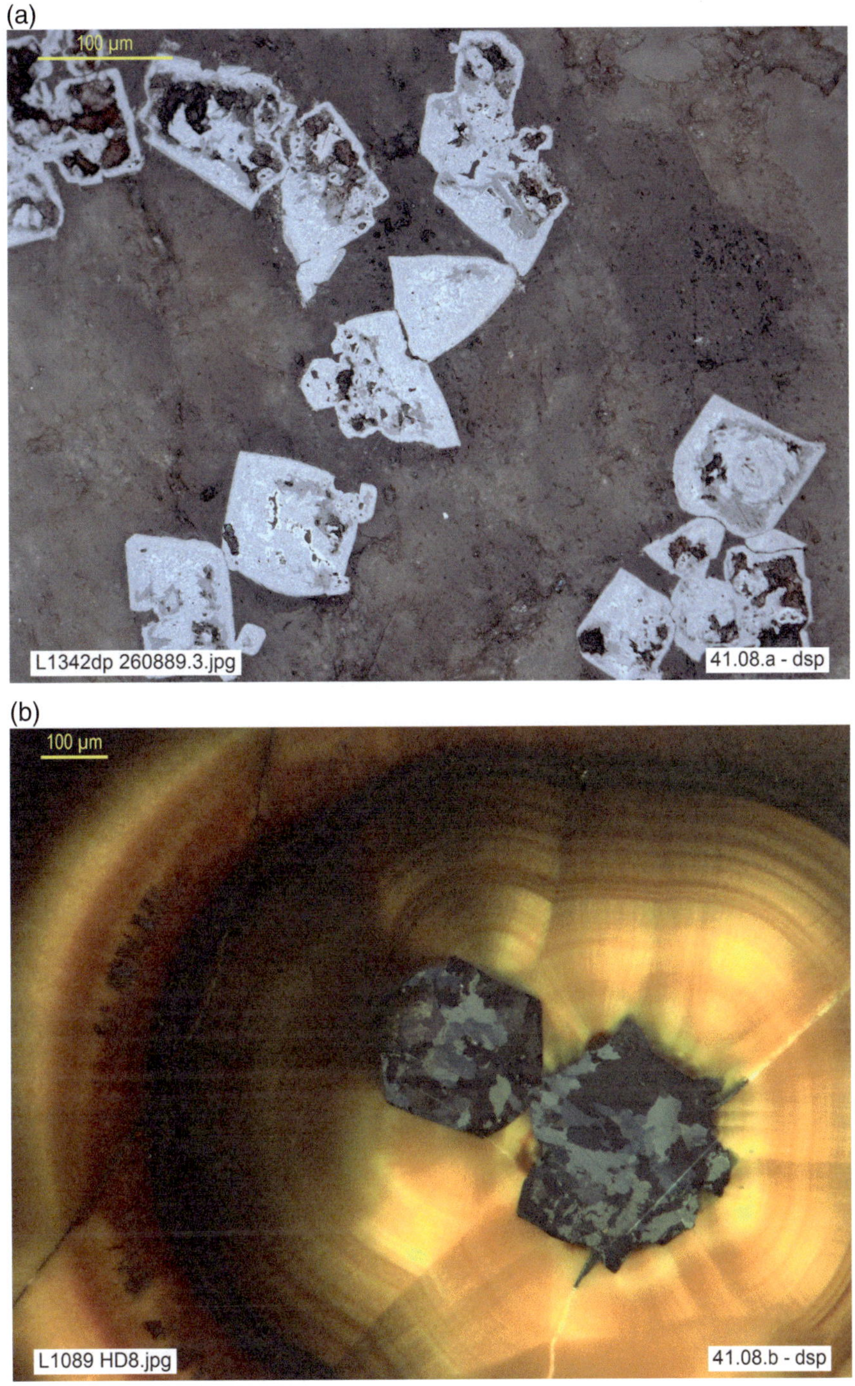

Fig. 1.10 **a** Cubic forms of limonite pseudomorphic of pyrite. M260889_3, Rodalquilar, Almeria, Spain (**dsp**). **b** Botryoidal or layered radiating sphalerite aggregate (*schalenblende*), encrusting a core originally formed by two idiomorphic gratonite crystals (center). These are totally replaced by jordanite, as microcrystalline pseudomorphs inheriting the six-sided contours of the rombohedral gratonite sections. HD 8, Wiesloch, Baden-Württemberg, Germany (**d + p**). **c** Pyritization of *Rotliegendes* sandstone, mainly affecting matrix and porous, reactive rock fragments. While quartz sand grains (dark gray) are still in an incipient phase, lithic grains have been completely replaced by pyrite. Rub10.07, Spremberg *Kupferschiefer*, Brandenburg, Germany (**dsp**). **d** *Banderz* ore: impregnation of pyrite and sphalerite (gray) ± galena (bluish white) ± marcasite (greenish white). Pyritized microfossil (*Fusuline*: center) and disseminated py framboids and pyritized organic remnants. HD 6153, Rammelsberg, Niedersachsen, Germany (**osp**). **e** Paramorphic replacement of marcasite (greenish white) by pyrite (yellow). HD6765, Teufelsgrund, Schwarzwald, Germany (**osp**). **f** Skeletal crystals of Bi, encrusted by thin rim (*corona*) of white arsenides (rammelsbergite, safflorite and löllingite, intergrown), included in massive maucherite (faint pinkish hue). HD 1222, Eisleben, Germany (**dsp**)

(c)

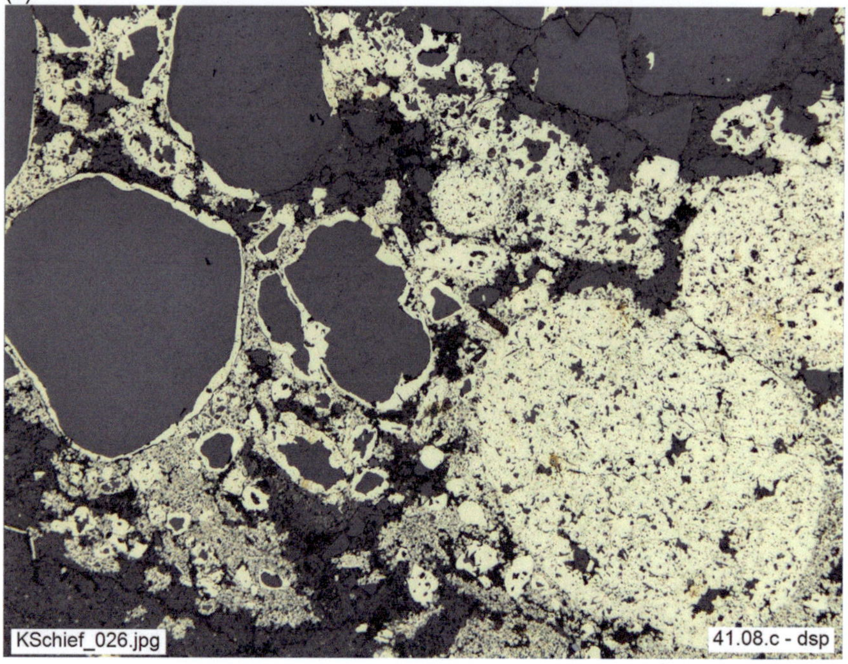

(d)

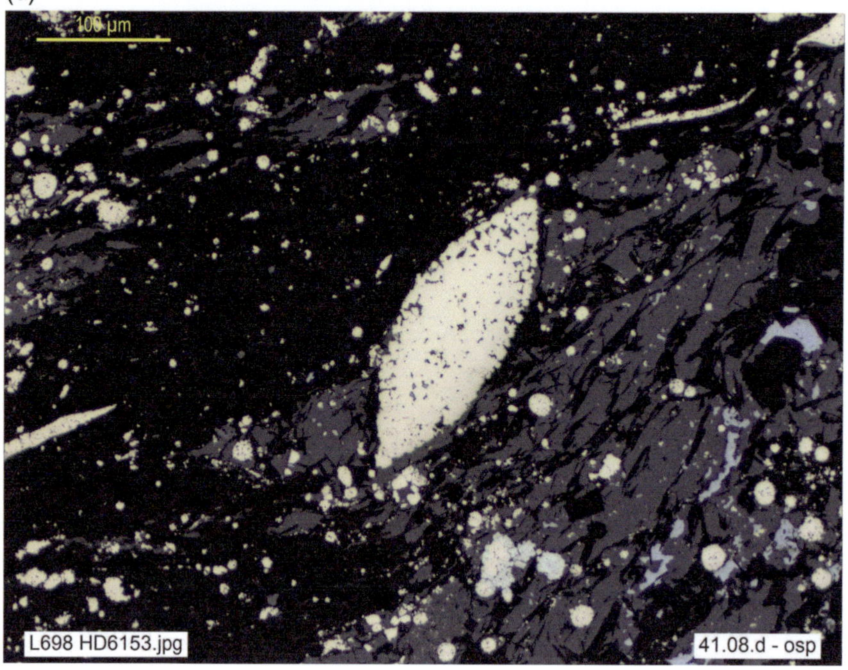

Fig. 1.10 (continued)

(e)

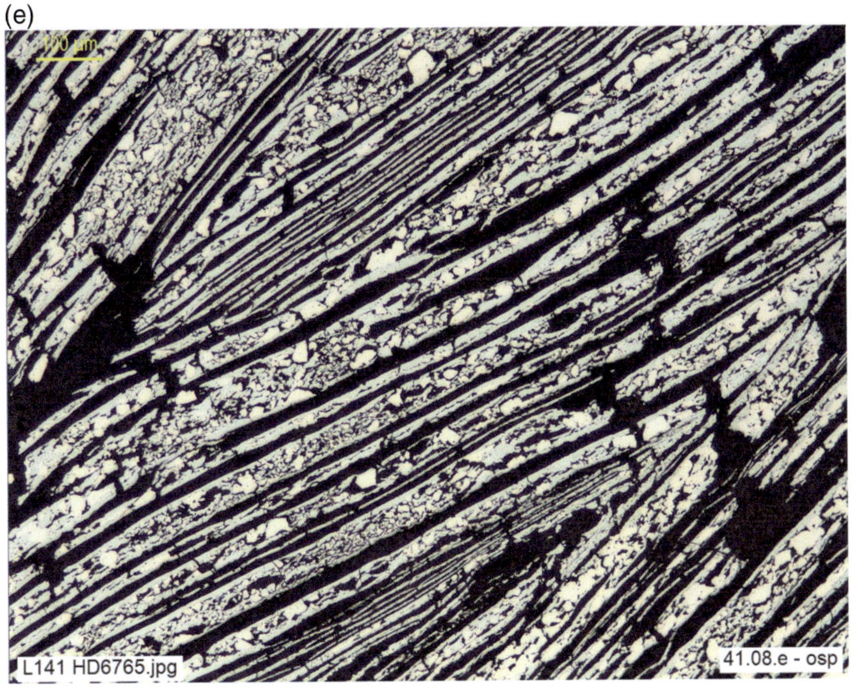

(f)

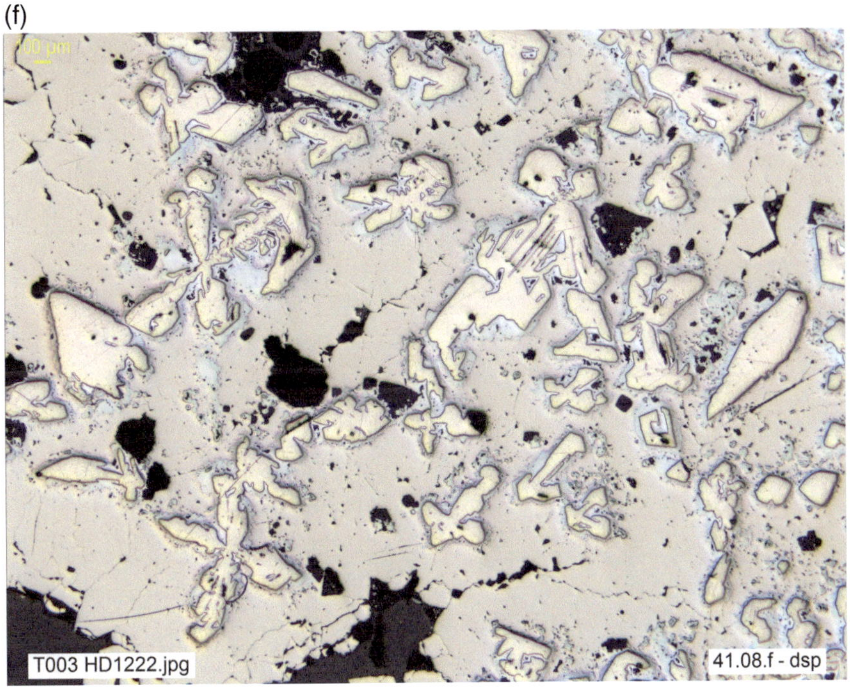

Fig. 1.10 (continued)

Grain size can be quantified by measuring with a graduated scale (or by automated image analysis techniques, as well as by screening meshes or sieves in the comminution circuits). However, a qualitative estimation is often sufficient, and in any case it should never be absent from microscopic study. As a guide to defining estimated particle size, the following grain size ranges are proposed: *submicroscopic, very fine, fine, medium, coarse and very coarse*, defined by the thresholds of 1 μm, 30 μm, 300 μm, 3 mm and 10 mm (Table 1.1). This proposal follows and simplifies that of Schneiderhöhn (1952) to facilitate communication through the objective expression of rapid estimates based on simple, direct observation (see, for example, Figs. 1.11 and 1.12). However, the terminology to be used in specific studies is often determined a priori by the nature of the problem (e.g. grinding sieves, specific scales for sedimentology).

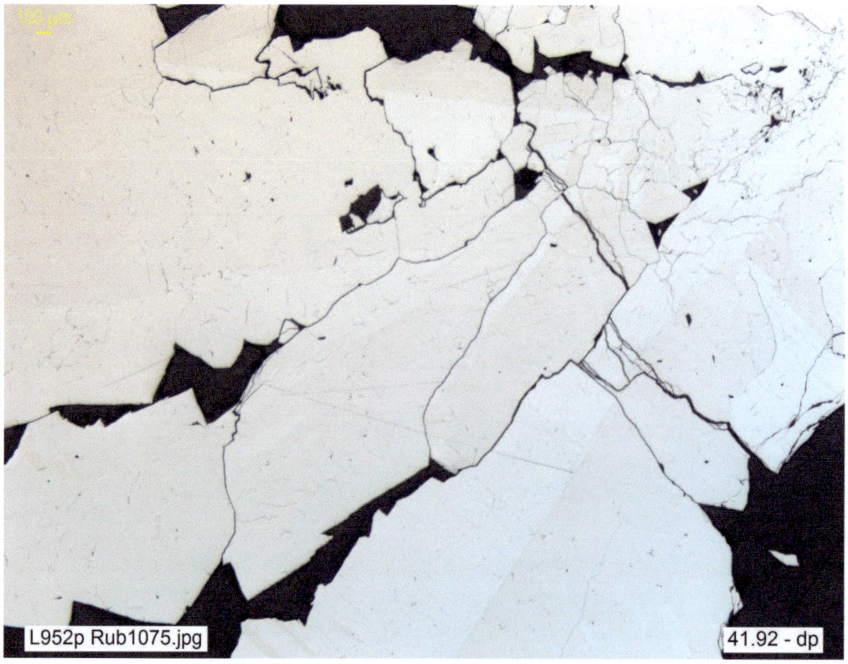

Fig. 1.11 Coarse phaneritic aggregate of marcasite. Rub1075, Friedrich der Grosse Mine, Germany (**dsp**)

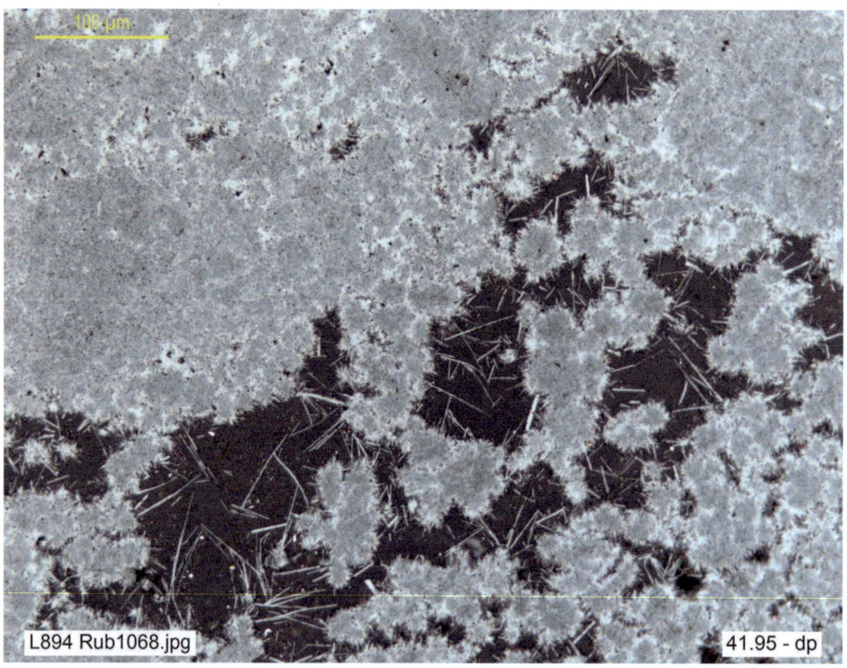

Fig. 1.12 Very-fine-grained limonitic ore. Finely intergrown lepidocrocite, goethite, hematite (and opal) in tiny fibrous-radial aggregates. Rub1068, Ober-Neissen, Nassau, Germany (**dsp**)

Table 1.2 Description of ore mineral aggregates-intergrowths

Crystallinity and grain size	Figure 1.13 Phaneritic Figure 1.14 Aphanitic/microcrystalline Figure 1.15 Cryptocrystalline/framboidal/botryoidal Amorphous	
	Equigranular Figure 1.16 Inequigranular (porphyritic, etc.)	
Compactness (degree of filling)	Figure 1.17 Compact	
	Figure 1.18 Porous	
	Figure 1.19 Vuggy	
Orientation *vs.* isotropy	Figure 1.20 Massive (unstructured)	
	Figure 1.21 Banded/layered	
	Figure 1.22 Oriented	
	Figure 1.23 Fibrous	
	Figure 1.24 Radial	
	Figure 1.25 Spherulitic	
	Figure 1.26 Spheroidal	
Type of grain-to-grain contact	Figure 1.27 Simple (smooth)	
	Complex	Figure 1.28 Jagged
		Figure 1.29 Concentric (corona, oolitic, etc.)
		Figure 1.30 Emulsion
		Figure 1.31 Other (symplectitic, etc.)
		Figure 1.32 Irregular (erratic)
Spatial arrangement of the components	Figure 1.33 Sandwich	
	Figure 1.34 Stockwork	
	Figure 1.35 Encrustation	
	Figure 1.36 Dissemination/impregnation	
	Figure 1.37 Interstitial	
	Figure 1.38 Intergranular film	

The textural description of **ores**, which are comprised of mineral **aggregates**, must characterize not only the individual grains but their spatial distribution and the intergrowths and relationships of the mineral species present, based on criteria such as those in Table 1.2. In general, the criteria and concepts applied are those of common rocks, whose mastery is assumed. The reader unfamiliar with these is referred to any basic petrography text, since only those relevant to common ores are discussed here.

Mineral aggregates can first be conveniently described by their crystallinity and grain size. However, grain size comprises not only the measure of particles' dimensions, as for individual grains; it also comprises new aspects, such as comparison of the size of the various component particles and definition of their intergrowths, as well as their degree of crystallization. Some specific terms characterize aggregates as a whole by their crystallinity and relative grain size. When the crystal size is submicroscopic or smaller than the resolving power of the microscope, they are referred to as **cryptocrystalline** aggregates (Fig. 1.15a); when the grains/crystals are seen individually under the microscope yet not with the naked eye, they are called **aphanitic** (Fig. 1.14 a)/**microcrystalline** (Fig. 1.14b), and if they are clearly visible to the naked eye they are called **phaneritic** (Fig. 1.13).

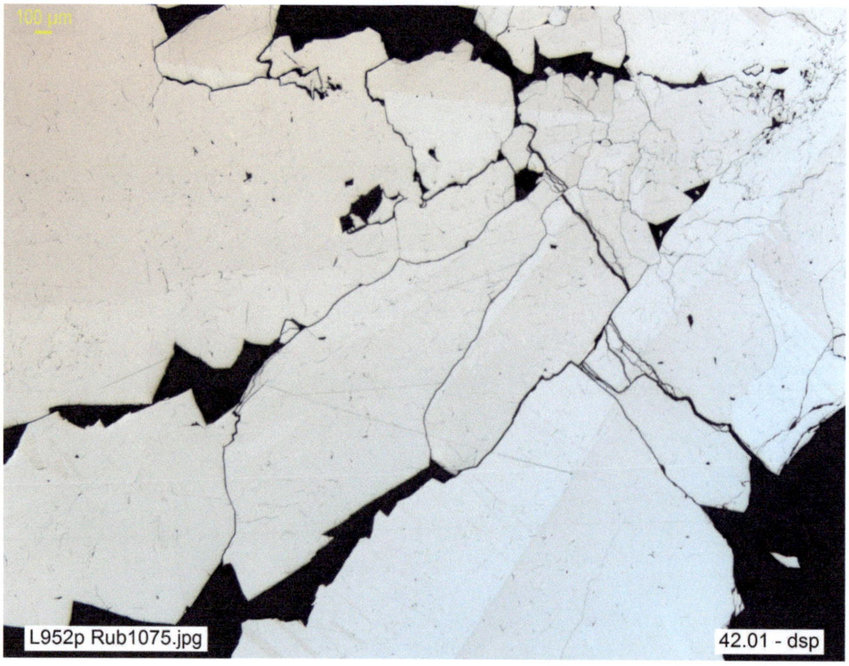

Fig. 1.13 Phaneritic marcasite aggregate: individual grains can be seen with the naked eye. Rub1075, Friedrich der Grosse Mine, Germany (**dsp**)

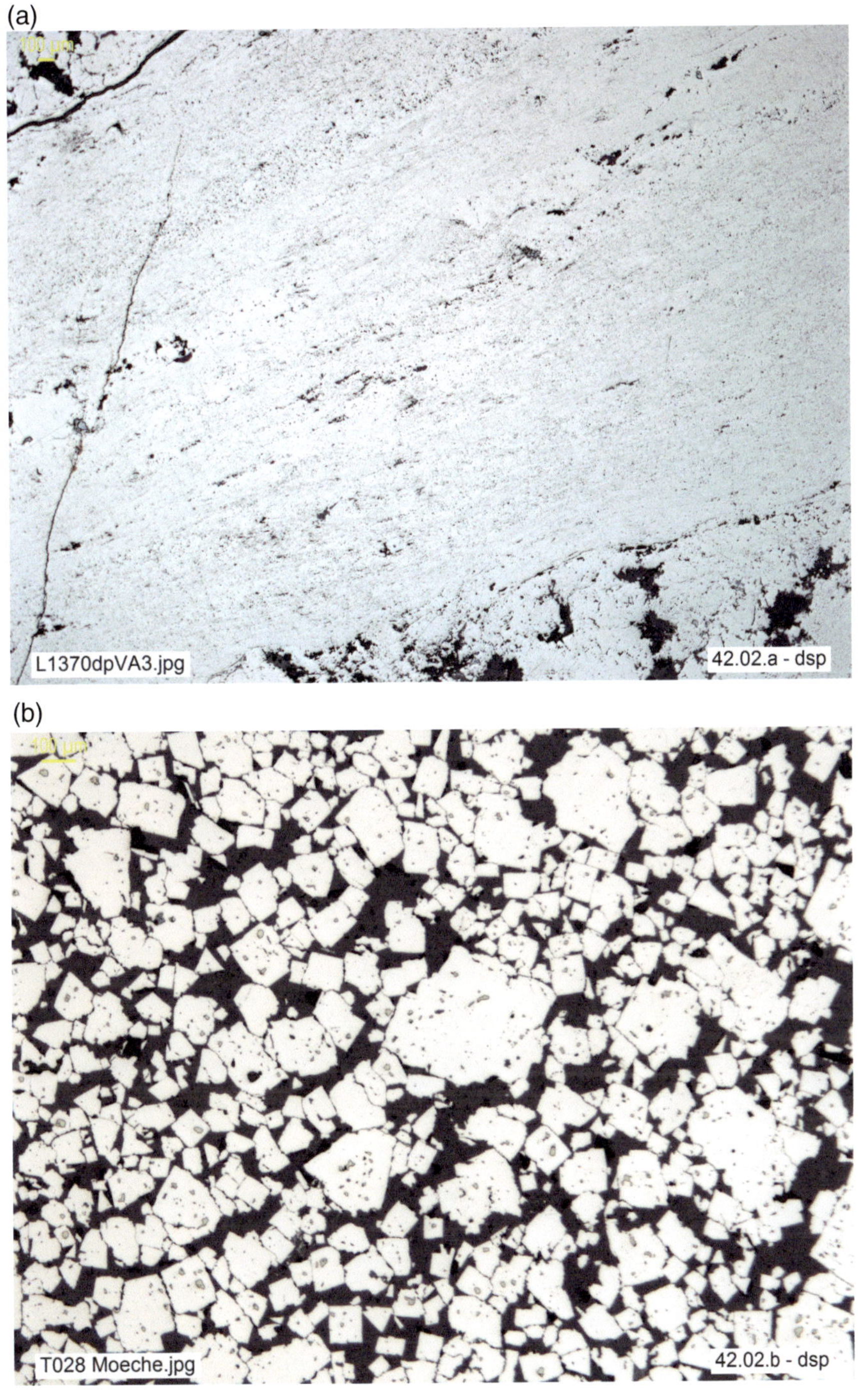

Fig. 1.14 **a** Aphanitic pyrite aggregate, with relict collomorph syn-sedimentary banding, and quartz-carbonate gangue, in massive sulfide ore. VA3-503.45, Masa Valverde, Huelva, Spain (**dsp**). **b** Massive microcrystalline aggregate of pyrite and quartz; inclusions of chalcopyrite and sphalerite in pyrite. Metamorphic massive sulfide ore. M031200.10, Piquitos II Mine, Moeche, Coruña, Spain (**dsp**)

Special mention is made of textures such as **framboidal** and **botryoidal**, clearly recognized by their morphology yet of ambiguous definition after their crystallinity, which can vary between crypto- and microcrystalline:

Framboids, so called for their raspberry-like appearance (Fig. 1.15b, c), are tiny spherical aggregates (usually of *very-fine* size, as defined in Table 1.1), typically consisting of minute crystallites of pyrite—or, more rarely, of other species: *vbgr* Fig. 1.61e—sometimes submicroscopic. They may comprise interstitial organic matter.

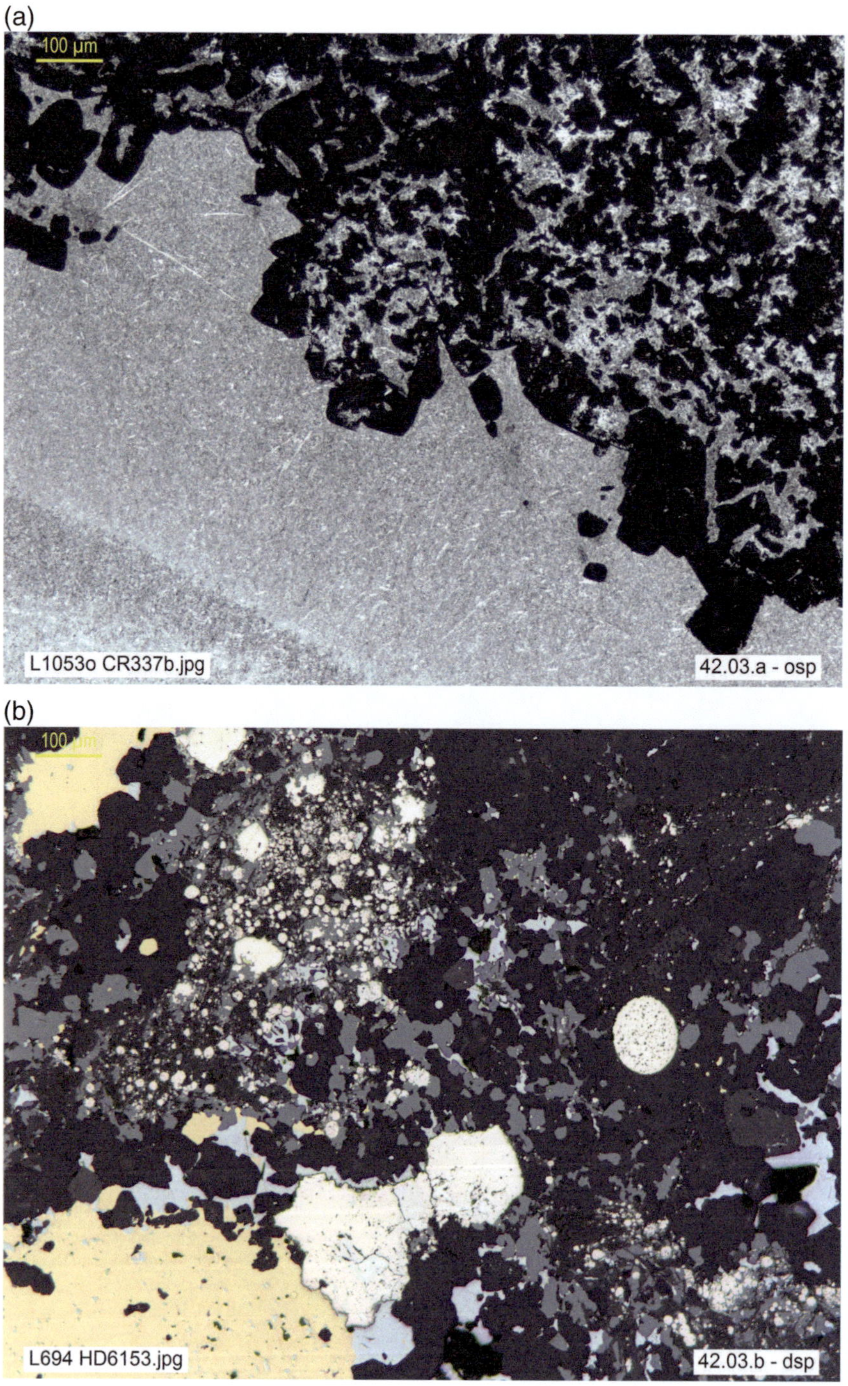

Fig. 1.15 A Cryptocrystalline manganese oxide ore (gray): cryptomelane intergrown with romanèchite and with gangue (black). CR 337.b, Veta Clemira, El Remanso, Santiago del Estero, Argentina (**osp**). **b** Pyrite framboids in massive sulfides, *Banderz* ore (chalcopyrite, pyrite, pyrite, sphalerite, marcasite, galena and gangue); the exceptionally large framboid (center right) shows the internal structure. HD6153, Rammelsberg, Niedersachsen, Germany (**dsp**). **c** Framboidal pyrite in massive sulfide ore: py, sphalerite, with minor galena, tetrahedrite, chalcopyrite and gangue. The original porous structure of the framboids (clearly visible in the upper right corner) is gradually overprinted by coalescence and recrystallization (bottom left). HD6153, Rammelsberg Banderz, Niedersachsen, Germany (**osp**). **d** Botryoidal sphalerite (*schalenblende*) encrustation on dendritic aggregate of gratonite, mostly pseudomorphosed by jordanite. HD 8, Wiesloch, Baden-Württemberg, Germany (dsp)

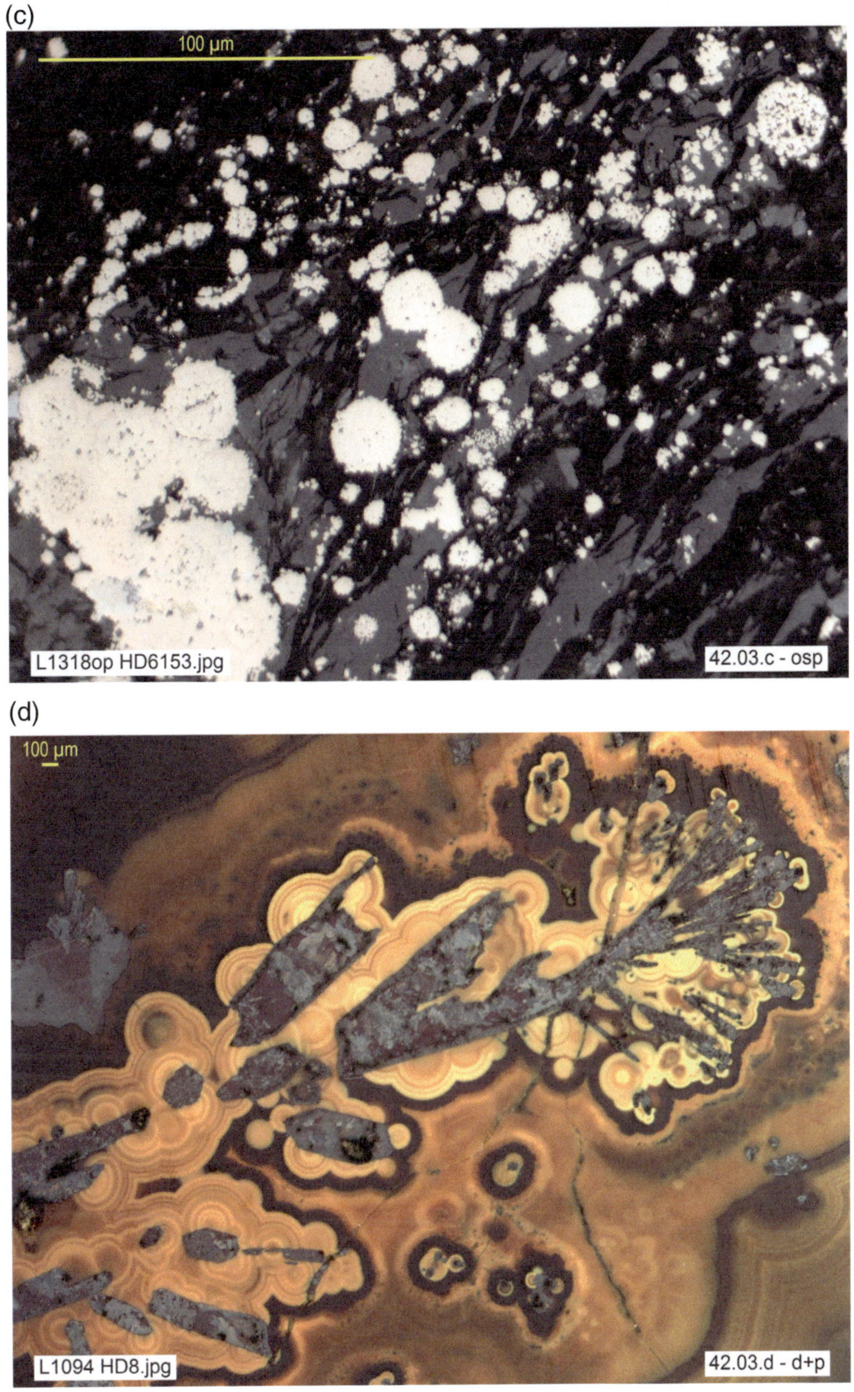

Fig. 1.15 (continued)

Botryoidal textures occur in sulfides such as wurtzite and sphalerite (*schalenblende*: Fig. 1.15d) or pyrite (*melnicovite*), or in oxides such as limonite and psilomelane. These textures are characterized by the juxtaposition of fine spherical corpuscles (*botryoids*), like a bunch of grapes. The internal structure of these corpuscles may vary from submicroscopic (apparently amorphous) to **spherulitic** (spherical forms constituted internally of radiating acicular crystallites, growing from the centre like spokes of a sphere, cf. Fig. 1.25a–c). Overall, the external surface of these aggregates shows a typical kidney-shaped appearance, hence they are also called **reniform**. When the corpuscles lack internal structure (*spheroids*), they are generally referred to as **spheroidal** textures (cf. Fig. 1.26a–c).

Exceptionally, **amorphous** products such as bauxite or limonite can be found, although closer study often reveals that they contain cryptocrystalline components; neither bauxite nor limonite are mineral species accepted by the IMA, *International Mineralogical Association*. Nevertheless, these terms are maintained because they correspond to natural substances important to industry. A particular case is **metamict** materials, minerals that were naturally transformed into an amorphous material by bombardment due to decay of their own or neighbouring radioactive elements, which has destroyed their original lattice.

The relative grain sizes of the components allow to distinguish **equigranular** aggregates (grains of equal or similar size, e.g. Figs. 1.14b and 1.39a), and **inequigranular** or **heterogranular** aggregates (grains of different sizes: Fig. 1.16a), among which **porphyritic** aggregates are a particular case (coarse grains disseminated in a finer-grained matrix: Fig. 1.16b).

Fig. 1.16 **a** Pyritic porphyroclasts in brecciated metapyritite, cemented by chalcopyrite and gangue. Py100r, Moeche, Coruña, Spain (**dsp**). **b** Magnetite idioblasts in hematite matrix, whose micro/crypto-crystalline granulometry is evidenced in (+p) by the "salt and pepper" texture. CR 371, BIF Minas Gerais, Brazil (**d + p**)

Fig. 1.17 Massive, compact, ore of pyrrhotite, chalcopyrite, sphalerite, arsenopyrite, quartz. 170800_1, Sulitjelma, Norway (**dsp**)

The textures also condition some physical properties of the ores, for instance the specific gravity of the ROM ores, which depends not only on the nature of the minerals but on the compactness of the aggregates, or whether they have more or less pores or cavities. A distinction is made between a **compact** texture, in which all the space is occupied by minerals (Fig. 1.17), a **porous** texture, in which there are empty pores (Fig. 1.18a–c), a **spongy** texture (extremely abundant pores or voids) and, when the voids are coarse (*vugs*) and clearly visible to the naked eye (size $\geq$ 3 mm), a **vuggy** texture (Fig. 1.19a–b).

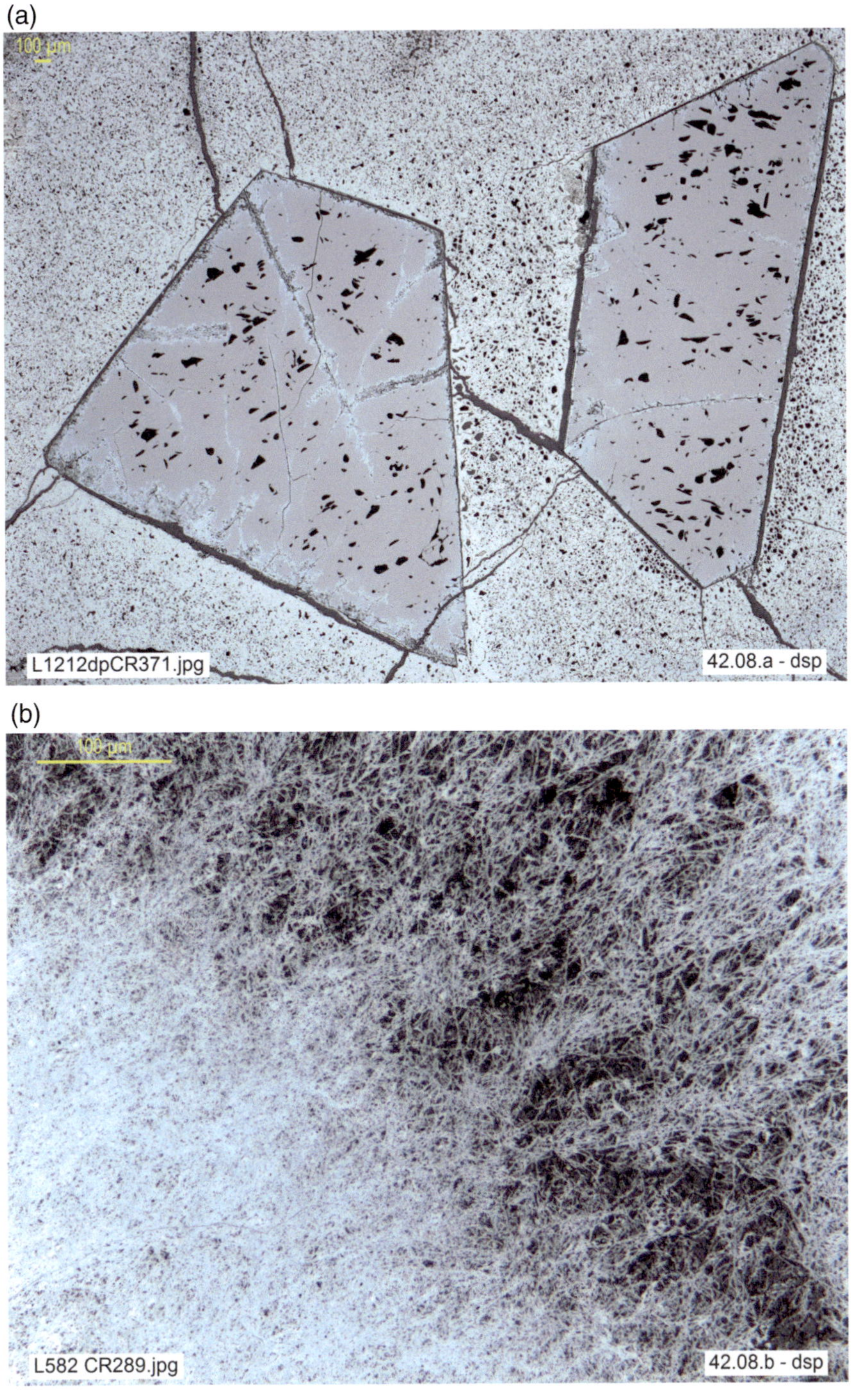

Fig. 1.18 **a** Porous, slightly weathered hematite and magnetite ore (BIF), with incipient replacement of magnetite (idioblasts) by hematite (matrix). CR 371, Minas Gerais, Brazil (**dsp**). **b** Porous ore resulting from fibrous texture of psilomelana. CR 289, Amalienhöhe Mine, Waldalgesheim, Bingerbrück, Germany (**dsp**). **c** Arsenopyrite-quartz lode microbreccia, with intense weathering resulting in high porosity and almost total replacement of arsenopyrite by limonite. CR-316, Magros Mine, Beariz, Orense, Spain (**dsp**)

(c)

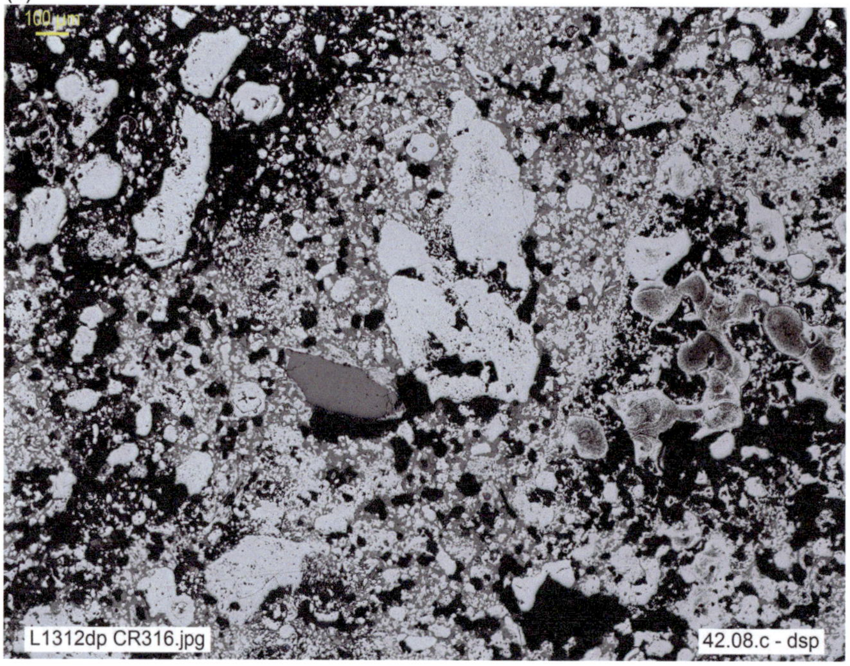

Fig. 1.18 (continued)

(a)

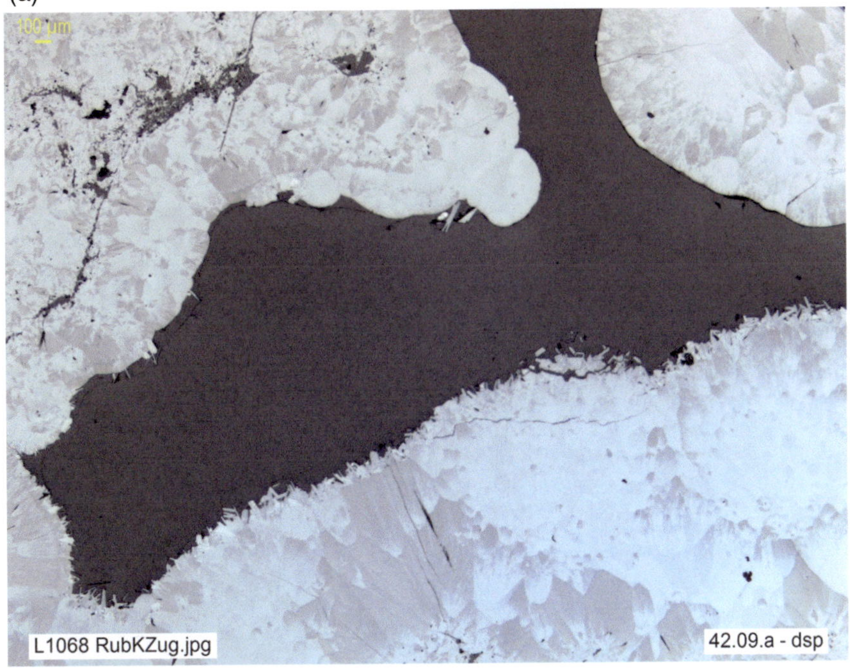

(b)

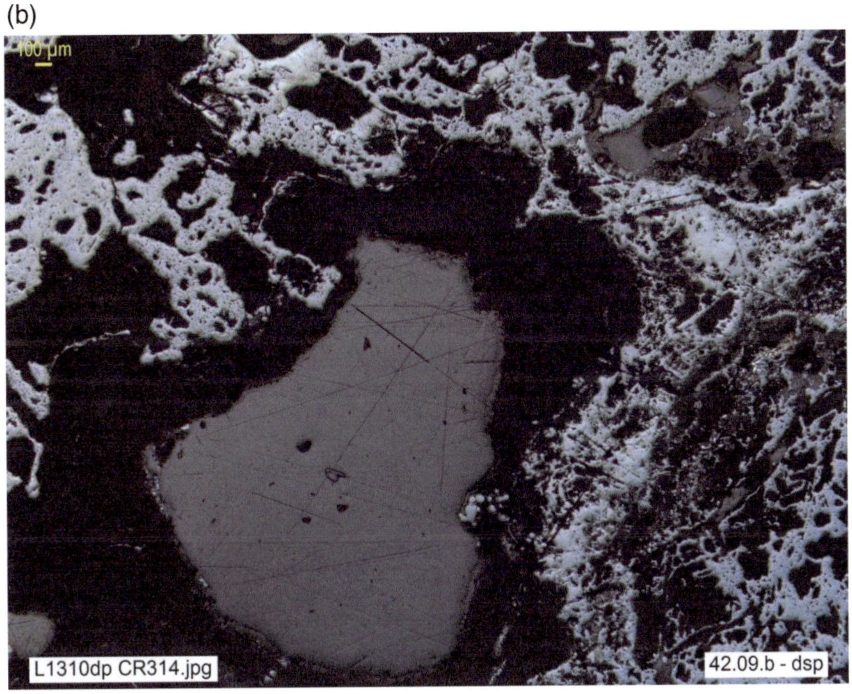

Fig. 1.19 **a** Fine border of lepidocrocite (acicular crystals, light gray), coating open geode (black) on botryoidal goethite (gray, in variable shades). Detail of vuggy texture. RubKzug, Kahlberger Zug, Olpe, Sauerland, Germany (**dsp**). **b** Vuggy ore of limonite (goethite, hematite) and psilomelane, with cavity (void, center) filled by *epoxy* resin (gray, scratched). CR-314, Ojos Negros, Teruel, Spain (**dsp**)

Although minerals, because of their crystal structure, are themselves anisotropic, aggregates may be isotropic on a larger scale when there is no preferred orientation of the grains as a whole: in these cases the aggregates or rocks are said to be **massive** (Fig. 1.20a, b). By contrast, **oriented** aggregates (Fig. 1.22a–d) comprise minerals with a preferential orientation (noticeable especially in minerals of elongated, prismatic or acicular, and flattened, tabular, lamellar or leafy habits). An intermediate case is that of **banded** textures, defined by the spatial distribution of components in bands or layers of various composition (and usually color), but not necessarily oriented (Fig. 1.21a–d).

(a)

(b)

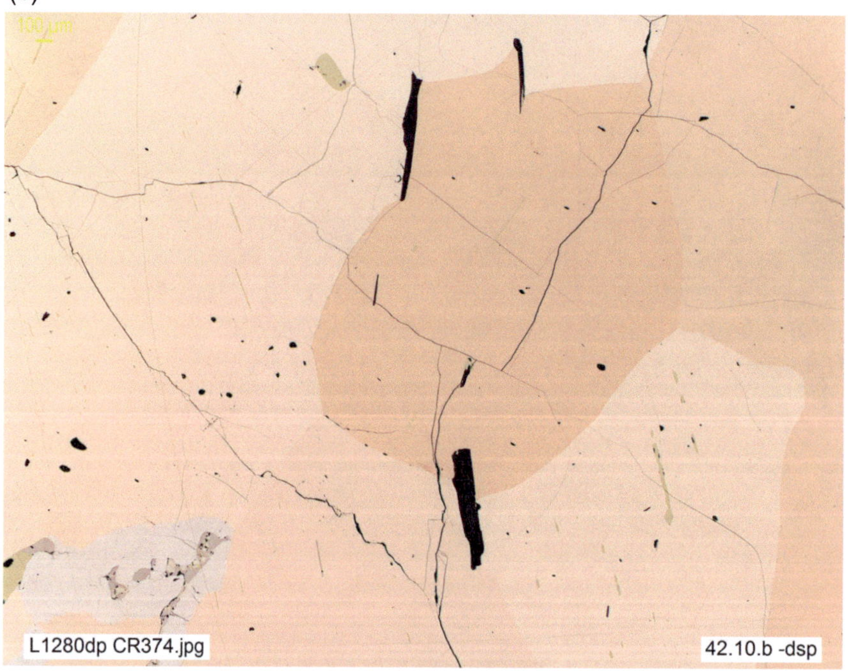

Fig. 1.20 **a** Massive chalcopyrite, pyrite and sphalerite ore. M190496_1, Neves Corvo Mine, Portugal (**dsp**). **b** Massive nickelite ore (with maucherite, pyrrhotite, chalcopyrite). CR-374, Frood Mine, Sudbury, Canada (**dsp**)

(a)

(b)

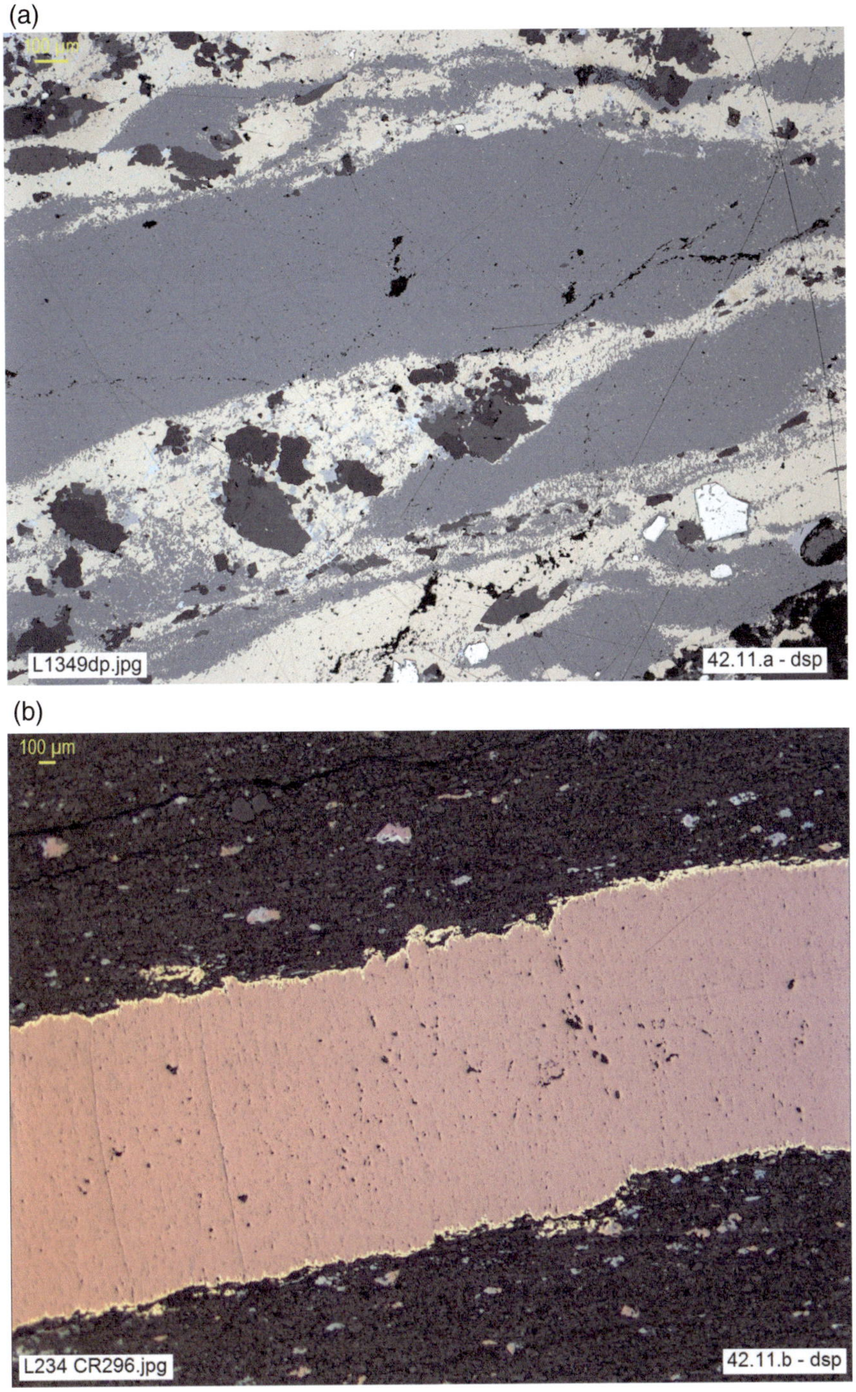

Fig. 1.21 **a** Tectonic banding on sulfide-rich vein filling: lenticular ductile shear bands of chalcopyrite, galena and sphalerite, with microclasts of pyrite, quartz and carbonates. SV_1, Rio Gibranzos, Caceres, Spain (**dsp**). **b** Bornite microband and disseminated copper sulfides in metapelite. CR296, Kupferschiefer, Poland (**dsp**). **c** Banded massive sulfide (*kiesiges Braunerz*): sphalerite, galena, pyrrhotite (pyrite). HD6775, Rammelsberg, Germany (**dsp**). **d** Mylonitic banding on massive sulfide *feeder* (*stockwork*), with pyrite (brittle, showing incipient pressure solution effects) and chalcopyrite (ductile). VA3-590.90, Masa Valverde, Huelva, Spain (**dsp**)

(c)

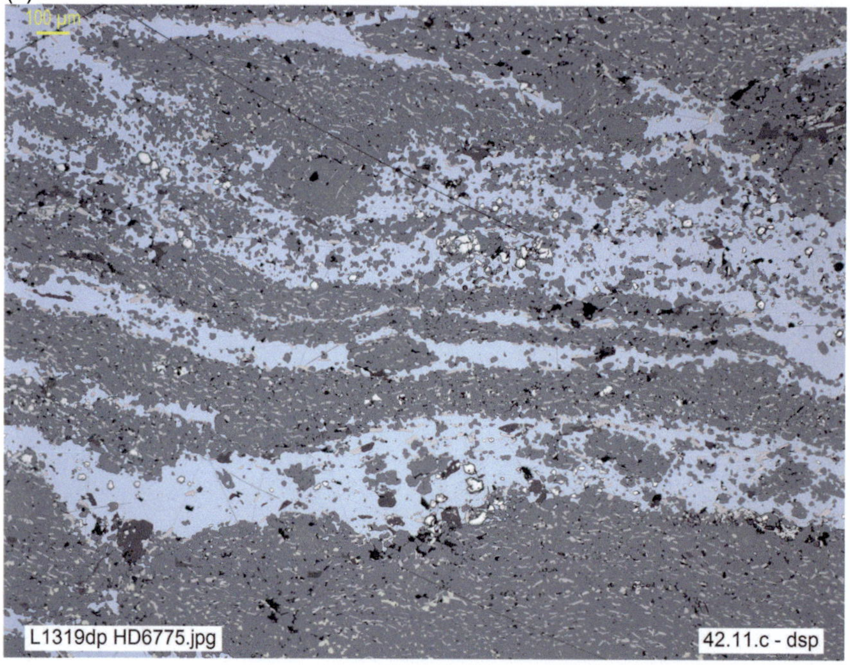

(d)

Fig. 1.21 (continued)

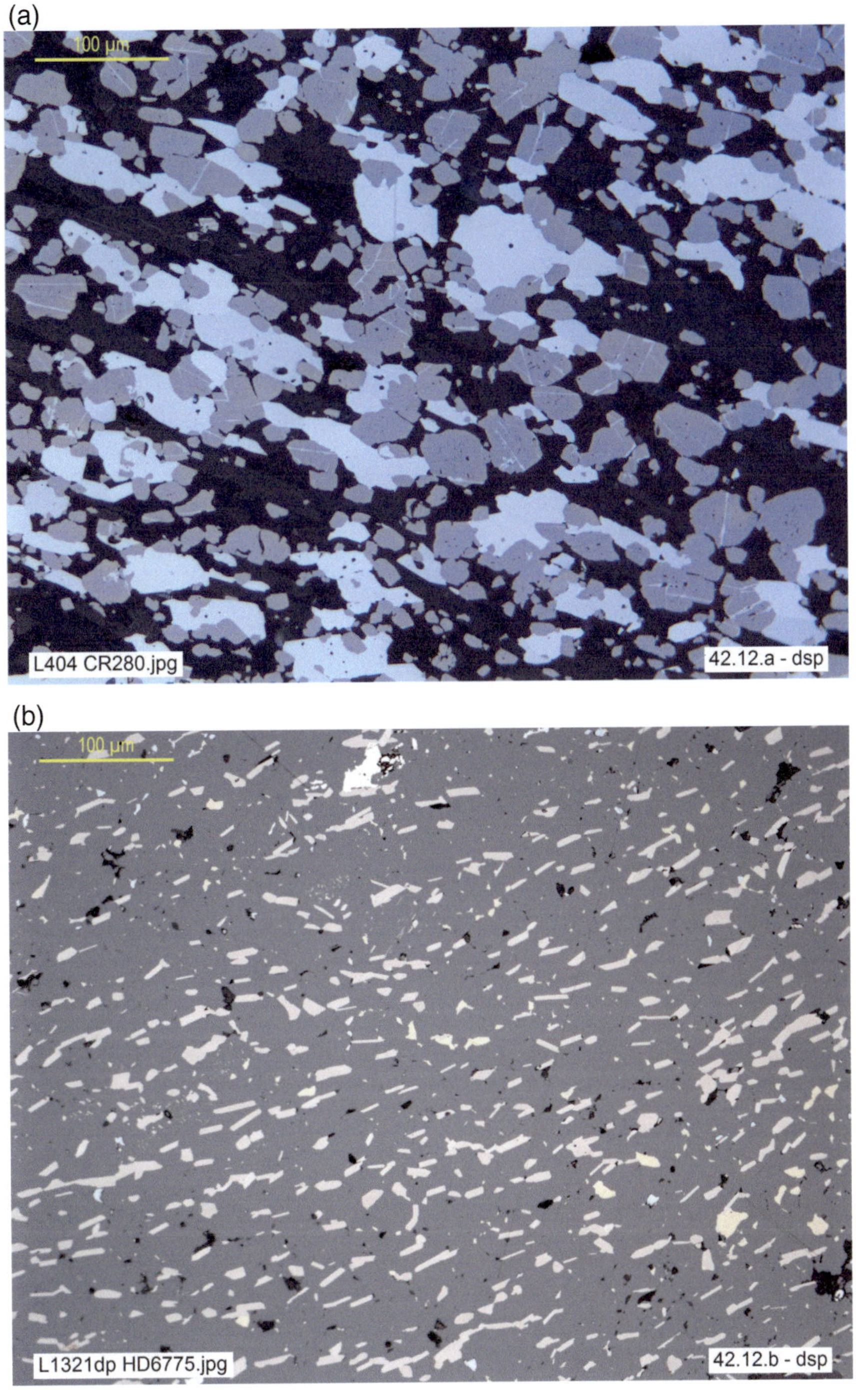

Fig. 1.22 **a** Metamorphic magnetite-hematite ore (Banded Iron Formation), oriented by parallel arrangement of elongate minerals (hematite and phyllosilicate-amphibole gangue). CR280, *BIF* (Banded Iron Formation), Harare, Zimbabwe (**dsp**). **b** Orientation of pyrrhotite in sphalerite band (*kiesiges Braunerz*). HD6775, Rammelsberg, Germany (**dsp**). **c** Pyrite orientation by pressure-solution and solution-transfer, resulting in elongated aggregates parallel to the schistosity (marked by the orientation of silicate inclusions in pyrite). VA14-377.80, FPI_Masa Valverde, Huelva, Spain (**dsp**). **d** Orientation parallel to banding in SedEx massive sulfides: sphalerite, galena, pyrrhotite, pyrite. HD6775 (*kiesiges Braunerz*), Rammelsberg, Germany (**dsp**)

(c)

(d)

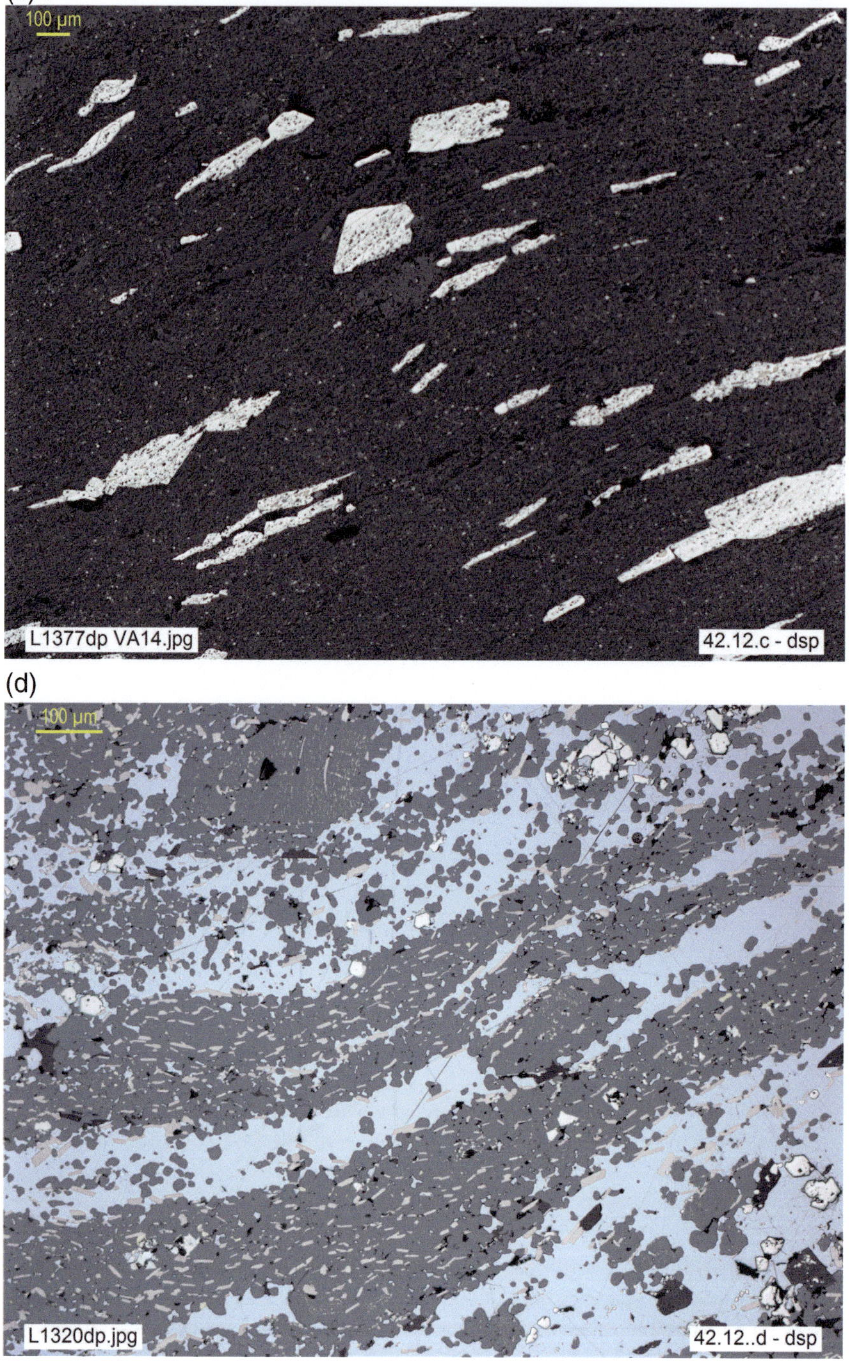

Fig. 1.22 (continued)

Fibrous aggregates (Fig. 1.23a, b) are characterized by the juxtaposition of minerals of fibrous habit, which usually form bundles or bands of parallel fibres or fill veinlets. Elongated crystal growth may radiate from a point, giving rise to **radial** aggregates; these are often constituted by fine fibres growing from a centre, like the spokes of a sphere: **fibrous-radial** (Fig. 1.24a–f) and **spherulitic** (Fig. 1.25a–c) textures. Spherulites are typically limited to the outside by a net spherical surface, while in radial and fibrous-radial aggregates the external shape is defined by the terminations of the different fibres, so it is usually more irregular. This is relevant to their interpretation, since only spherical shapes can be explained as the effect of surface tension, typical of primary colloidal deposition (*colloform or colomorphic textures*), although they may later show a radial structure as a secondary feature (later modification, due to recrystallisation); otherwise, radial and fibro-radiated structures may be primary. The spherical forms lacking any internal structure are designated as **spheroidal** (Fig. 1.26a–c). These types are transitional to the botryoidal ones already described.

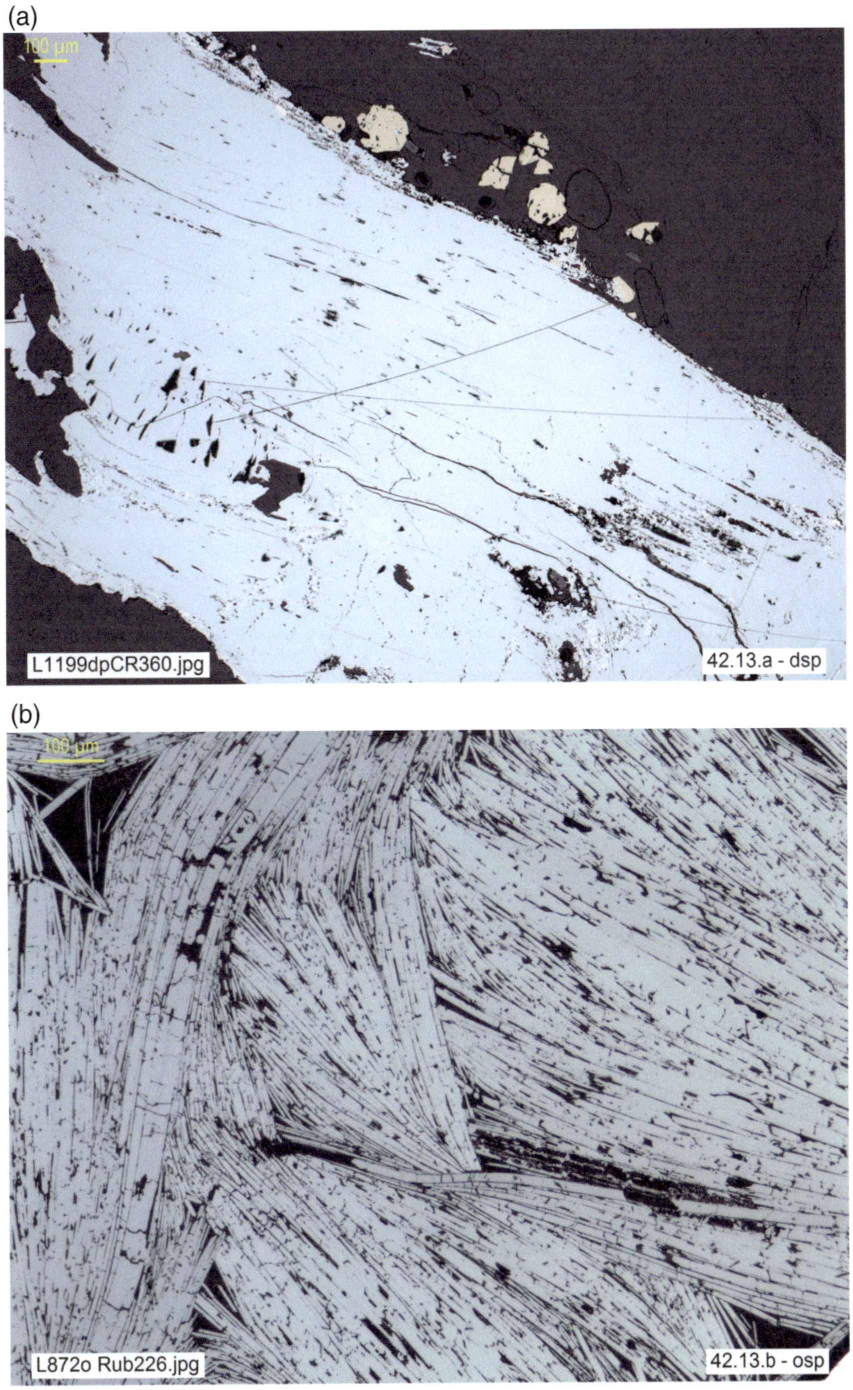

Fig. 1.23 **a** Fibrous aggregate of cosalite, with inclusions of galena (*triangular pits*, bottom left) and sparse pyrite, chalcopyrite and Bi. CR360, Cariboo Au Mine, Wells, Brit. Columbia, Canada (**dsp**). **b** Fibrous aggregates of hematite, partially folded. Rub226, St. Martin am Silberberg, bei Huettenberg, Kaernten, Austria (**dsp**)

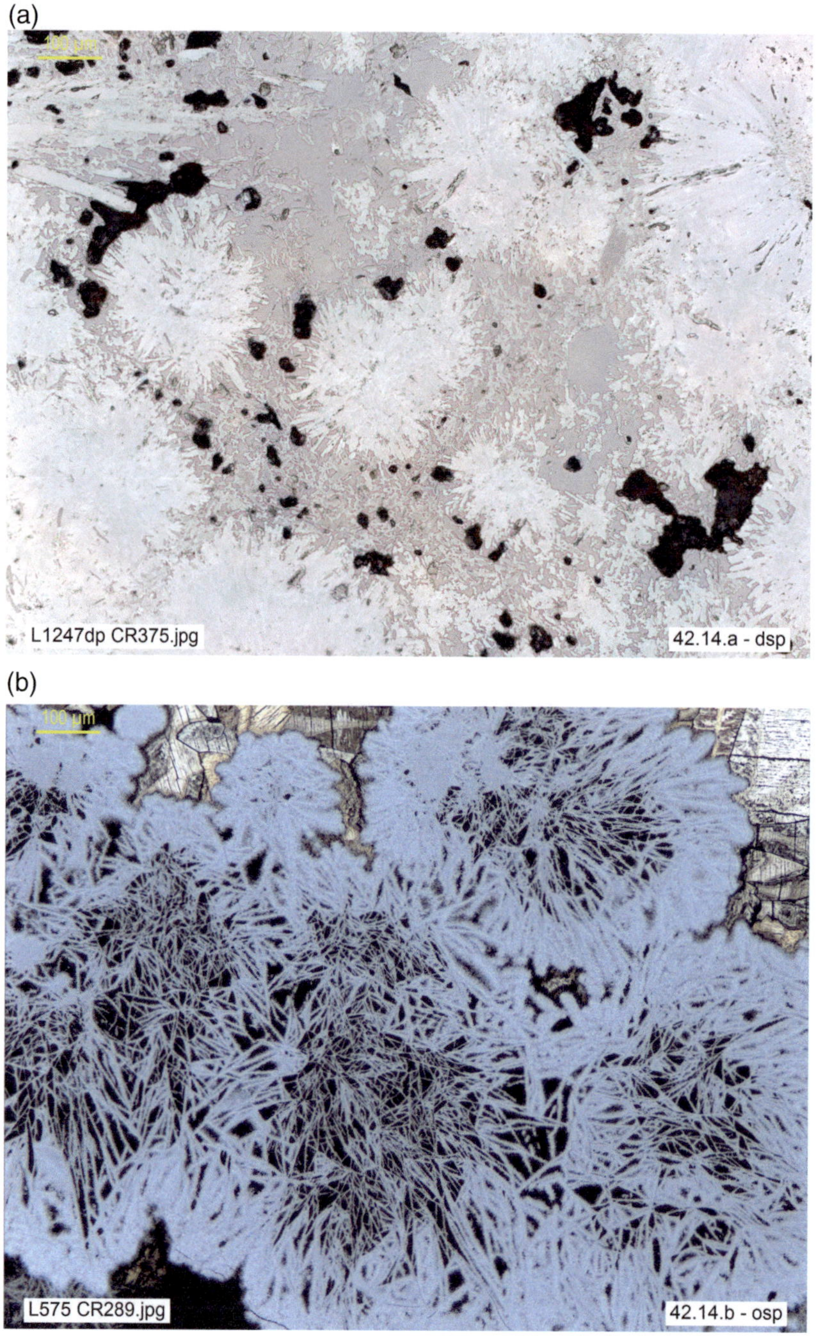

Fig. 1.24 **a** Fibrous-radial aggregates of orpiment (yellowish-white, bireflective), intergrown with erythrin (light pink). CR 375, Cobalt, Ontario, Canada (**dsp**). **b** Transition from radiating to spheroidal psilomelane aggregates, with residual pyrolusite (top). CR 289, Amalienhohe Mine, Waldalgesheim, Bingerbrück, Germany (**osp**). **c**, **d** (**osp** and **o + p**, resp.). Radial aggregates of safflorite (white, anisotropic), with skutterudite (white, isotropic) and bismuth ochre (gray) in quartz gangue. TF Pe31, Schneeberg, Saxony, Germany. **e**, **f** (**osp** and **o + p**, resp.). Radial aggregates of löllingite (white, *palmated texture*), on relict band of Bi (bright white, very scratched). Rub1070, Hüttenberger Erzrevier, Lölling, Schacht Hauptlager, Kärnten, Austria

Fig. 1.24 (continued)

(e)

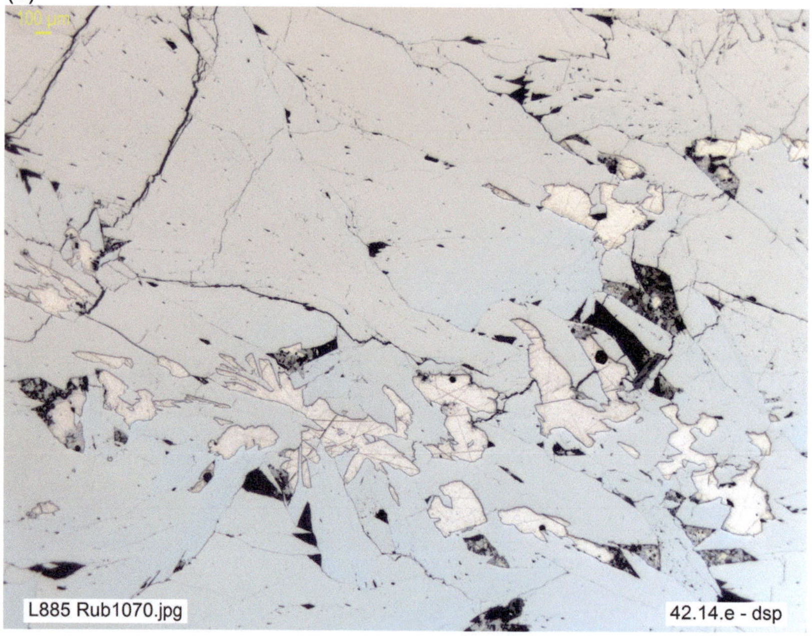

(f)

Fig. 1.24 (continued)

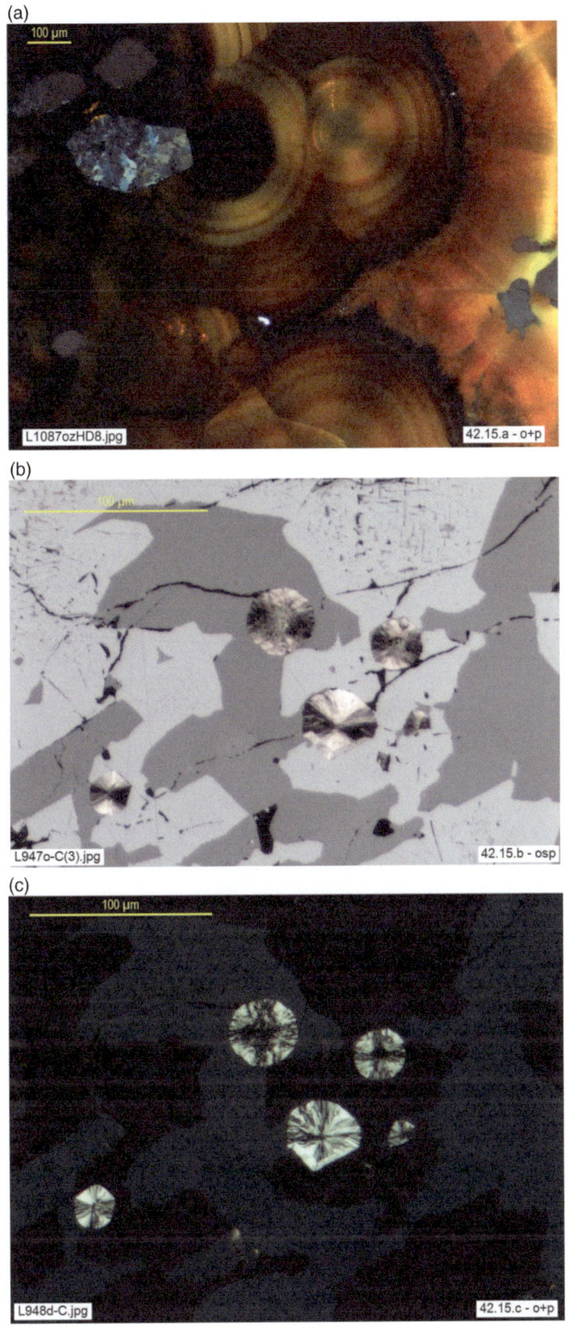

Fig. 1.25 **a** Coalescent spherulites of sphalerite (*schalenblende*), with inclusions of gratonite (in extinction) partly replaced by jordanite (colored: anisotropic). HD 8, Wiesloch (Baden-Württemberg, Germany (**osp**). **b**, **c** (**osp** and **o + p**, resp.). Graphite spherulites (*retortengraphit*) in titanomagnetite (isotropic) and ilmenite (purplish gray, anisotropic) ore. RubFsk sample, Foskor, Palabora, South Africa

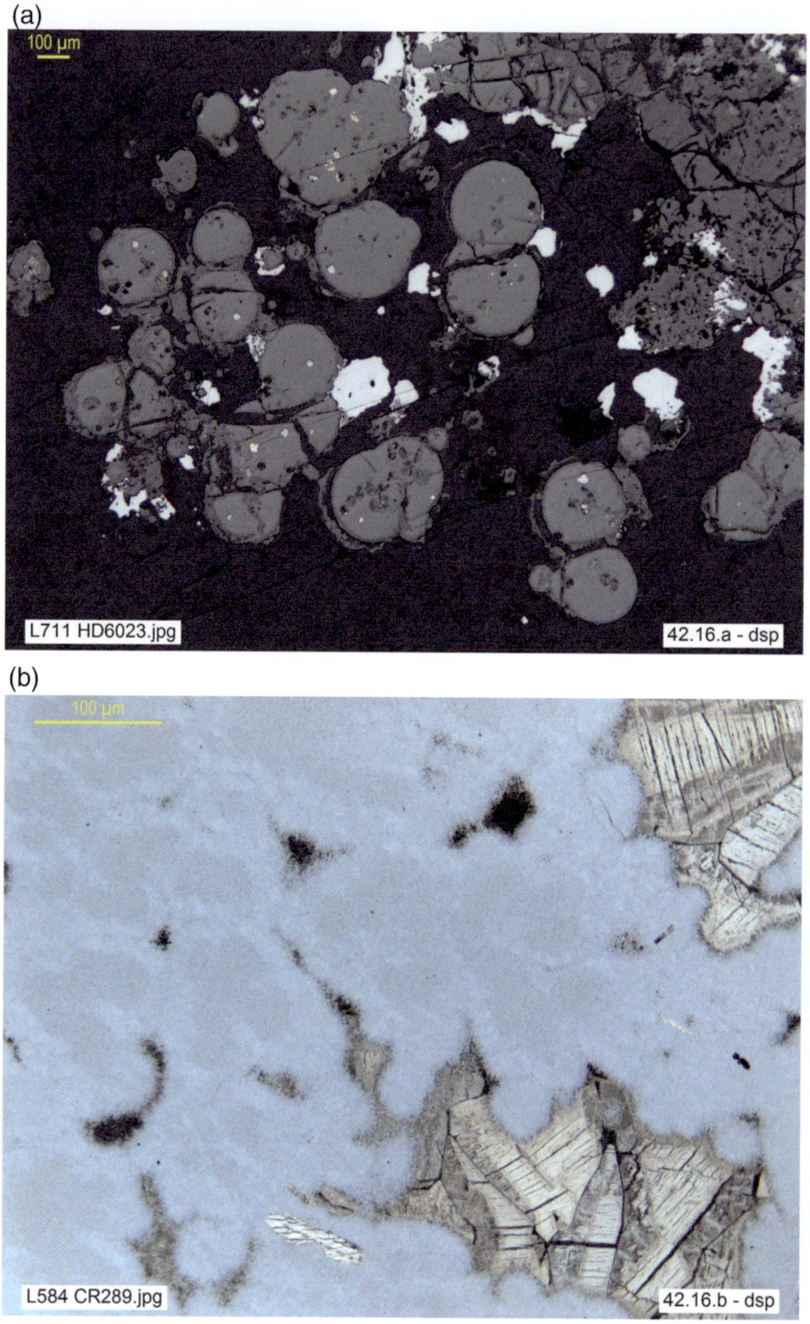

Fig. 1.26 **a** Massive, spheroidal uraninite ore with galena (white). HD6023, Wölsendorf, Bavaria, Germany (**dsp**). **b** Spheroidal psilomelane, secondary to pyrolusite (interstitial relicts). Detail of Fig. 1.27c. CR289, Amalienhohe Mine, Waldalgesheim, Bingerbrück, Germany (**dsp**). **c** Replacement of pyrolusite by psilomelane, which is evolving from porous, fibrous, radial textures (periphery) to more compact, massive, spheroidal textures (core). Compare to Fig. 1.26b, which shows in detail the massive core zone. CR289, Amalienhohe, Germany (**dsp**)

(c)

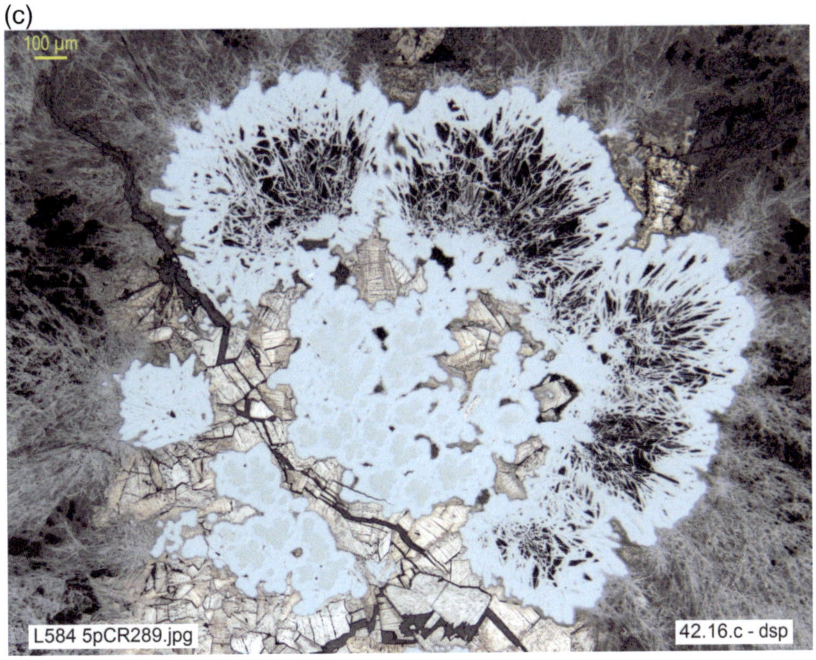

Fig. 1.26 (continued)

The type of contact between grains is a key textural property, both for genetic interpretation and for industrial processes. These applications are discussed below (*cf.* § 1.3); here we dwell only on their morphological description. Although rectilinear or smooth contacts (designated as **simple contacts**) do exist between the components of an ore (Fig. 1.27a–b), intricate or interpenetrated contacts (designated as **complex contacts**) are very frequent and are given names according to their morphology. **Jagged** contacts are interpenetrated, *saw-toothed* or lobed (Fig. 1.28a); a particular case is the contacts in **caries** or indented on a small scale, starting from grain boundaries (Fig. 1.28b, c). The **concentric** types (Fig. 1.29a–c) are defined by superimposed concentric microlayers, often circular, oval or elongated. If it is a single layer overlying a pre-existing core, it is called a **corona**, while several rhythmically repeated layers characterize a **rhythmic** texture, and in the case of **oolites** they usually contain a detrital core.

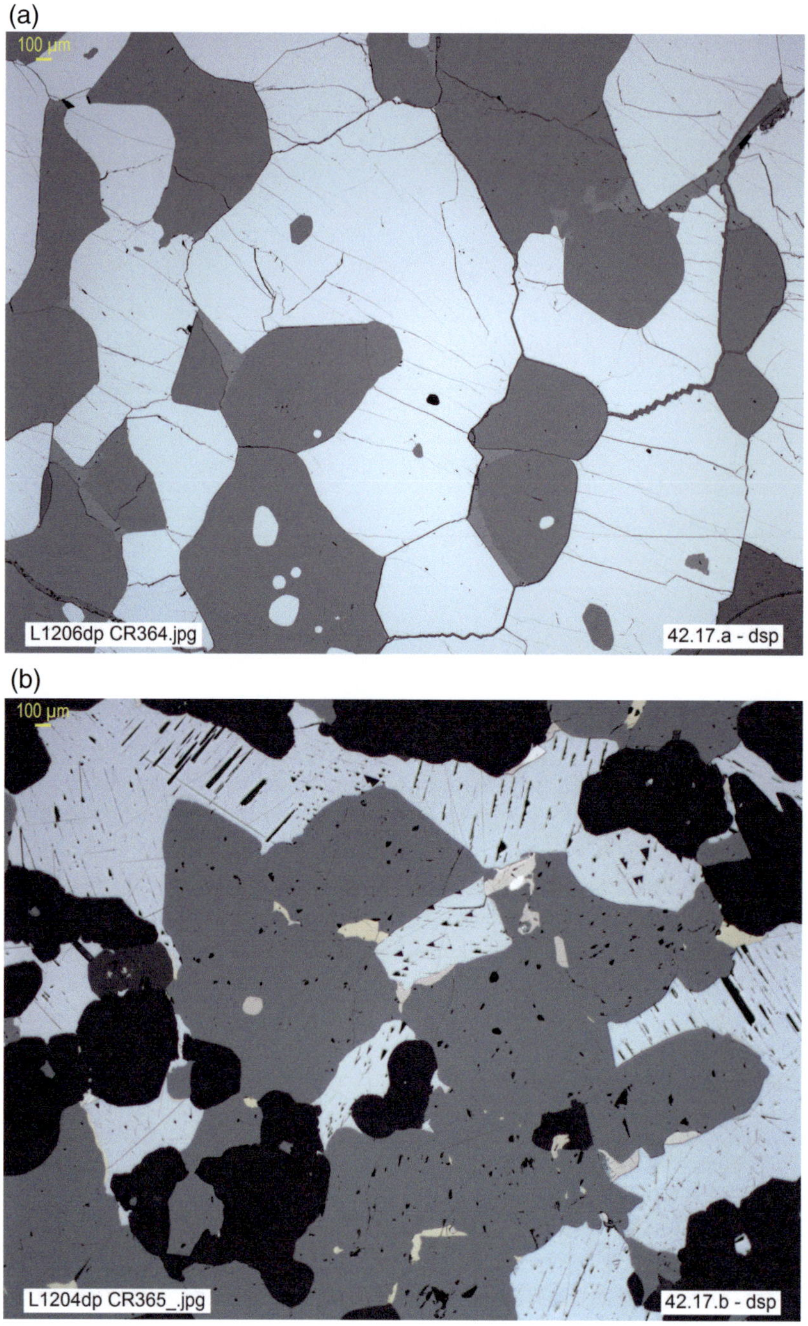

Fig. 1.27 **a** Simple contacts between granoblastic franklinite (light gray) and gangue (gray). CR 364, Franklin Furnace, New Jersey, USA (**dsp**). **b** Smooth contacts, suggestive of mineral equilibrium, in metamorphosed ore: sphalerite (gray), galena (light gray), pyrrhotite (pinkish brown), chalcopyrite (yellow), arsenopyrite (white) and gangue (dark gray). CR365, Zn Corp Mine, Broken Hill; NSW, Australia (**dsp**)

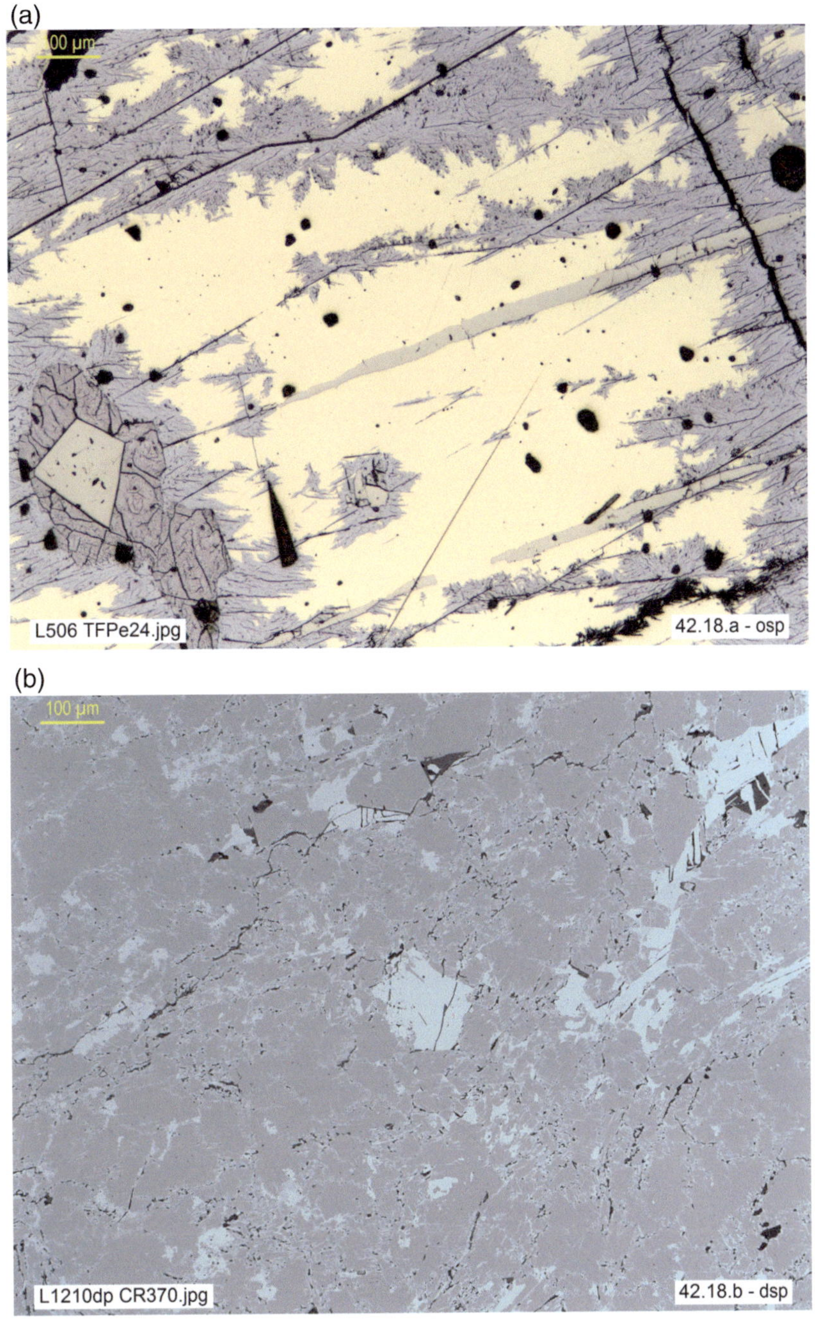

Fig. 1.28 **a** Jagged contacts between millerite, *ml* (yellow), and secondary polydymite (light pinkish gray), replacing ml. Relicts of pyrite (pale yellowish white, compared to millerite) and violarite (dark pinkish gray) in *corona* on pyrite. TFPe24, Dry Nickel Mine, Bindura, Zimbabwe (**osp**). **b** Magnetite with incipient martitization (alteration to hematite), which progresses following microcracks, reticular planes and grain boundaries in mt; these acquire an irregular contour (beginning of *caries texture*). Minor maghemite and goethite. CR 370, Mina Recanzoni, Alidos, Mexico (**dsp**). **c** Development of *caries* texture on magnetite idioblast: alteration to hematite starting from the grain boundaries on, progressing in favor of microcracks and reticular planes, resulting in an irregular, indented outline (compare to Fig. 1.28b, which shows an earlier stage). In the outer, microcrystalline mass, the alteration is almost total. CR371, BIF Minas Gerais, Brazil (**dsp**)

(c)

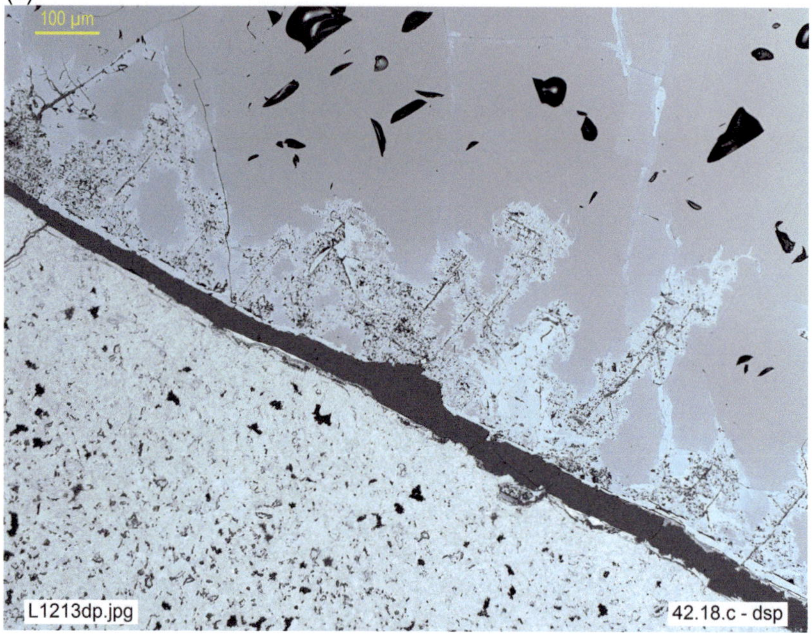

Fig. 1.28 (continued)

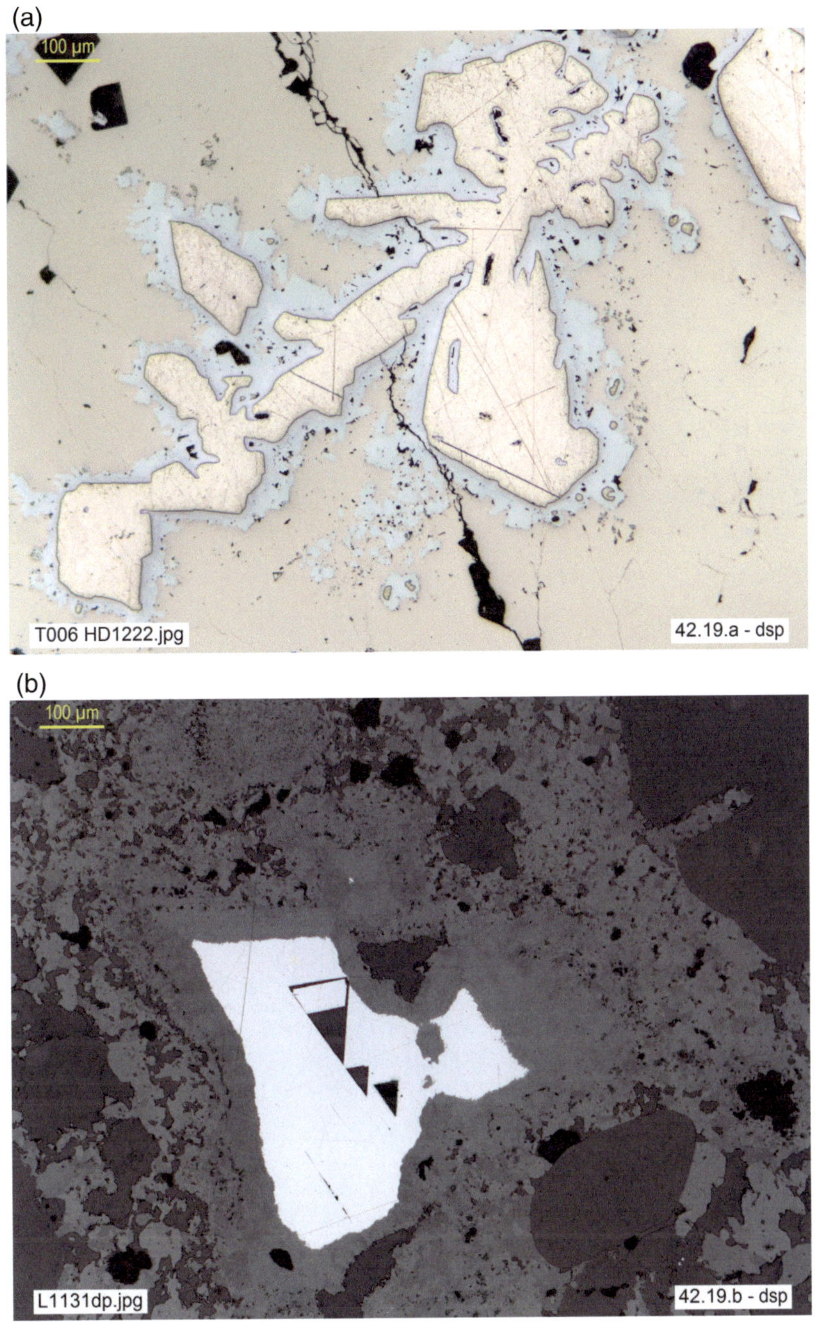

Fig. 1.29 **a** *Corona* encrustation of white arsenides (rammelsbergite + safflorite+ löllingite, finely intergrown) on native bismuth, all included in massive maucherite. HD 1222, Eisleben, Germany (**dsp**). **b** Alteration of galena (white) to concentric layers of anglesite and cerussite (both gray). CR347, Webb's Consols Mine, New England, Australia (**dsp**). **c** Oolitic iron ore: magnetite, maghemite, goethite and gangue. CR325, Gora Djebilet, Tindouf, Algeria (**osp**)

(c)

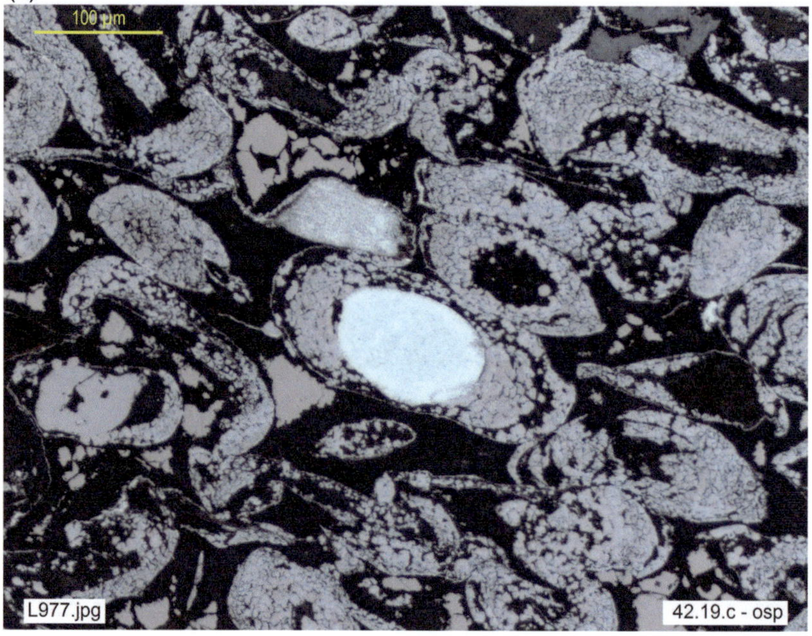

Fig. 1.29 (continued)

The **emulsion** texture (Fig. 1.30a–c) comprises a multitude of tiny inclusions of a *guest* component dispersed in a *host* component, like the two phases of an emulsion (the name caries, which has also been applied to it, is not recommended, as it is more appropriate when the contacts of the two phases are irregular and marginal, as just explained and shown in Fig. 1.28b, c). When the geometry of the intergrown ores acquires particular characteristics due to their shape, orientation or spatial arrangement, the emulsion textures can be transitional to **other** types which, because of their potential significance, receive specific names (*myrmektitic*, *symplectitic*, *reticulate or mesh, lattice*, *boxwork*, etc., *cf.* Fig. 1.31).

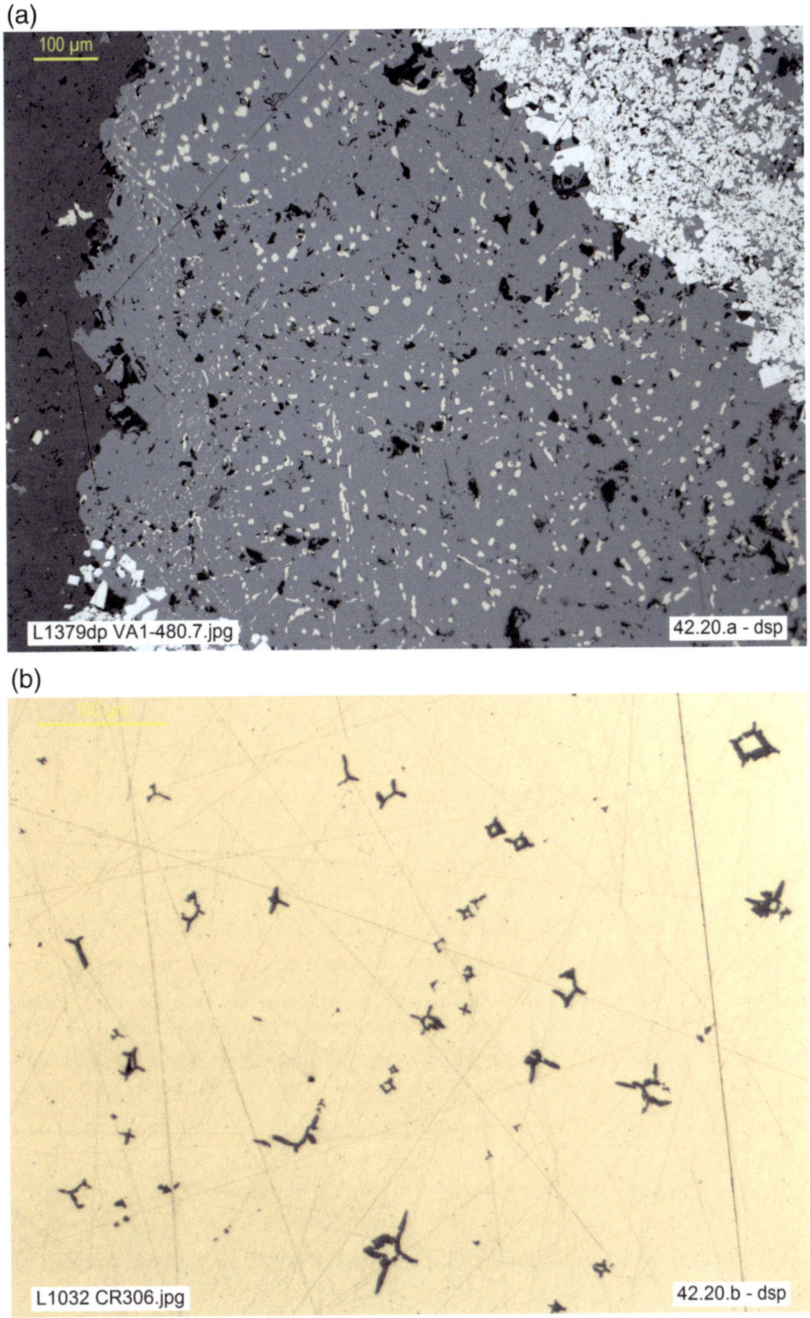

Fig. 1.30 **a** Chalcopyrite emulsion in sphalerite. VA1-480.70, FPI_Masa Valverde, Spain, DDH3 (**dsp**). **b** Emulsion of sphalerite *stars*, marking crystalline directions in chalcopyrite. CR 306, As Sombras Mine, Lobios, Orense, Spain (**dsp**). **c** Emulsion of galena (white) in pyrargyrite (bluish gray). CR 330, Sta. Cecilia Mine, Hiendelaencina, Guadalajara, Spain (**dsp**)

(c)

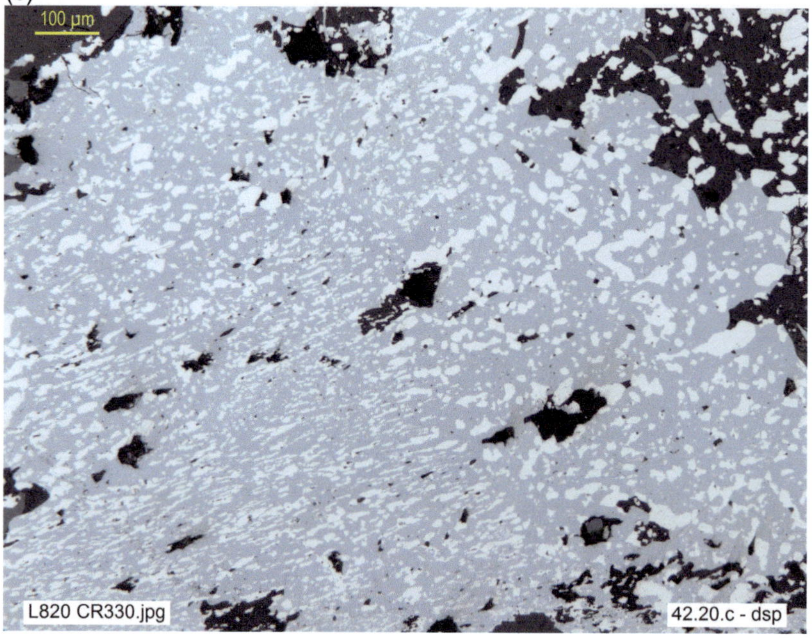

Fig. 1.30 (continued)

The **myrmekitic** texture is characterized by the intimate intergrowth of two mineral phases forming small elongated corpuscles with gently curved edges (like worms, hence the name); this texture was originally defined for silicates (quartz and feldspars) in igneous rocks, so it has been suggested to name it **symplectitic**, in the case of ores (Fig. 1.31a–c); the name *eutectic* is sometimes applied to it, but this is not appropriate as a descriptive term, because of its genetic connotation.

Schwartz (1951) groups with myrmektitic the *graphic, micrographic or subgraphic* textures, considered as variants and characterized by intergrowths of angular, wedge-shaped contours (like the characters of *cuneiform* writing, a name also applied to the graphic texture, Fig. 1.31d). However, from the formal point of view, taking into account their typical rectilinear contacts, these textures could be included in the group of simple contacts.

Other important variants occur when the geometry of the intergrowth is conditioned by the crystalline lattice of one of the phases, which occurs frequently in magnetite, as shown in Fig. 1.31f, and in many other ores (sphalerite, chalcopyrite, etc.), as well as in their supergene oxidation products (*boxwork*, *pigeonhole*, *honeycomb*, etc., textures of limonite: Fig. 1.31g). They are referred to, in general, as **reticulate** textures or intergrowths. This is a very wide group that includes well-known types such as *Widmanstätten* (hematite lamellae included in magnetite, forming a lattice that mimics the octahedral //111 directions of the latter: Fig. 1.31e), and many others, for instance the *cloth-like* texture (Ramdohr 1980) characteristic of ulvöspinel (or ulvite) in magnetite, defined by a lattice of tiny, oriented //100 magnetite lamellae resembling the weave of a fabric (*cf.* Fig. ulv 5).

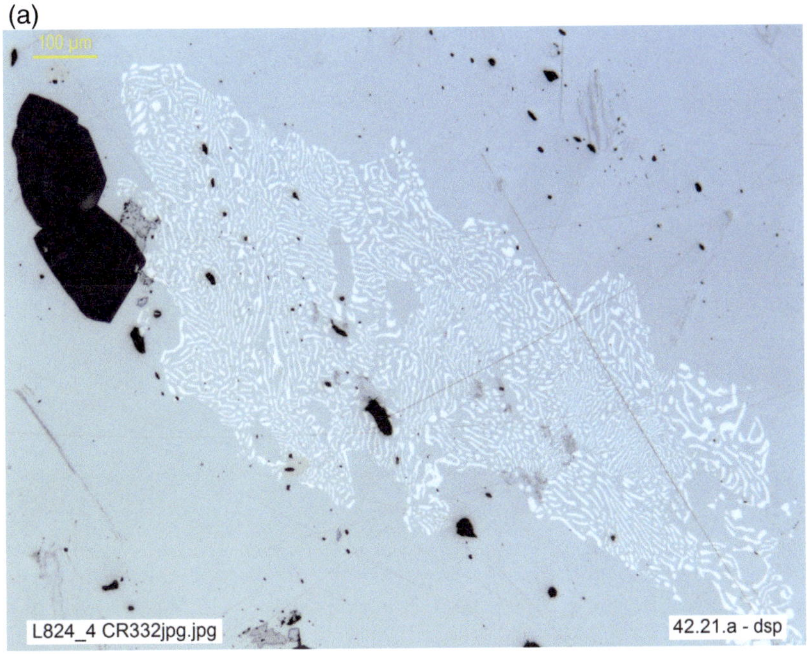

Fig. 1.31 **a** Symplectitic intergrowth of galena in chalcocite. CR 332, Vrancice, Pribram, Czech Republic (**dsp**). **b** Symplectitic (or myrmekitic) intergrowth of proustite and polybasite. CR 327, Freiberg, Saxony, Germany (**o + p**). **c** Symplectitic intergrowth of native silver (white) in allargentum (light gray). CR 362, Nipissing Mine, Cobalt, Ont, Canada (**dsp**). **d**, **e** (**dsp** and **d + p**, resp.). Micrographic texture of disseminated titanomagnetite in basalt. Polarization (*d + p* image) evidences the octahedral magnetite lattice, distinctly shown by the (111) arrangement of thin ilmenite and martite (ht) lamellae, included in mt (*Widmanstätten texture*), not apparent in the (*dsp*) setting (Fig. 1.31d). BL-070513, basaltic pegmatoid, Lanzarote, Canary Islands. **f** Reticulate texture in magnetite, defined by tiny spinel cubes and lamellae (dark gray) aligned with octahedral directions of mt (//111, sometimes associated with unmixed il); acicular segregations of il (pinkish gray), horizontal and much longer, but more scarce. CR253, L. Berta, Quebec, Canada (**o + p**). **g** *Boxwork* texture in indigenous limonite, pseudomorphic of pyrite microclast. 260889.3, Rodalquilar, Almeria, Spain (**dsp**)

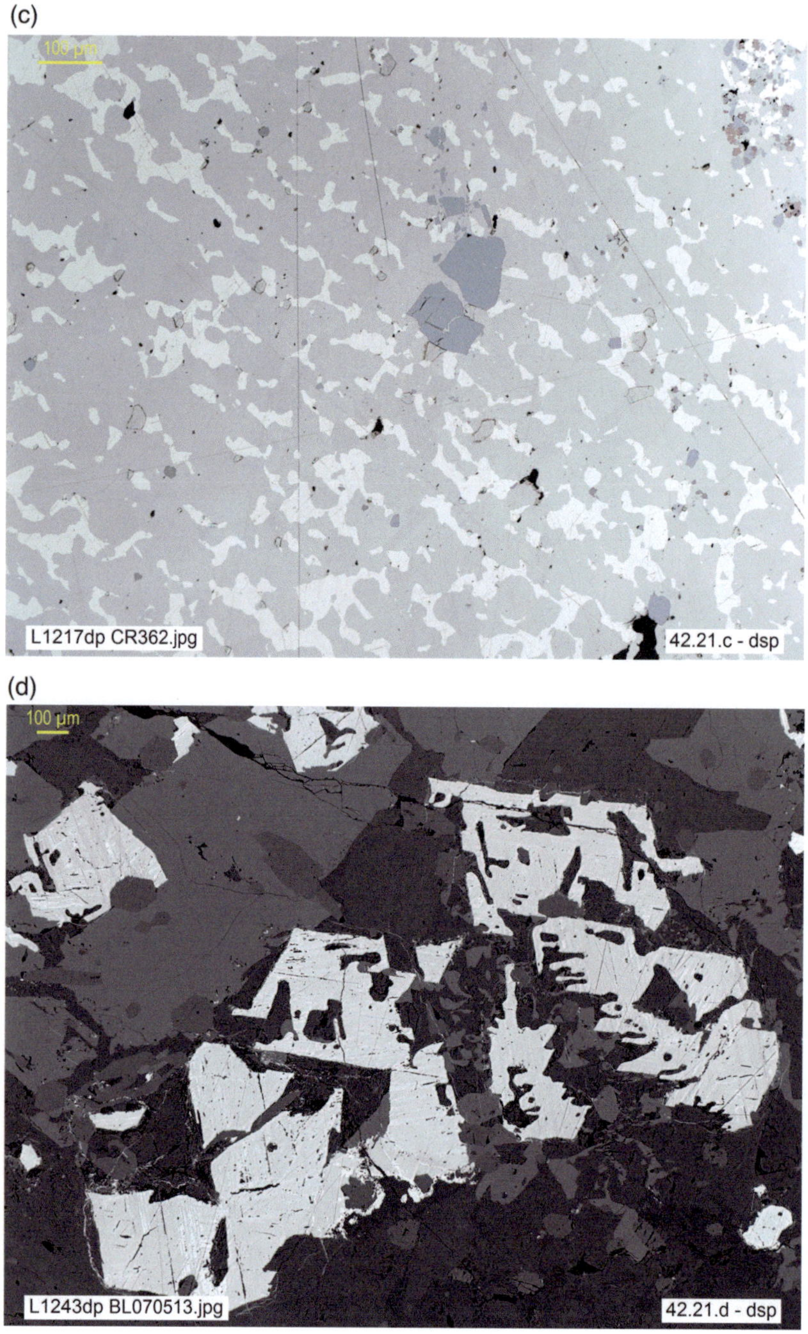

Fig. 1.31 (continued)

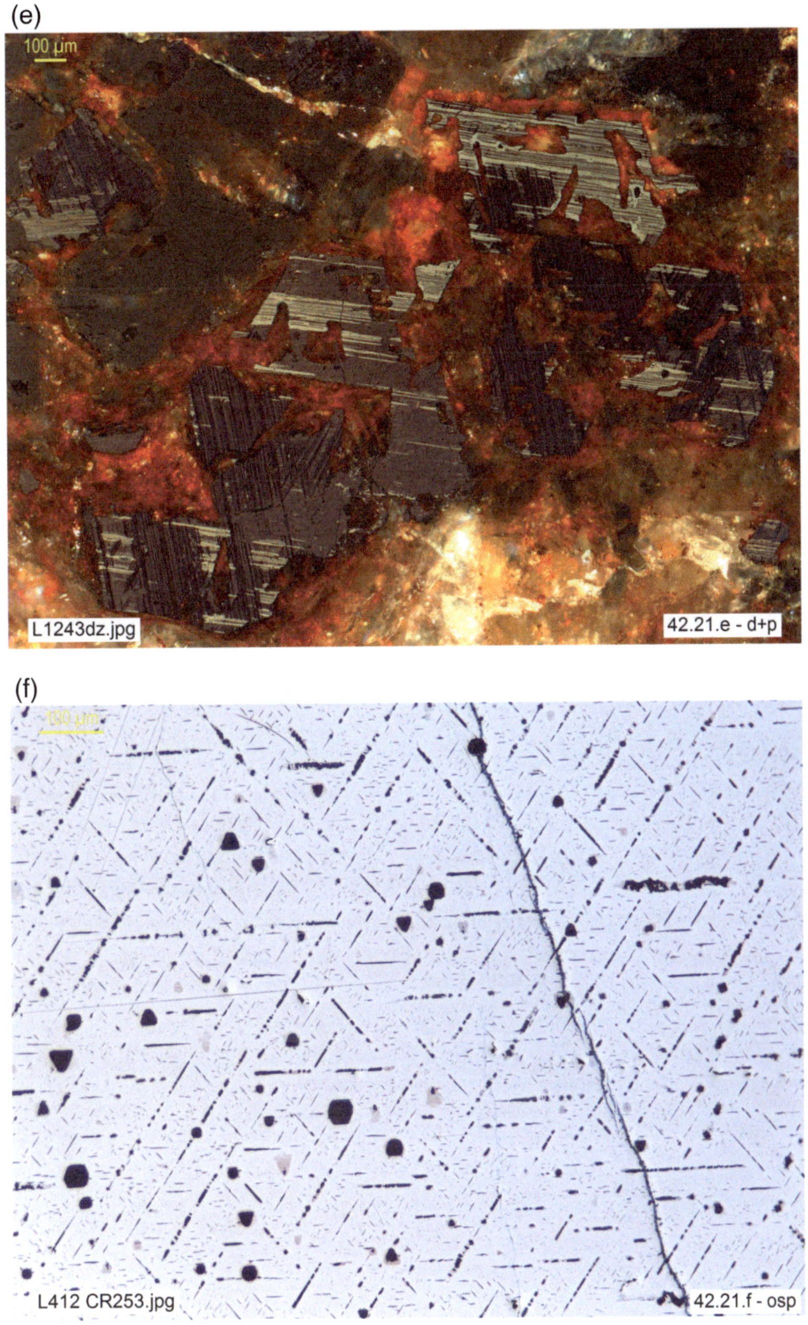

Fig. 1.31 (continued)

(g)

Fig. 1.31 (continued)

Finally, contacts whose complex morphology cannot be reduced to any of the previous classes are grouped as **irregular** contacts (Fig. 1.32a–c) and may require a specific description.

The distribution of the ore components in space is also of interest for both genetic interpretation and industrial characterization. The possible arrangements are highly varied but, leaving aside the **massive** morphologies (those with no differential features), most fall into one of the six basic types of spatial arrangement in Table 1.2. ***Sandwich*** morphology (Fig. 1.33a, b) is defined, as the name suggests, as a sandwich of a tabular mass or sheet of one component between two of another component; it is typical of, but not exclusive to, veins or veinlets at various scales.

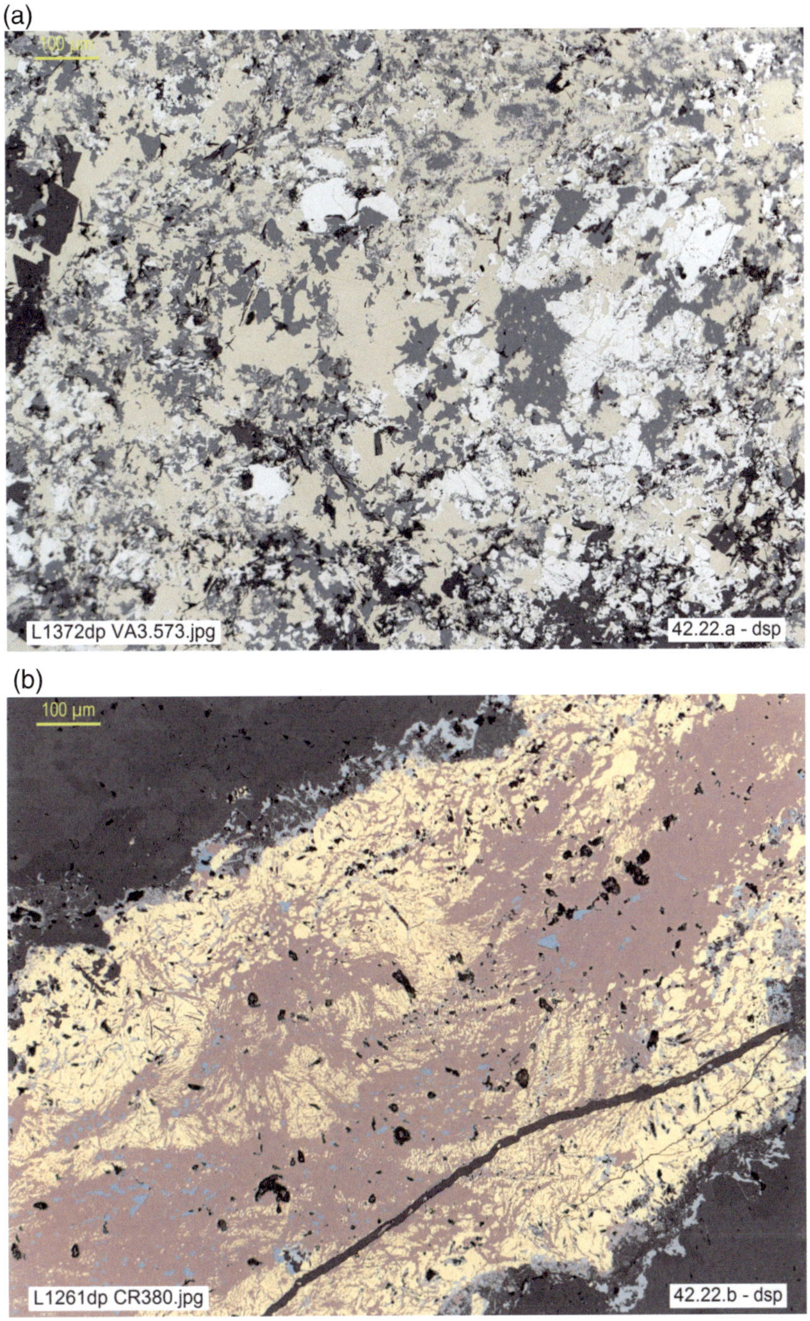

Fig. 1.32 **a** Intergrowths with irregular contacts in massive sulfides (*complex sulfide* ore, with chalcopyrite, pyrite, sphalerite). VA3-573.10, Masa Valverde, drill hole 3, Huelva, Spain (**dsp**). **b** Irregular contacts of chalcopyrite (yellow) with bornite (pink). Minor tetrahedrite, digenite and covellite. CR-380, Tsumeb Mine, Namibia (**dsp**). **c** Stanniferous massive sulfide ore: irregular contacts of intergrown chalcopyrite, pyrite, sphalerite and cassiterite. 190496.1, Neves Corvo Mine, Algarve, Portugal (**dsp**)

(c)

Fig. 1.32 (continued)

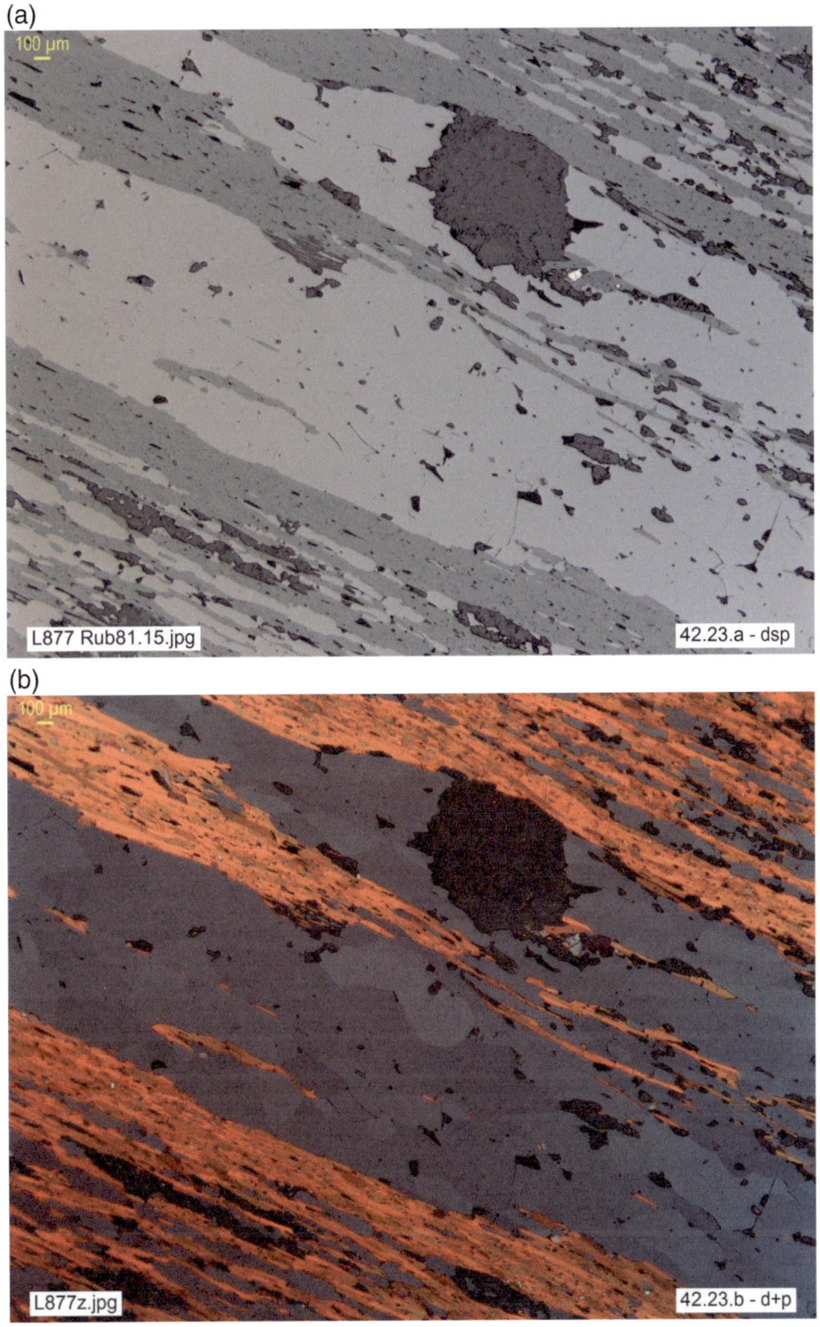

Fig. 1.33 a, b (**dsp** and **d + p**, resp.). Intercalation of magnetite (very light gray, isotropic) and vonsenite (gray, strongly anisotropic with reddish polarization tints) microbands, defining a *sandwich* texture. Rub81.15, Burguillos del Cerro, Badajoz, Spain

Stockwork is characterized by a network of veinlets criss-crossing in all directions (Fig. 1.34a–c); usually, the ore cementing the veinlets is also disseminated in the stockwork host rock. It is important to note that, in its original meaning, the term stockwork was applied to the described morphology, yet at an orebody or deposit scale (millions of tons). Evidently, in such cases a microscopic section alone cannot represent the geometry of the whole orebody. As seen in Fig. 1.34c, a sample of a few centimetres across can only show a tiny portion from either a vein infill, disseminated ore or massive replacement. The actual texture is only evident from observations at the appropriate scale. In the text following, *when referring to textural characterization the term stockwork is restricted to the small scale*.

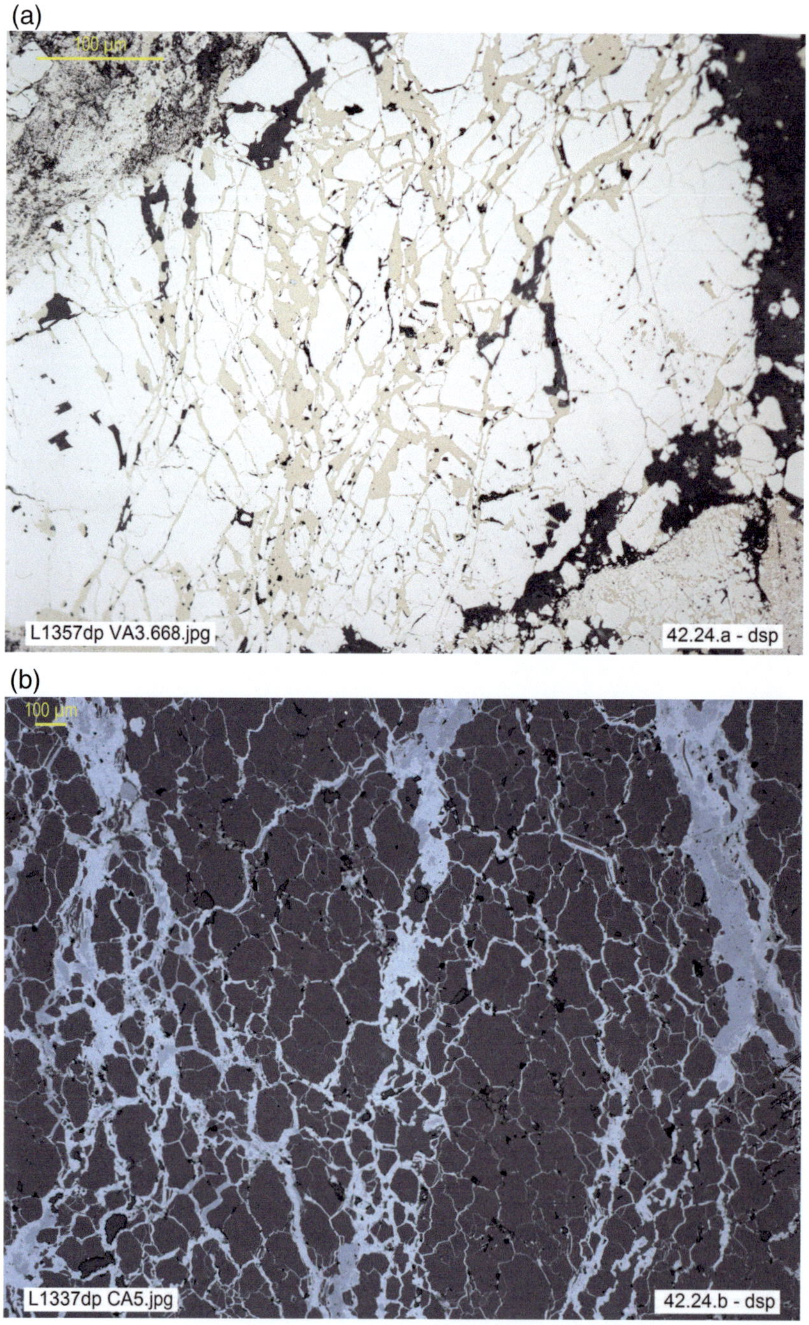

Fig. 1.34 **a** Microstockwork of pyrite (host) and chalcopyrite (microcracks). VA3-668.80, Masa Valverde, drill hole 3, Huelva, Spain (**dsp**). **b** Crackled quartz (microbreccia) with limonite cement, incipient stage of microstockwork development (indigenous limonite replaces original pyrite infill). CA-5, Cantal, Lomo de Bas, Murcia, Spain (**dsp**). **c** *Laminated feeder* (pyrite with pressure solution strain, chalcopyrite and silicates), corresponding to a former massive sulfide stockwork vein. As is typical in real deposits, the dimensions implied in the process exceed the microscopic scale. The structure must be reconstructed by field data. VA3.590, Masa Valverde, Huelva, Spain (**dsp**)

(c)

Fig. 1.34 (continued)

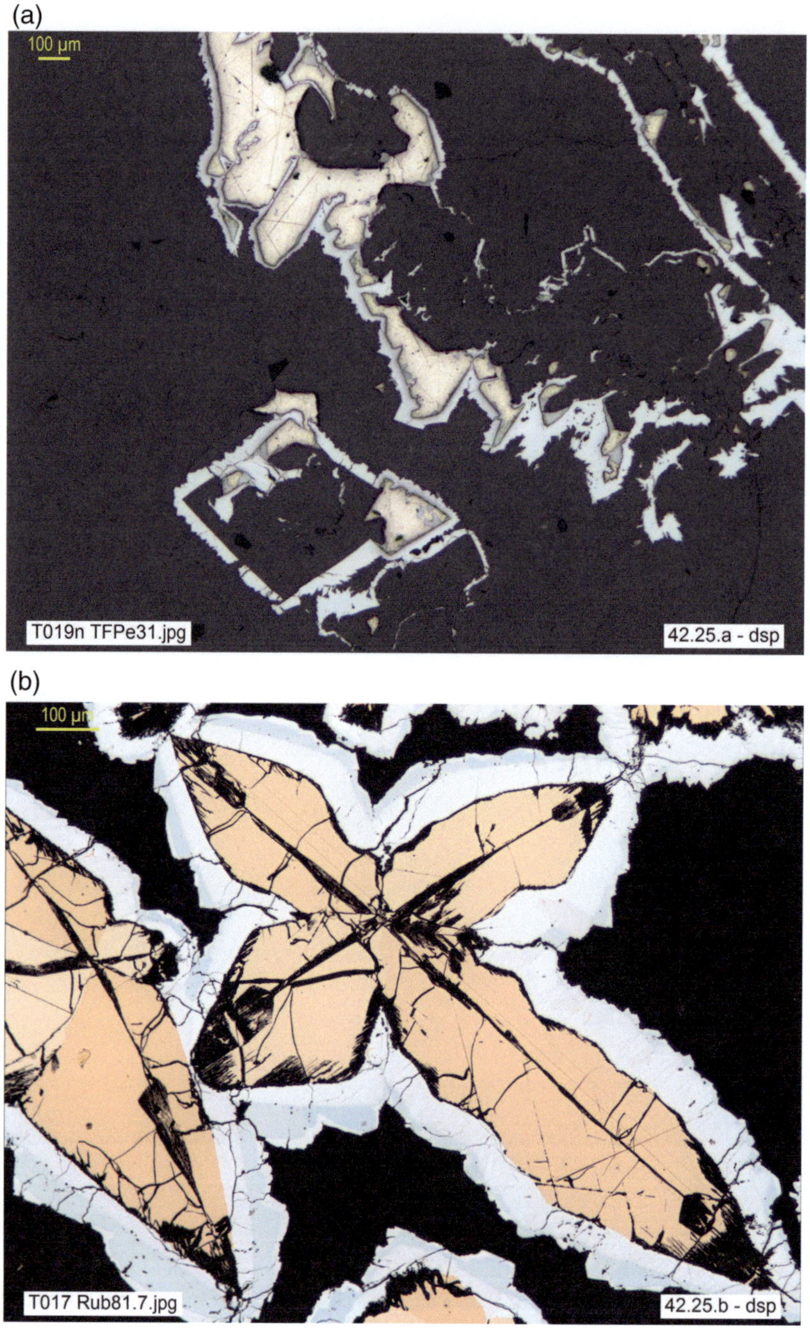

Fig. 1.35 **a** Encrustation of safflorite and skutterudite (white rim) on native Bi (yellowish white, bright), bismuthinite (gray) and quartz (partly pseudomorphic of rhombohedral calcite). TF Pe31, Schneeberg, Saxony, Germany (**dsp**). **b** White arsenides (rammelsbergite, skutterudite and safflorite) encrusting on nickeline skeletal aggregates (pink). Rub81.7, Talmessi, Anarek, Iran (**dsp**). **c** Sphalerite encrustation (schalenblende variety, botryoidal, very visible by its abundant yellowish-brown internal reflections) on dentritic relicts of jordanite pseudomorphic after gratonite. HD 8, Wiesloch, Baden-Württemberg, Germany (**dsp**). **d** Embedding of organic remains: bone texture fossilized by pyrite. Rub.sn. Brilon, Nord-Rhein-Westphalen, Germany (**dsp**)

(c)

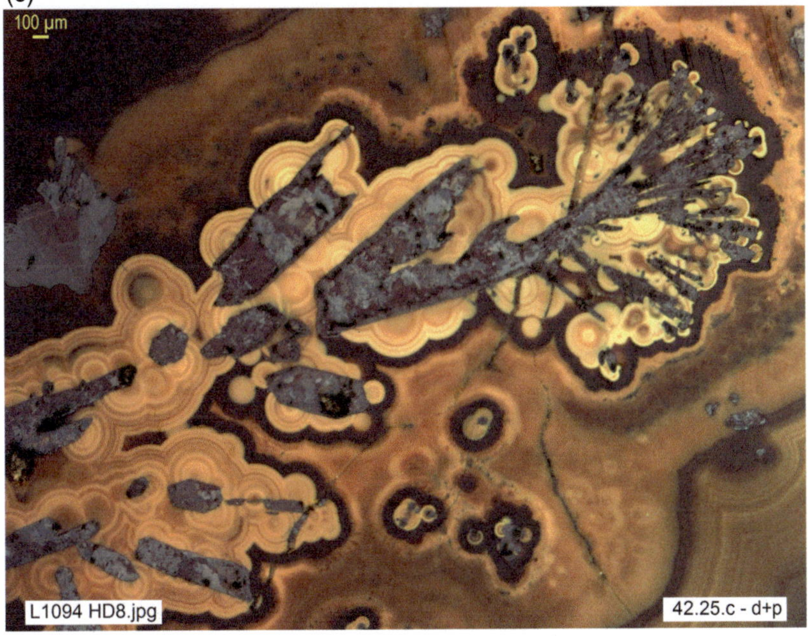

(d)

Fig. 1.35 (continued)

A mineral phase deposited as a thin rim or coating on pre-existing materials or voids (on grains, aggregates, clasts, pores, cavities: Fig. 1.35a–d) is said to form an **encrustation**. The dispersion of mineral particles more or less at random throughout the volume of the rock is called a **dissemination**. When there is evidence that such dispersion is due to fluid percolation and pore filling, it may be referred to as **impregnation**, but this concept anticipates genetic connotations and, for a general case, dissemination is to be preferred (Fig. 1.36a–g). A mineral phase is said to be **interstitial** when it infills the interstices between the other phases present (Fig. 1.37a–c). It is called an **intergranular film** (Fig. 1.38) if it occurs as a thin film along the contacts between grains of another phase, like a very narrow rim marking the grain contours.

(a)

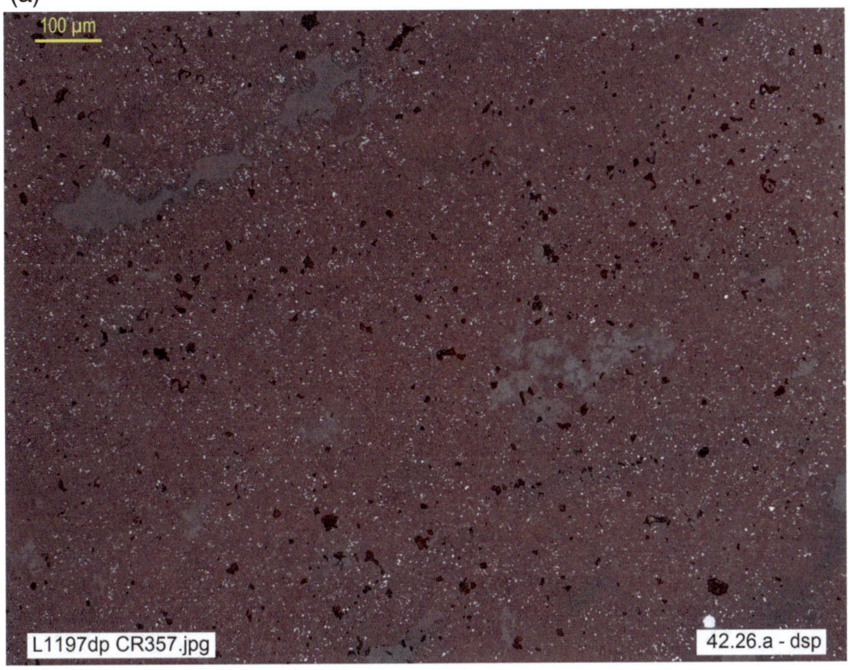

(b)

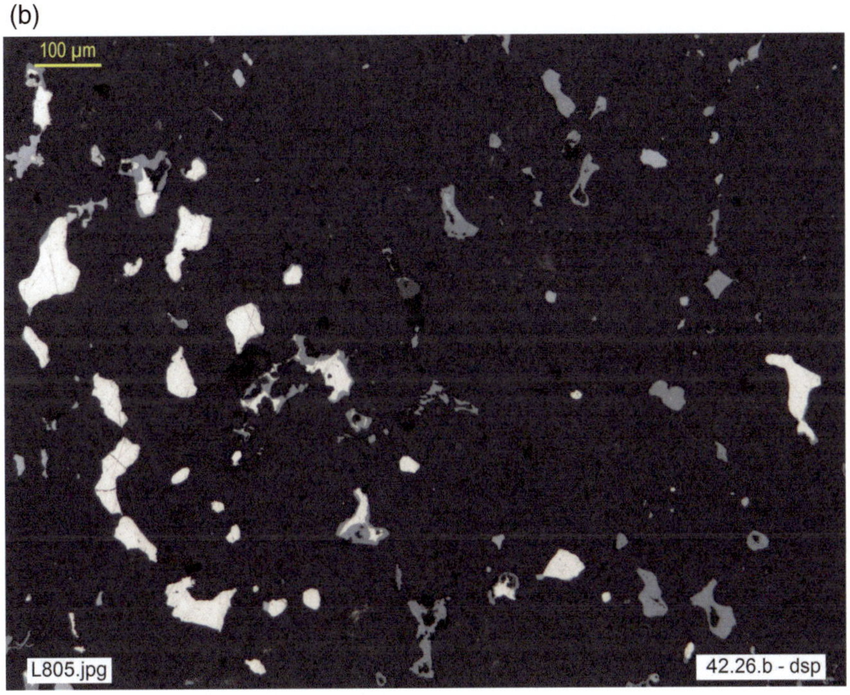

Fig. 1.36 **a** Fine-grained cinnabar impregnation (white dots) on fluorite. CR357, El Realito Mine, Guanajuato, Mexico (**dsp**). **b** Dissemination of native silver (white), acanthite and polybasite in quartz vein. CR329.4, Imiter, Morocco (**dsp**). **c** Vein infill of azurite and malachite (gray, bireflective) with disseminated chalcocite (light gray, with a very faint blueish hue) and covellite (blue). CR 349, Grand Gulch, Coconino County, Arizona, USA (**dsp**). **d** Impregnation of native tellurium and sylvanite in a vuggy chalcedony vein. CR390, Teine Mine, Milsuyama-Sanko Vein, Hokkaido, Japan (**dsp**). **e** Banded copper ore mineralization (disseminated digenite, bornite, covellite) in *Kupferschiefer*. CR 296, Lubin Kupferschiefer, Poland (**osp**). f. Impregnation of allargentum, $Ag_{1-x}Sb_x$, with unmixed (lighter) amalgam (likely eugenite, $Ag_{11}Hg_2$), partially coated by lo (lower centre). CR324, Bouismas Mine, Bou Azzer, Morocco (**dsp**). g. Pyrite dissemination (idioblastic) in slate. VA14-79.75, Masa Valverde, Huelva, Spain (**dsp**)

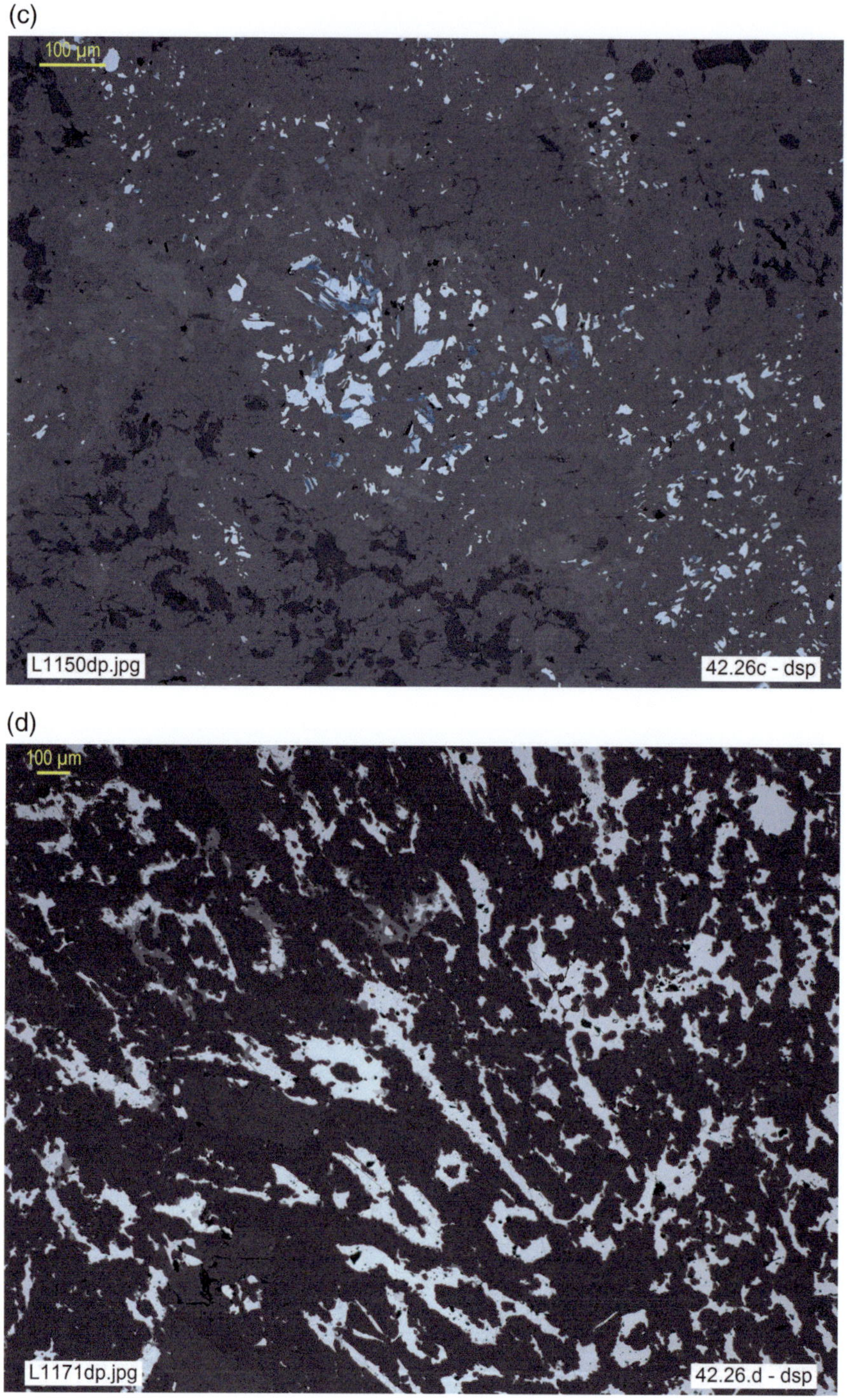

Fig. 1.36 (continued)

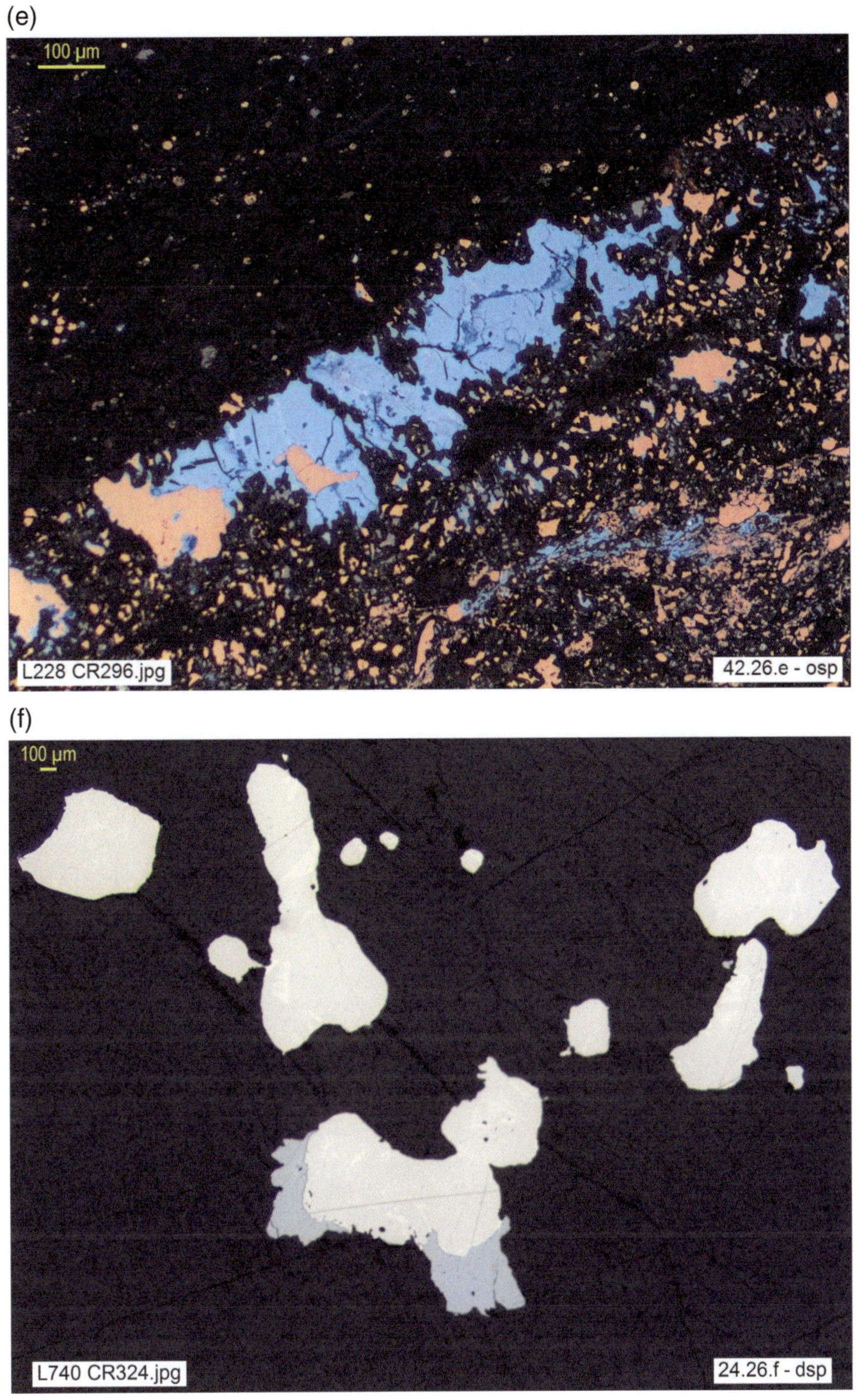

Fig. 1.36 (continued)

(g)

Fig. 1.36 (continued)

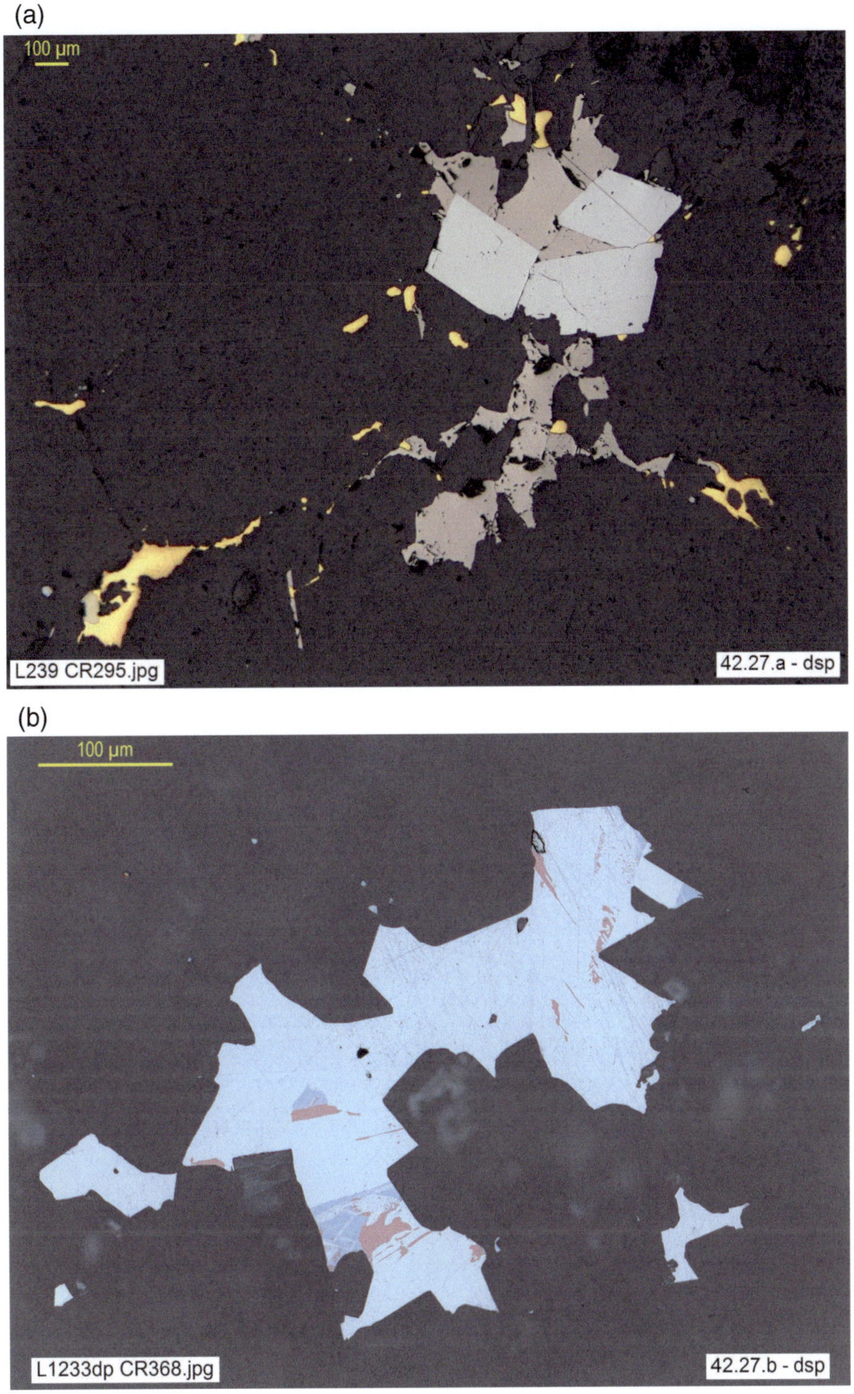

Fig. 1.37 **a** Interstitial native gold (yellow, bright) and pyrrhotite (pinkish brown), in quartz (black)-arsenopyrite (white) vein. CR-295, La Rinconada, Puno, Peru (**dsp**). **b** Interstitial chalcocite, digenite and bornite in quartz vein. CR 368, Crescent Mine, Olympic P., Washington, USA (**dsp**). **c** Cassiterite (gray, idiomorphic) vein filling, with late, interstitial sulfides (pyrite, bornite, etc.) and quartz. Rub607, Geevor Mine, Cornwall, UK (**dsp**)

(c)

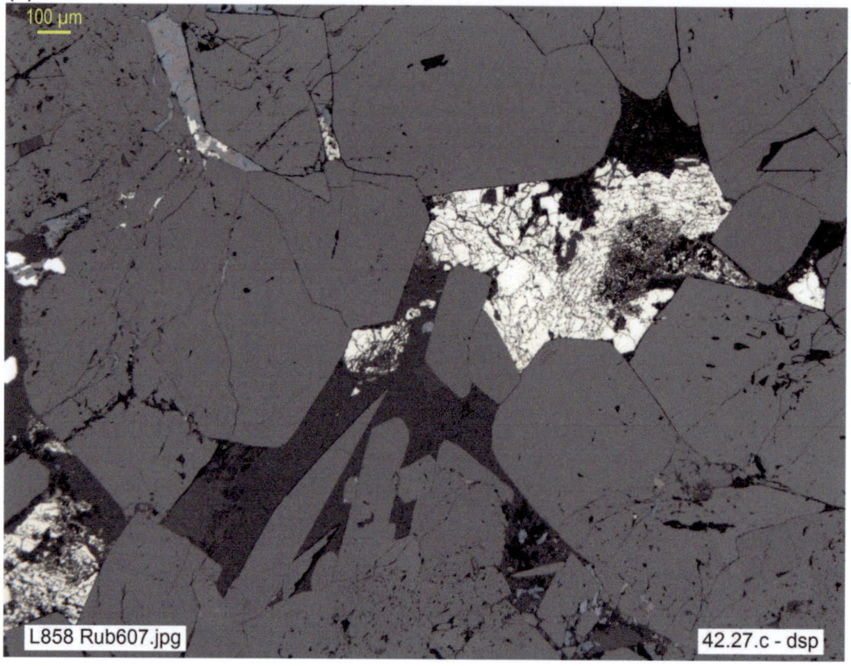

Fig. 1.37 (continued)

Fig. 1.38 Realgar impregnation, interstitial and along intergranular films, in carbonate gangue from an argentiferous lode. CR 379, Cobalt, Ontario, Canada (**dsp**)

1.2.2 Genetic Classification

Genetic classification (Table 1.3) attempts to order the textures systematically in relation to the processes that produce them. It is the first step to interpreting them and understanding the origin of the ore. Unlike in the descriptive classification, interpretation plays an important role here, but rigorous observation and discussion should still allow a tight control of subjectivity so that reasonable conclusions can be reached that are based on criteria consistent with both geological and experimental data. The succinct exposition of these criteria is the subject of this section. It must be approached with humble realism, aware that, although textural interpretation is usually a valuable source of information, it is equally true that many textures are ambiguous and of questionable interpretation, as noted earlier. It is important to bear in mind that microscopic information must be complemented with observations at other scales, particularly in hand specimens, as Taylor (2009) clearly shows. The integration of all these observations leads to the correct interpretation.

Given the ease with which metallic ores register in their textures the changing conditions of their environment throughout geological history, it is pertinent to distinguish first of all two classes of textures: those in which the original character is evident (**primary** textures) and those that have been subsequently transformed by thermal, dynamic or physico-chemical processes (**modified** textures). The first class corresponds to the *primary precipitation textures* of Ramdohr (1980), while the second class comprises the *transformation textures* (replacement, exsolution, decomposition, etc.) together with the *deformation* textures of Schneiderhöhn (1952) and Ramdohr (1980). The reason for this grouping is that it is not always possible to separate thermal, dynamic and physico-chemical re-equilibration processes in nature—think, for example, of metamorphism.

The primary textures are arranged in types, taking into account the fundamental concentration processes in the generation of ore deposits. **Cumulate** textures (Fig. 1.39a–c) correspond to the gravitational accumulation towards the base of the magmatic chamber of the most heavy, early minerals (crystal settling). These are due to fractional crystallisation in ultramafic magmas, and are typical of stratiform chromite deposits (Bushveld, R. South Africa, Fig. 1.39a). Nevertheless, they are also found in podiform deposits (Fig. 1.39c) and in titano-magnetite or ilmenite-hematite deposits related to gabbro-anorthositic complexes. They may contain sulfides, usually also segregated from the melt by liquid-phase immiscibility, in the interstices of the cumulates (**intercumulus**, Fig. 1.39b).

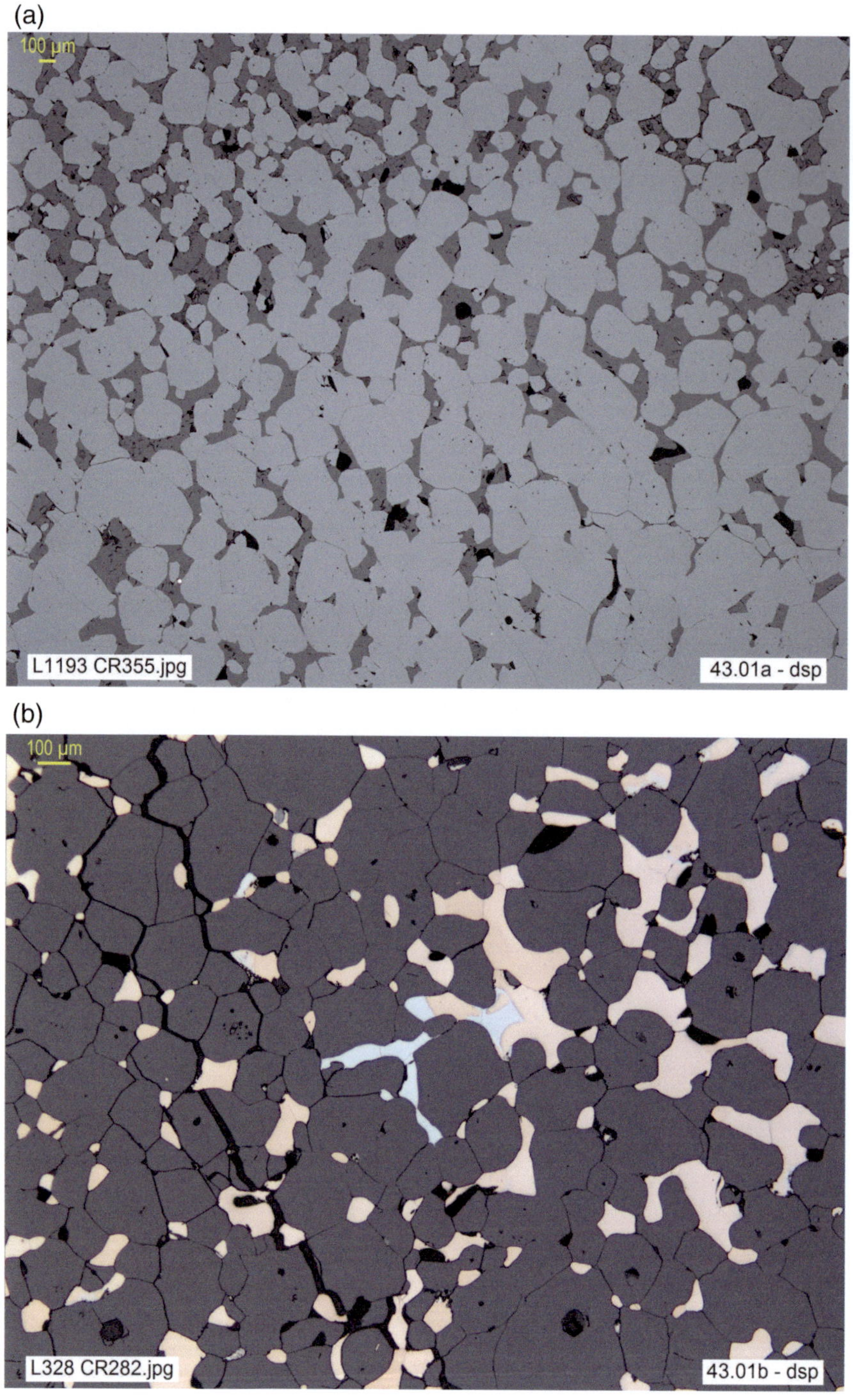

Fig. 1.39 **a** *Adcumulate* texture in chromitite: accumulation of chromite with scarce *intercumulus* (interstitial) silicates and scattered minute pentlandite particles. CR 355, Prootboom Mine, Bushveld, RS Africa (**dsp**). **b** Cumulate textures in Cr-Ni ore: accumulated chromite (*cumulus*) with interstitial nickeline and gersdorffite (*intercumulus*). CR 282, La Gallega Mine, S^a Ronda, Málaga, Spain (**dsp**). **c** Cumulitic-banded texture: alternating bands of chromite and silicates (olivine and serpentine). Rub20, Chalkidiki, Greece (**dsp**)

(c)

Fig. 1.39 (continued)

Table 1.3 Genetic classification of ore textures

Class	Processes and resulting textures		
Primary textures	Figure 1.39 Cumulates (cumulus, intercumulus, adcumulate, etc.)		
	Figure 1.40 Vein and cavity filling		
	Figure 1.41 Sequential		
	Figure 1.42 Colloidal		
	Figure 1.43 Detrital (various, including oolitic)		
Modified textures	Figure 1.44 Re-equilibration	Annealing, exsolution, decay, etc	
		Crystallization of gels	
	Figure 1.45 Metamorphism	Metamorphic (varied)	
	Figure 1.46 Remobilization	Various (infill, dissemination, etc.)	
	Replacement	Figure 1.47 Reaction rims, jagged contacts	
		Figure 1.48 Relicts	
		Figure 1.49 Ghosts	
		Figure 1.50 Supergene	Second. sulfide zone
			Oxidation zone
	Figure 1.51 Infiltration, weathering	Infiltrative (var. infill and replacement)	
	Figure 1.52 Deformation	Brittle: breccias/cataclasites	
		Ductile: mylonitic textures	
	Figure 1.53 Fluid-assisted strain	Pressure solution Solution transfer Pressure shadows	

Cavity-infill textures (Fig. 1.40a–j) are due to complete or partial filling of fractures, veins, cracks, geodes, pores or voids by a mineral phase precipitated from a fluid. Here, the concept of a *void* means any hollow space in the rock, such as the empty interstices between rock fragments in a clast-supported breccia, while the concept of a *fluid* excludes, for example, the matrix of a matrix-supported breccia. The vein and cavity infill textures are typically related to hydrothermal processes (i.e. processes involving hot aqueous fluids of any origin), so are also often referred to as **hydrothermal infill textures**.

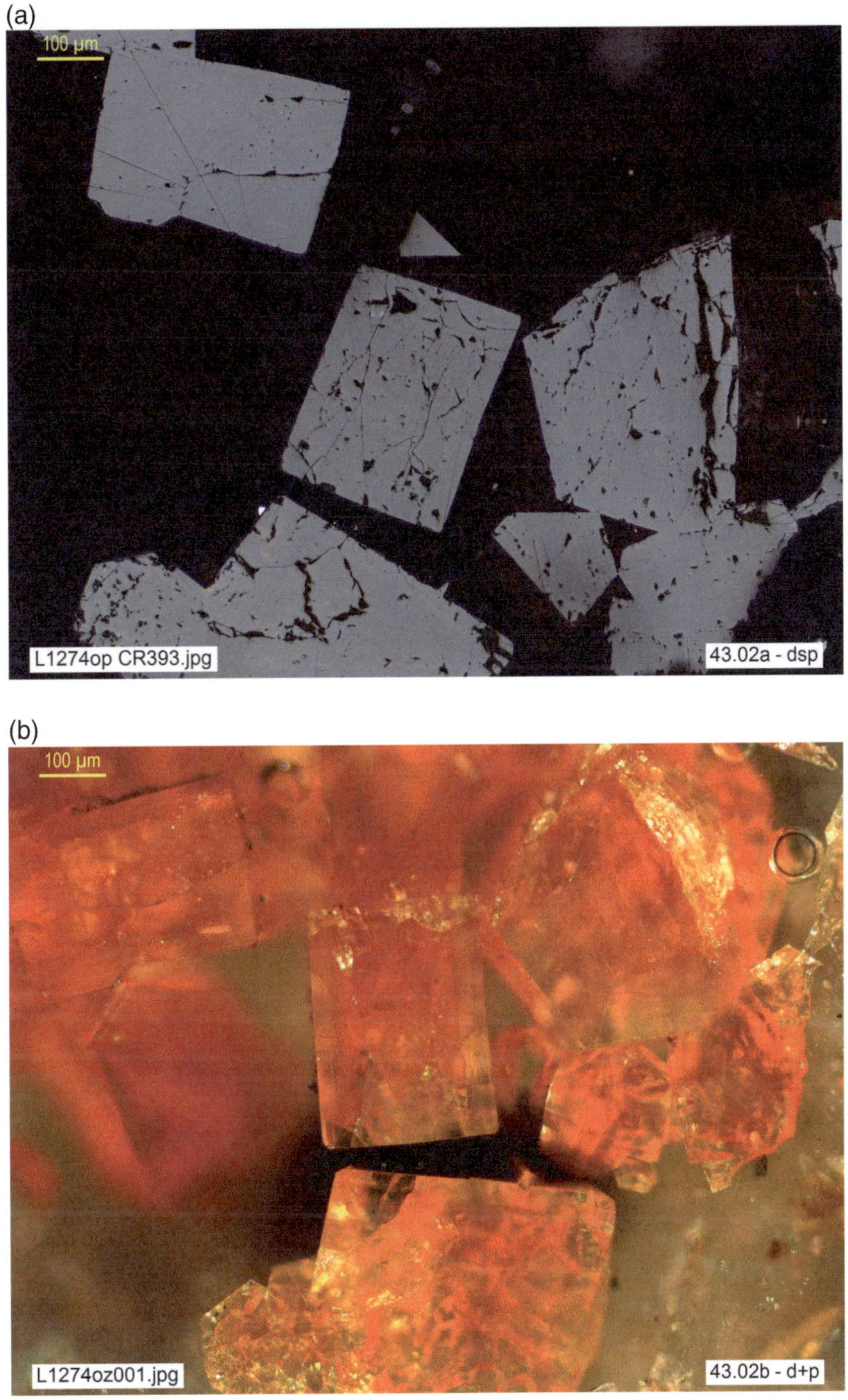

Fig. 1.40 a, b (**osp** and **o + p**, resp.). Deposit of crystalline vanadinite (gray, red internal reflections) in druse on quartz. CR 393, Globe, Arizona, USA. **c** Vein filling of cassiterite (light gray, idiomorphic, bireflective) and quartz (dark gray, xenomorphic, interstitial, top left), in a typical *cockscomb* intergrowth. TF Pn9, Montanchez, Caceres, Spain (**dsp**). **d** Massive, holocrystalline hydrothermal infill of safflorite-löllingite (light bluish gray) and cobaltite (light pinkish gray), with interstitial freibergite (gray), native Ag (white) and gangue (very dark gray). CR 375, Cobalt, Ontario, Canada (**osp**). **e**, **f** (**dsp** and **d + p**, resp). Massive hydrothermal filling of intergrown umangite (Cu_3Se_2, purplish-brown to light gray, pleochroic and strongly anisotropic) and carbonate (gray, anisotropic, internal reflections, twins), with minor ferroselite (FeS_2, white, bright). Rub583, San Francisco Mine, La Rioja, Argentina. **g** Hydrothermal filling of löllingite-skutterudite (gray) brecciated and cemented by native Ag (white) and carbonate (dark gray). CR 384, Cobalt, Ont, Canada (**dsp**). **h** Hydrothermal filling of sphalerite, chalcopyrite, pyrite, sparse galena, quartz and carbonate, with strong tectonic imprint: banded structure of the ore, sphalerite microlenticles and microclasts of pyrite and gangue. SV1 Rgib, Rio Gibranzos, Caceres, Spain (**dsp**). **i** Deposition of allargentum (gray) and native Ag (white) as breccia infill, cementing and impregnating carbonate gangue clasts (very dark gray). CR 362, Nipissing Mine, Cobalt, Ont, Canada (**osp**). **j** Posthumous infill, by downward infiltration (oxidation zone), of chrysocolla (bluish-green **IR**, *internal reflections*), goethite-cuprite (red IR), azurite (blue ir) and hematite (acicular). CR 356, San Antonio Mine, Sonora, Mexico (**dsp**)

(c)

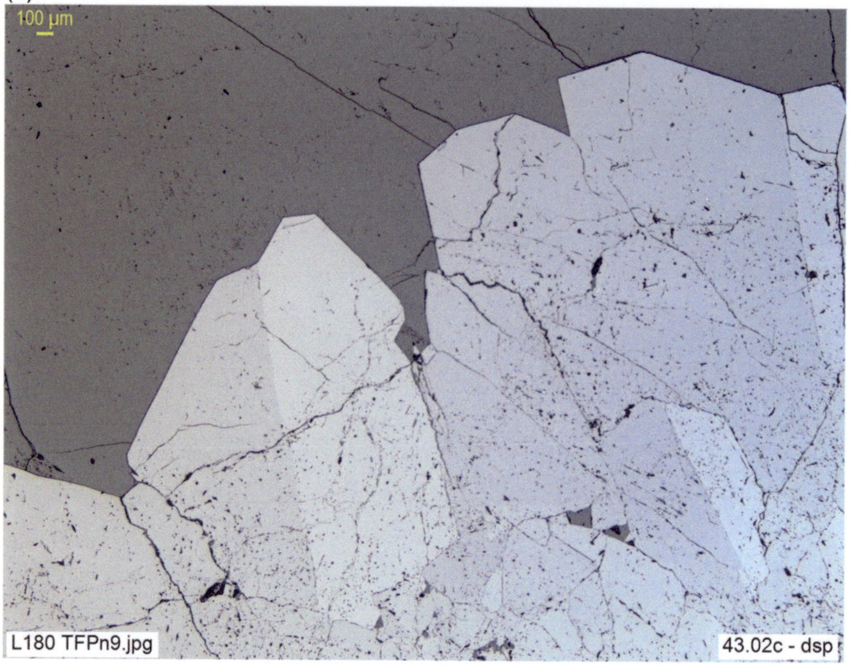

(d)

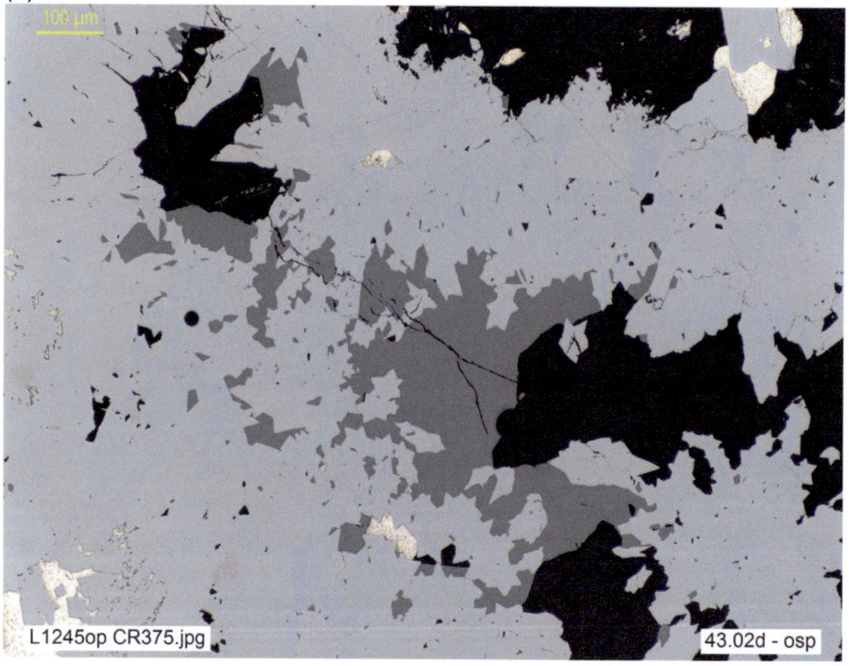

Fig. 1.40 (continued)

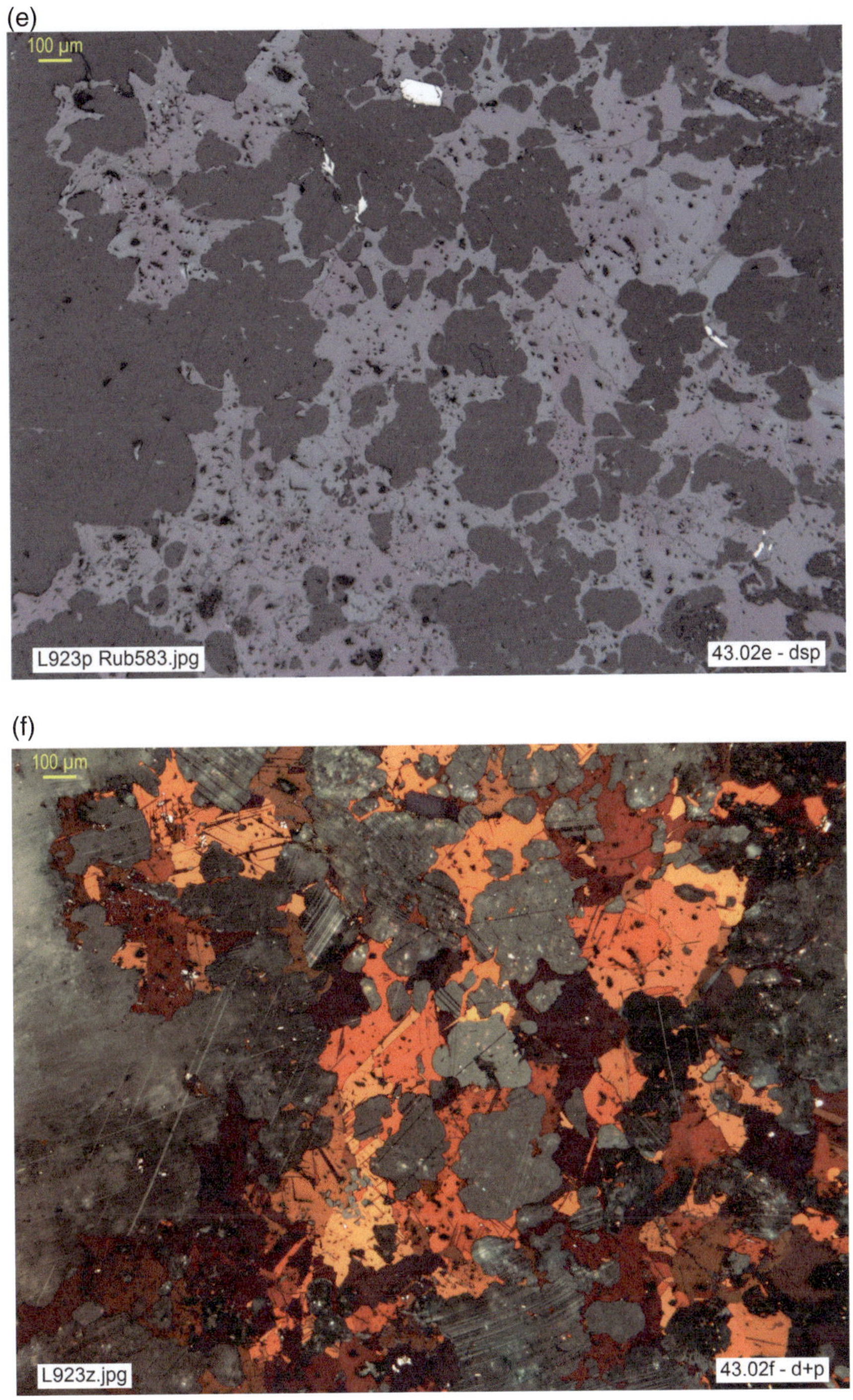

Fig. 1.40 (continued)

(g)

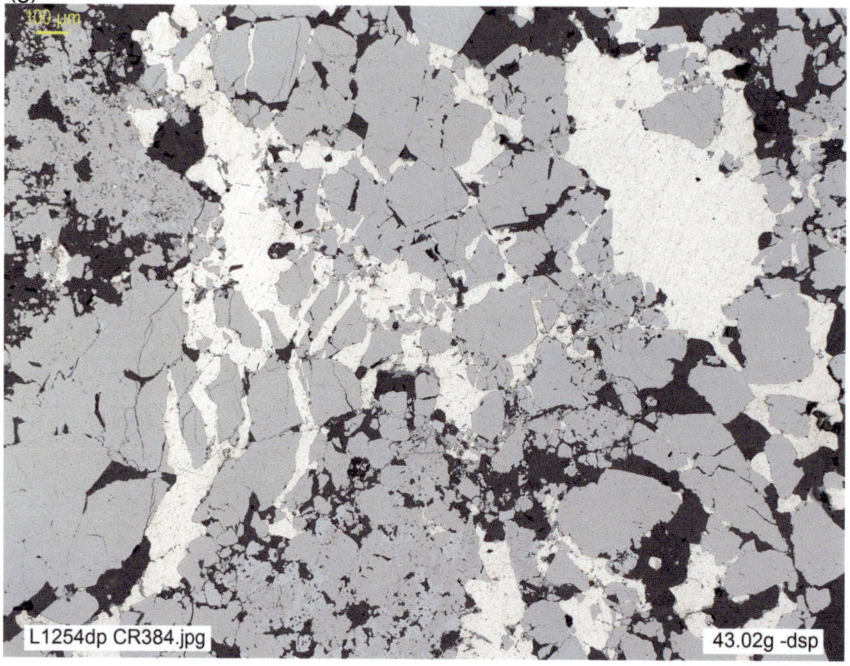

(h)

Fig. 1.40 (continued)

(i)

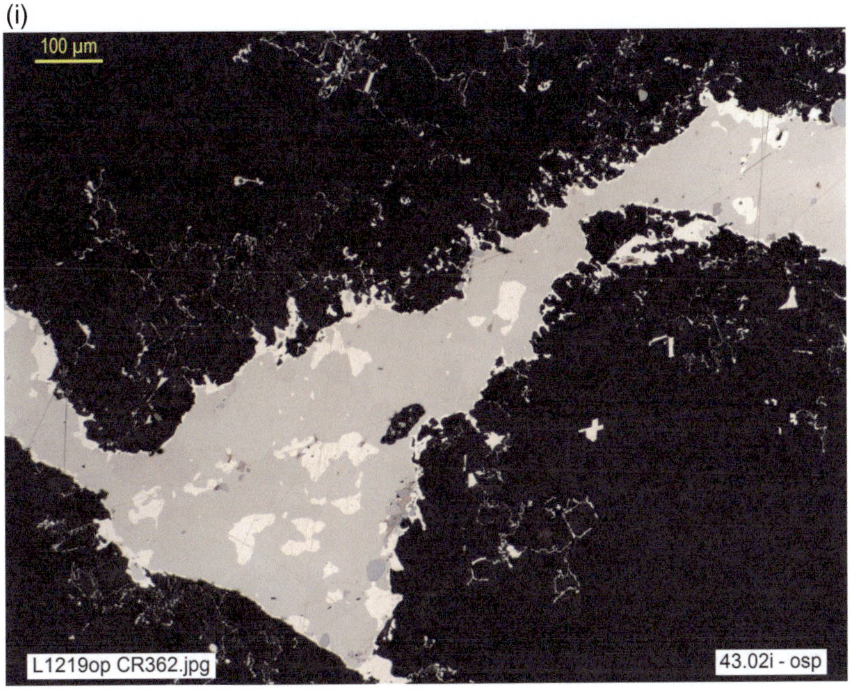

(j)

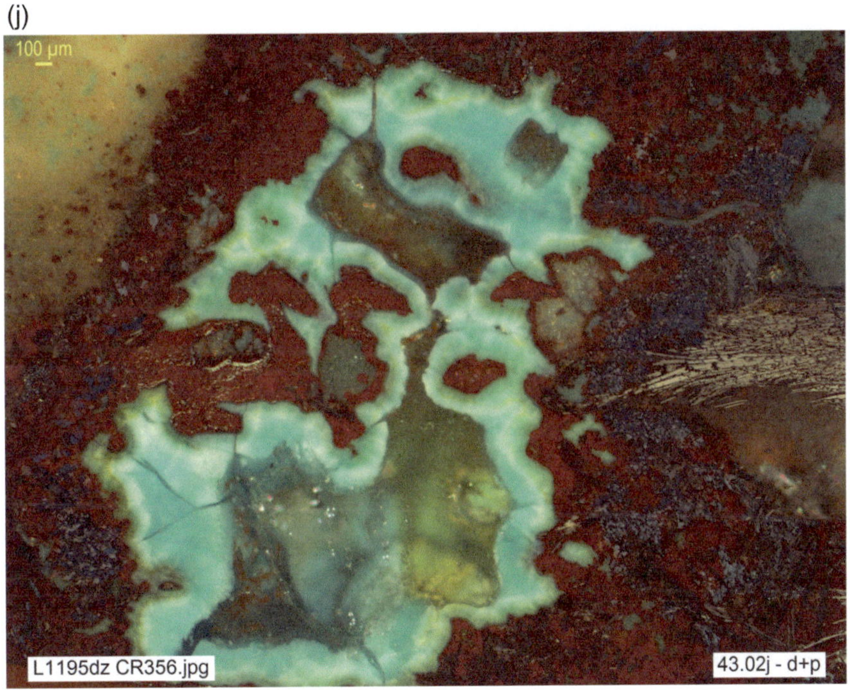

Fig. 1.40 (continued)

Although the filling process is, conceptually, easy to understand, the resulting textures can be varied and complex, not least because of the enormous range of possible scales, from kilometre-long and decametre-thick lodes to thin veinlets and microbreccias or microcracks on a micrometric scale.

In this context, banded and sequential textures are frequent, as may be zoned crystals sometimes, suggesting an evolution in the fluid composition, temperature, and so on. Colloidal textures can also be formed, either as posthumous manifestations of a waning hydrothermal activity or as the result of marked disequilibrium conditions that propitiate massive, instant (*flash*) mineral precipitation, so fast that ordered crystallization is not possible even at relatively high temperatures.

Hydrothermal infill may also comprise multiple textures that have already been defined by their morphology, only with geometric criteria (Table 1.2). Some receive particular names. Thus, for example, the *cockade* texture—due to deposition of successive layers on clasts from the host rock or an earlier vein filling—is formally equivalent to the *corona* texture (Table 1.2, Concentric), but on a mesoscopic scale (hand sample or outcrop). The *symmetrical banding* depositional texture, displaying the same pattern on both sides of the lode and, particularly, the *layer cake* texture so frequent in low-sulfidation epithermal concentrations, are, formally, *sandwich* textures (Table 1.2, Fig. 1.33), also at mesoscopic scale. *Cockscomb* (Fig. 1.40c) and *palisade* textures can be considered as encrustation variants (Table 1.2, Fig. 1.35) comprising one or more holocrystalline layers, each composed of crystals growing parallel to each other and perpendicular to the layering, that is, to the *salband* or vein margin. In these cases, the interpretation as vein infill is evident, following the usual petrographic criteria.

If the geometrical relations among the minerals that compose the ore allow to establish its crystallization order (*sequence*), they are said to define a **sequential** *texture* (Fig. 1.41a–h). These textures are frequently, but not exclusively, the result of hydrothermal filling. Actually, they can be both hypogene and supergene (*vbgr.* chemical sedimentary deposits, and also supergene encrustation processes: *cf.* Figure 1.40j). The criteria for establishing the sequential order are those usual in geology, commonly of geometrical type, as Figs. 1.40 and 1.41 show. In due course (§ 1.3) they are discussed further, in the interpretative analysis.

Fig. 1.41 **a** Vein infill, with sequential deposition of: (1) arsenopyrite (white relicts, bottom right) and sphalerite (gray), followed by: (2) stannite (light olive gray) and (3) galena (grayish white) and quartz (dark gray). Jaggged reaction contacts fix the sequence (1–2) and, together with spatial relationships (or brecciation-cementation episodes), determine the sequence (2–3). Rub483/77, Davidschacht, Saxony, Germany (**dsp**). **b** Hydrothermal filling of *white arsenides* (safflorite, löllingite, skutterudite, bluish gray to gray) and cobaltite (slightly pinkish gray), followed by argentopyrite (pinkish brown) and, finally, native silver (very bright white), which fills pores and corrodes the previous ores. CR 375, Cobalt, Ontario, Canada (**dsp**). **c** Sequential deposition of: (1) quartz (dark gray, euhedral), (2) skutterudite (grayish white, relief on Bi, relict, mainly in lower central zone), and (3) native Bi (yellowish white, bright) and bismuthinite (gray, very thin *corona* on skutterudite, barely visible). Order of deposition evidenced by spatial relationships, with successive episodes of microcracking, followed by cementation and corrosion indicated by reaction rims. C-15, Loma La Pizarra, Villanueva de Cordoba, Spain (**dsp**). **d** Sequence of crystallization in stanniferous lode, evidenced by spatial relationships: (1) cassiterite (gray, euhedral), slightly cracked; (2) sulfides (pyrite, covellite, bornite, digenite) and gangue, filling interstices, microgeodes and microcracks in cassiterite. Rub607, Geevor Mine, Cornwall, UK (**dsp**). **e** Sequential infill texture: (1) main episode: holocrystalline and euhedral alabandite (light gray, porous) and quartz (dark gray); (2) posthumous episode, sealing interstitial spaces and microgeodes: adularia (dark gray), carbonate (gray, bireflective) and stibnite (acicular inclusions, white, exclusively in adularia and carbonate). CR 344, Inakuraishyi Mine, Hokkaido, Japan (**dsp**). **f** Early cryptocrystalline aggregate of psilomelane and probable braunite (colomorph, in dark brownish-gray to gray bands), cracked and cemented by pyrolusite (yellowish white), which includes relicts of ramsdellite (grayish). CR-382, Iron Dyke Mine, Plumas County, California, USA (**osp**). **g** Succession of episodes of hydrothermal infill and diffusion in the host rock (schist with disseminated acicular rutile): (1) fracturing and early precipitation of arsenopyrite (white), carbonate and quartz (vein selvedges), accompanied by (2) metasomatic impregnation of the host rock by diffusion of arsenopyrite from the vein edges (crystals growing in continuity across the vein border) and followed by (3) precipitation of quartz, sphalerite (gray) and traces of galena (light gray) inside the vein. Diffusion of arsenopyrite by replacement of the host is evidenced by microinclusions of rutile (acicular, oriented parallel to the schistosity), inherited by arsenopyrite. PET.2. Inf409, Rio Gibranzos, Cáceres, Spain (**dsp**). **h** Successive cycles of fracturing and hydrothermal filling: arsenopyrite (white, relief, top left) seals the first episode and appears in microclasts and relicts in galena (light gray, main ore); pyrite (yellow, relief) appears in two generations: an early, cataclastic one, coarse-sized but corroded by galena, and a late, fine-grained, euhedral one, cementing microcracks in sphalerite. The generation of sphalerite (dark gray, top right) is later than arsenopyrite (it corrodes it), but precedes pyrite2, while galena seals the last episode and cements the whole. Orogenic gold deposit. 220910.4B, Veta Esperanza (San Vicente), Minera Retamas, Pataz, Peru (**dsp**)

(a)

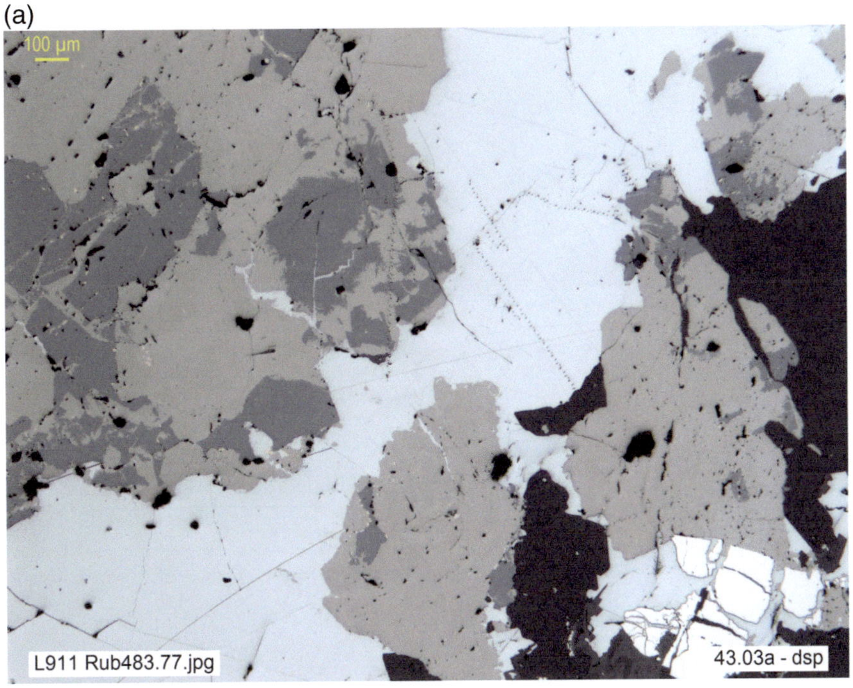
100 µm
L911 Rub483.77.jpg
43.03a - dsp

(b)

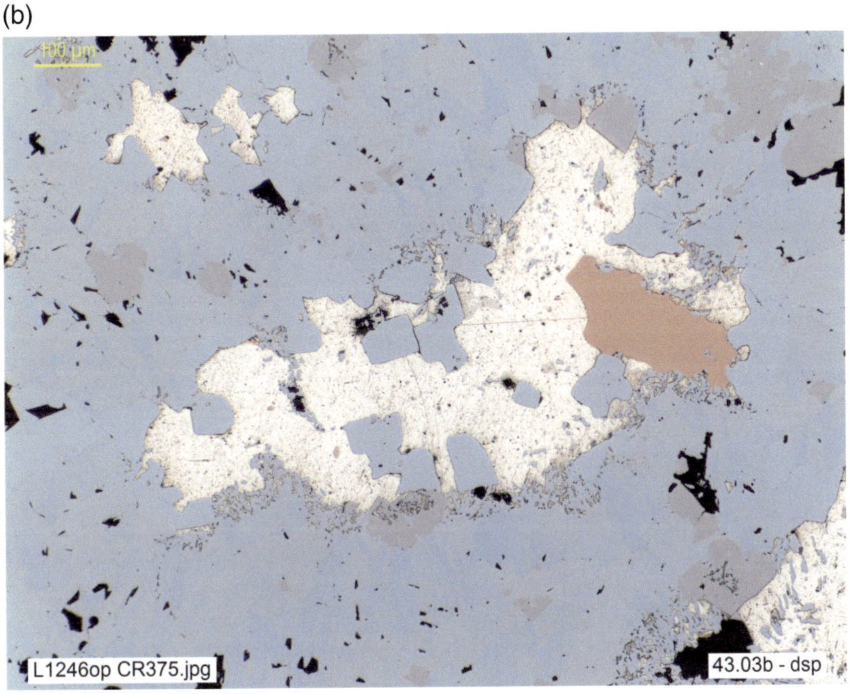
100 µm
L1246op CR375.jpg
43.03b - dsp

(c)

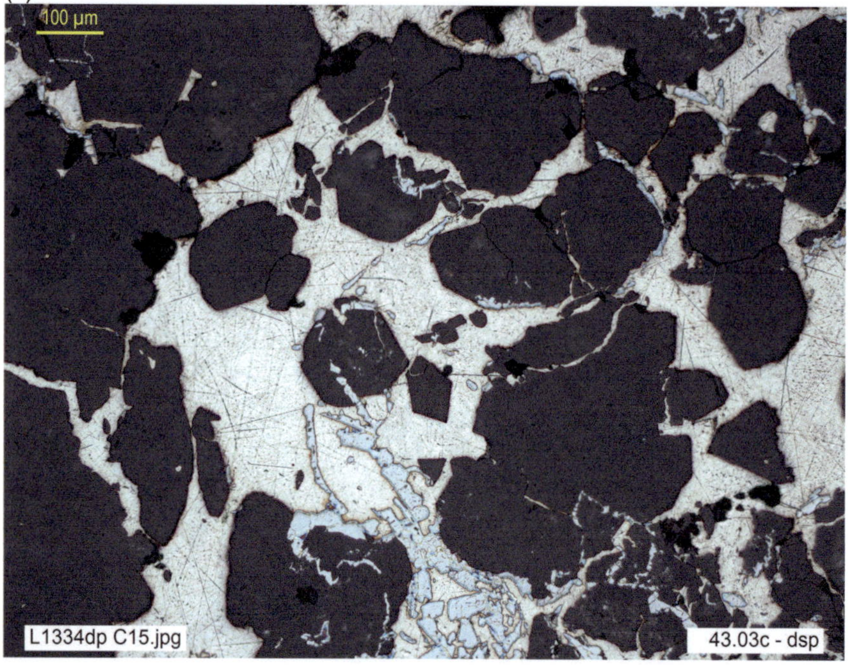

(d)

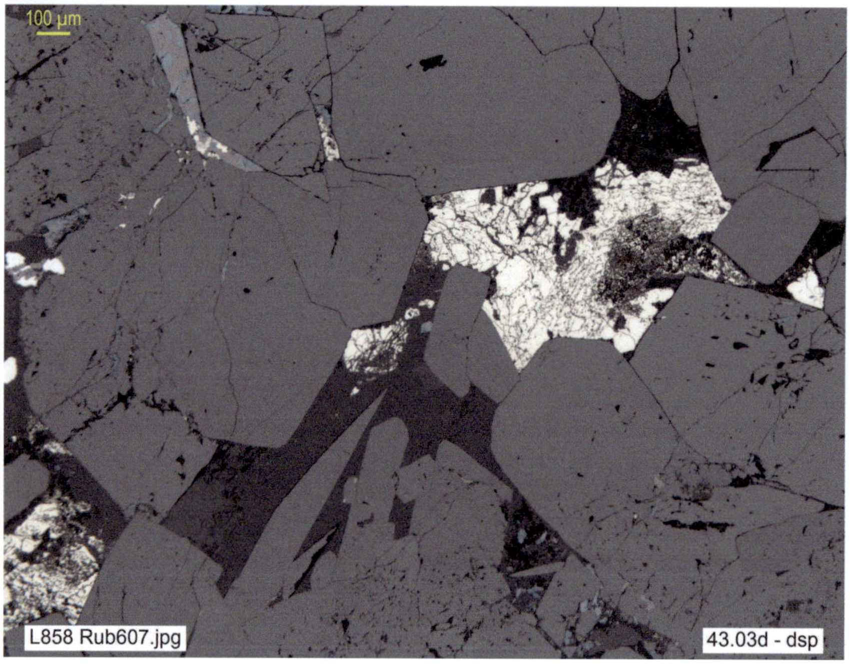

Fig. 1.41 (continued)

(e)

(f)

Fig. 1.41 (continued)

(g)

(h)

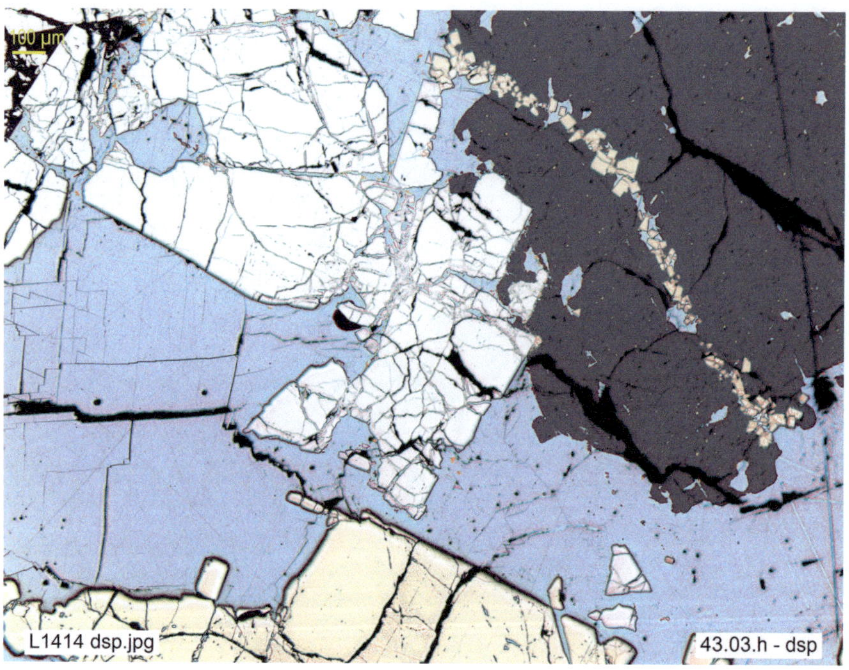

Fig. 1.41 (continued)

Colloidal or *colomorphic* textures are typically micro- to cryptocrystalline and are interpreted as the result of gel consolidation (Fig. 1.42a–h). Although the components tend to stabilize by increasing grain size through crystalline growth, they are generally characterized by a very fine, often sub-microscopic grain size, which can give the impression of an amorphous material. They may exhibit fine banding and striking morphologies, such as *colloform, fibrous-radiating, reniform* (kidney-shaped), *botryoidal* (Table 1.2, Fig. 1.15), and so on. They may have a massive, undifferentiated appearance, even if comprising several mineral phases, as often occurs in limonite: an aggregate of submicroscopic particles of goethite, lepidocrocite, jarosite, opal, hematite, and so on, which mostly cannot be individualized by the resolving power of the optical microscope.

(a)

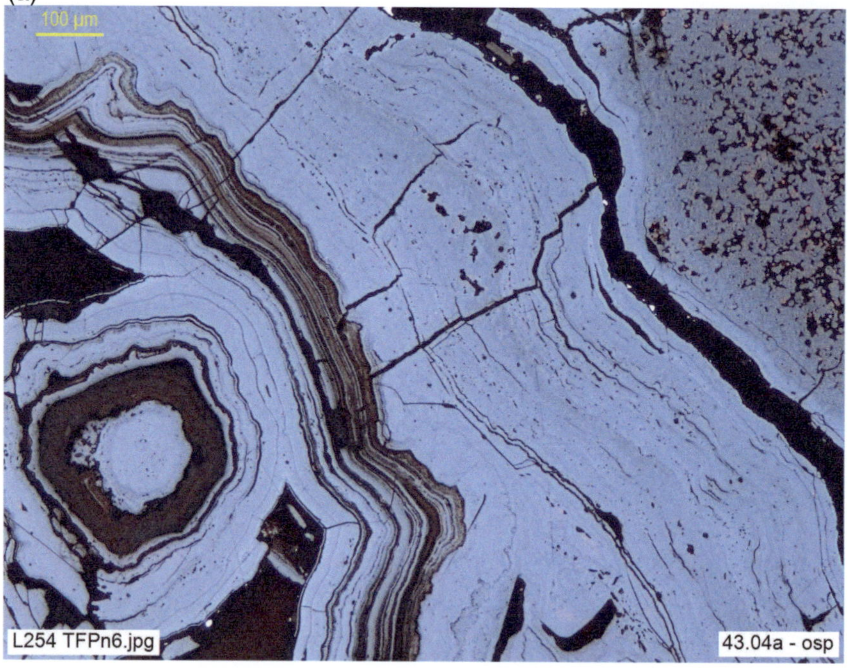

(b)

Fig. 1.42 **a**, **b** (**osp** and **o + p**, resp). Banded colloform aggregate of goethite (bluish gray, reddish-brown internal reflections) and limonite (brownish gray, massive and more yellowish IR). TF PN 6, Mina Sar, Santiago de Compostela, Spain. **c** Polymetallic massive sulfide ore with colomorphic aggregates (partially recrystallized) of pyrite and chalcopyrite, included in a microcrystalline mass of pyrite, sphalerite, silicates and chalcopyrite. VA3.529, Masa Valverde (drill hole 3), Huelva, Spain (**dsp**). **d** Relict syn-sedimentary colomorphic banding, preserving evidence of syn-sedimentary deformation: *slumping* shown by thin microfolded bands of pyrite, with subordinate chalcopyrite, galena, sphalerite and silicates (*cf* details in Fig. 1.42e). VA3-529.15, Masa Valverde (drill hole 3), Huelva, Spain (**dsp**). **e** Detail of Fig. 1.42d (lower right corner): microfolded bands and spheroids of porous pyrite and interlayered chalcopyrite, with scarce inclusions of sphalerite and galena (**osp**). **f** Botryoidal deposit of sphalerite (*schalenblende*), enhanced by its abundant and luminous yellowish-brown internal reflections, on dendritic relicts of gratonite pseudomorphosed by jordanite. HD 8, Wiesloch, Baden-Württemberg, Germany (**d + p**). **g** Colloform deposit of native As (white), partially recrystallized, fragmented and cemented by sulfides (galena and tetrahedrite, gray) and quartz (dark gray), which fill microcracks, infiltrate and partly replace the As bands. CR 397, Hiendelaencina (?), Guadalajara, Spain (**dsp**). **h** Ghost colomorphic texture in massive As (brownish gray) finely intergrown with stibarsen (*allemontite II*), with rhombohedral euhedral inclusion of löllingite (white). The fine intercalations of stibarsen (white) show the original concentric banded structure of the primitive As-Sb gel, whose unmixing explains the currently observed intergrowth. The dark tone of As is due to its typical fast tarnish, diagnostic to avoid confusion with stibarsen (unaltered, bright) and even to distinguish bands where one or the other dominates according to its brightness (cf Fig. sba1, 2, 5) to compare with the aspect of fresh or repolished As). CR 345, Engineer Gold Mine, Atlin District, BC, Canada (**dsp**)

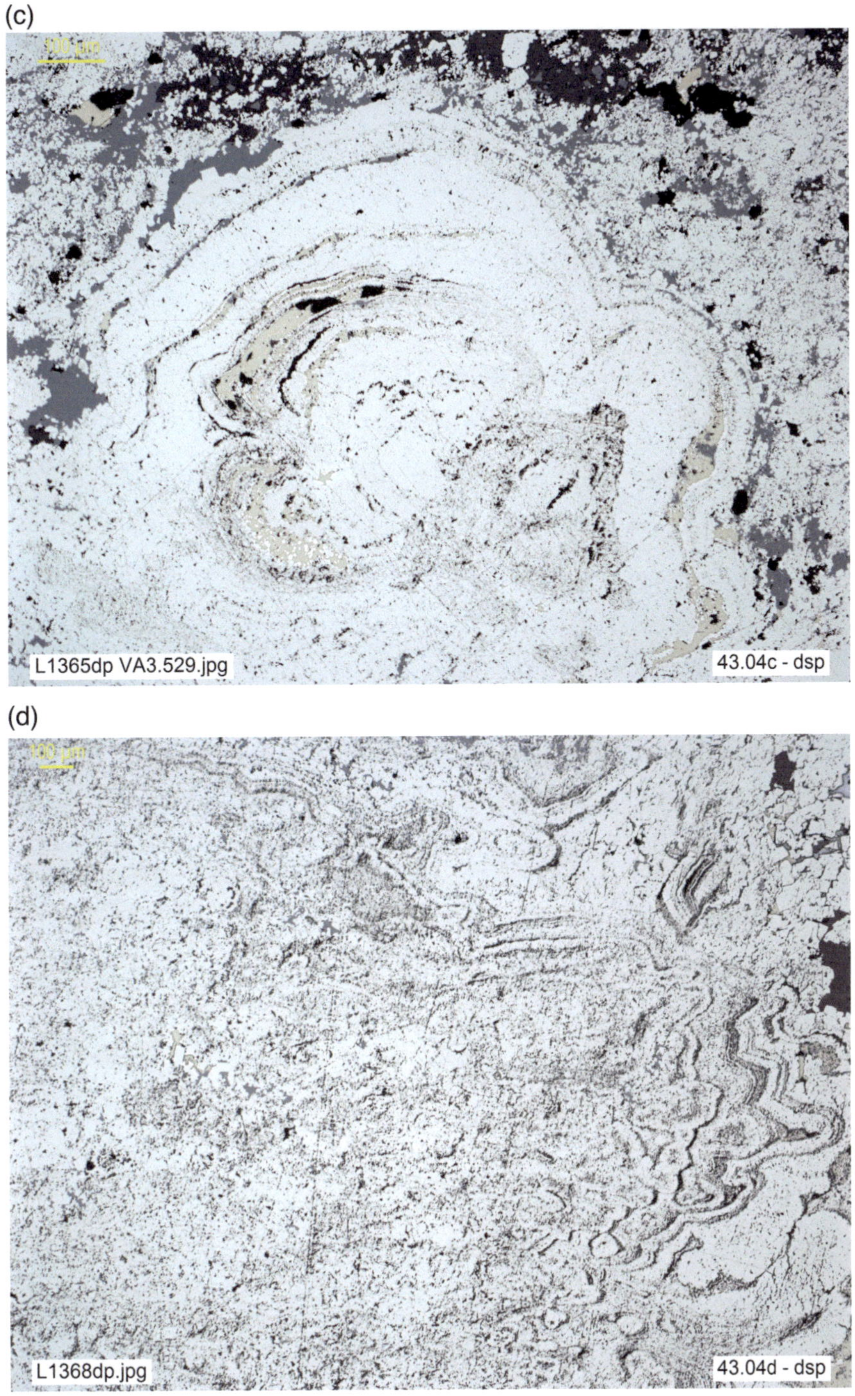

Fig. 1.42 (continued)

(e)

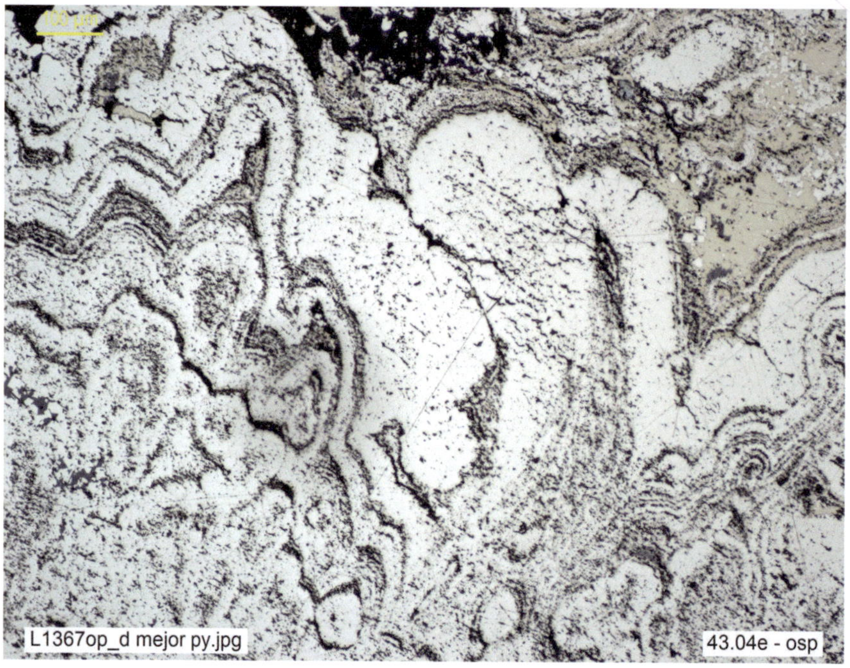

(f)

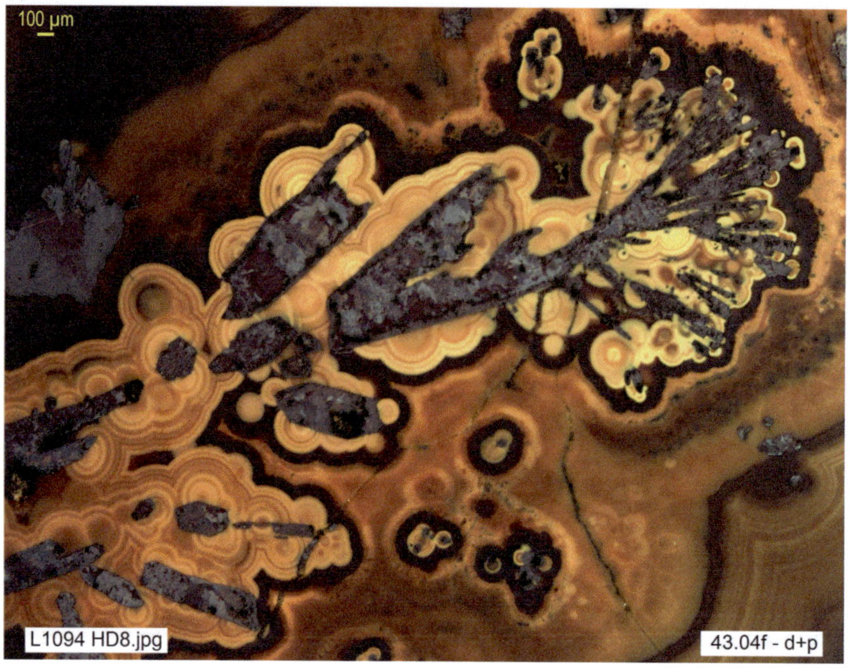

Fig. 1.42 (continued)

(g)

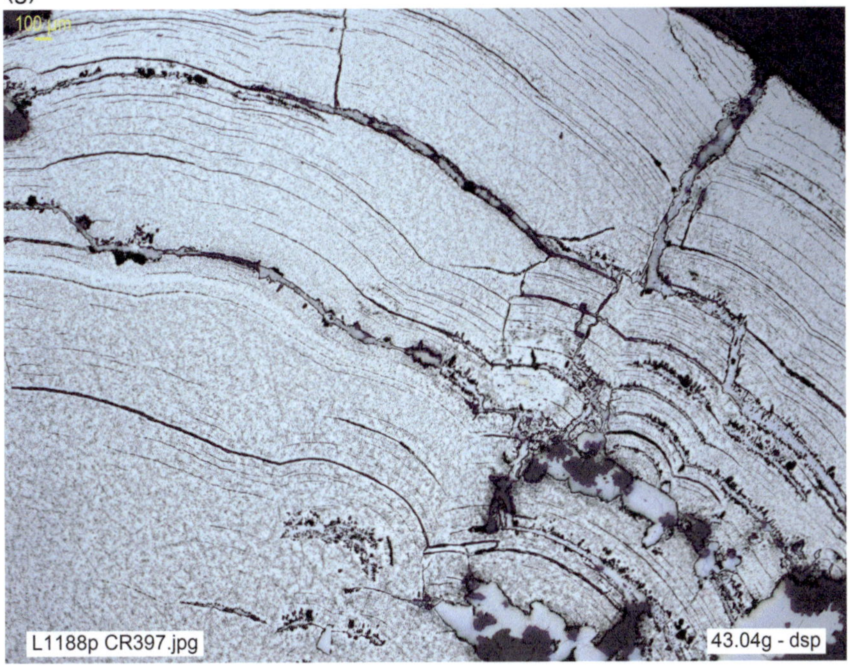

(h)

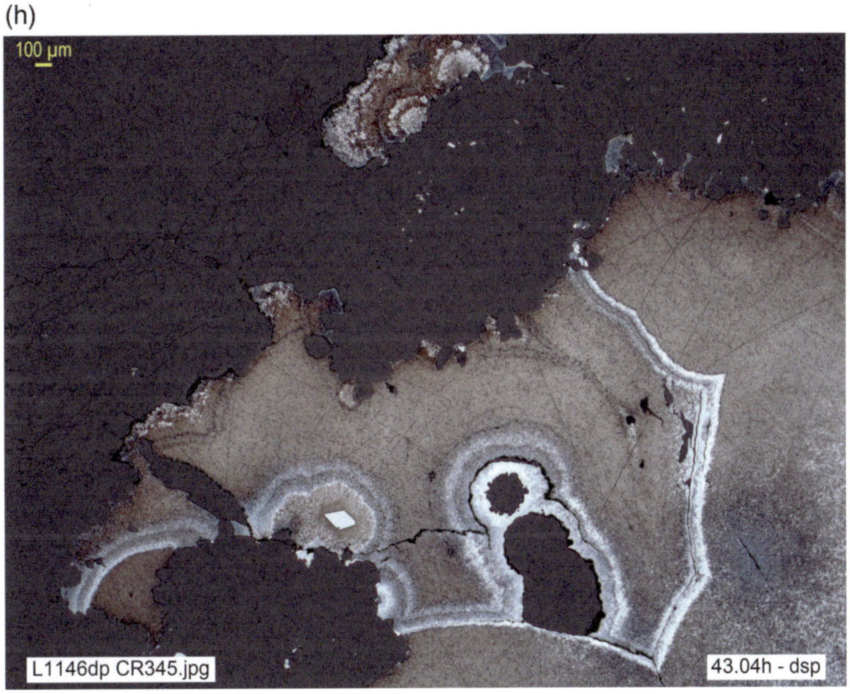

Fig. 1.42 (continued)

Detrital (Fig. 1.43a–g) or clastic textures are, as the name suggests, typical of the sedimentary cycle. They are common in ores that are resistant and stable in surface conditions, such as cassiterite, columbite, tantalite, gold (Fig. 1.43g) and platinum, traditionally beneficiated in placers or paleoplacers. *Oolitic* textures are associated with this group (Fig. 1.43e), although these can be considered as transitional to those of chemical precipitation: in fact, a typical oolite usually shows a detrital sand core covered by successive micro-layers of hematite, goethite, carbonates, and so on, extracted from the basin, giving the whole an oval external contour.

The recognition of detrital textures in sulfide ores has been relatively late, among other things because they are infrequent, given the low resistance of sulfides to weathering and transport processes in surficial conditions. However, the example of the auriferous conglomerates of the Witwatersrand (South Africa), which contain pyrite and uraninite sands in addition to gold, proves that they are also possible in sulfides, in this case, thanks to the Archean reducing atmosphere. In other cases, such as the volcanogenic massive sulfides, their preservation is explained by the restricted local conditions of the submarine basin, whose reducing environment makes possible the deposit and preservation either of sulfide gels and muds or of the cumulated clastic fragments (*sulfide mound*) derived from the collapse of the exhalative chimney (*black smoker*). However, pseudo-detrital textures can also occur in sulfides, due to replacement of detrital grains (Fig. 1.43f).

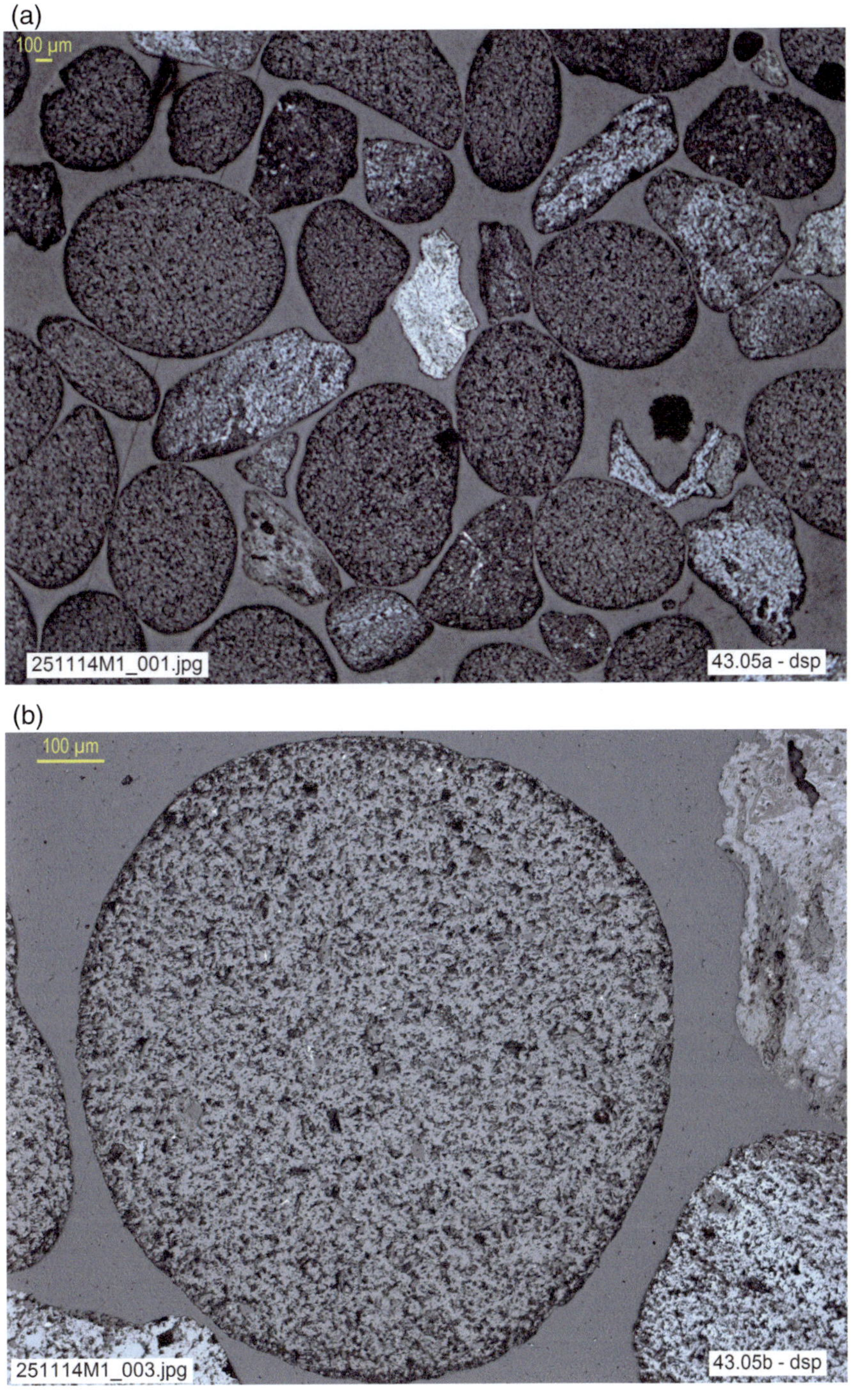

Fig. 1.43 **a** Detrital concentrate of monazite sands: monazite (gray, porous matrix), goethite-psilomelane (impregnation, lighter gray tones) and silicates (tiny inclusions, dark gray). Heavy fraction sample 251114.M1, Matamulas, Ciudad Real, Spain (**dsp**). **b**, **c** (**dsp** and **d** + **p**, resp). Detail of Fig. 1.43a: internal structure of nodule or monazite sand, with detrital inclusions (silicates, rutile) and incipient marginal zoning, discernible in (+p). **d** Witwatersrand auriferous metarenite (*gold ore*): dominantly detrital quartz sands (gray), with interstitial phyllosilicates, organic matter, pyrite, titanite, gold. The angular forms of detrital pyrite suggest a proximal source. Comparison of q and py grain sizes shows that detrital components are finer the higher their density. CR 367, Dagga Fontein Mine, Witwatersrand, South Africa, R. South Africa (**dsp**). **e** Oolitic sandstone with maghemite (white), magnetite (gray) goethite (slightly darker gray) and quartz (dark gray). CR 325, Gora Djebilet, Tindouf, Algeria (**dsp**). **f** Progressive replacement of *Rotliegendes* sands by pyrite: incipient process in the compact, stable quartz sands (just a thin py rim on the q grain), more advanced in the softer detrital matrix, and complete in the porous and reactive lithic fragments. It is evident that, unlike Fig. 1.43d, this pyrite is not detrital. Rub11, Spremberg, Brandenburg, Germany, *Kupferschiefer* (**dsp**). g. Heavy fraction of stream sediments, rich in detrital gold (nuggets), with minor zircon (euhedral crystal, dark gray), and covellite impregnation (dark blue). GR Sample 8.GR, Greenstone Belt of Minvoul, Gabon (**dsp**)

(c)

(d)

Fig. 1.43 (continued)

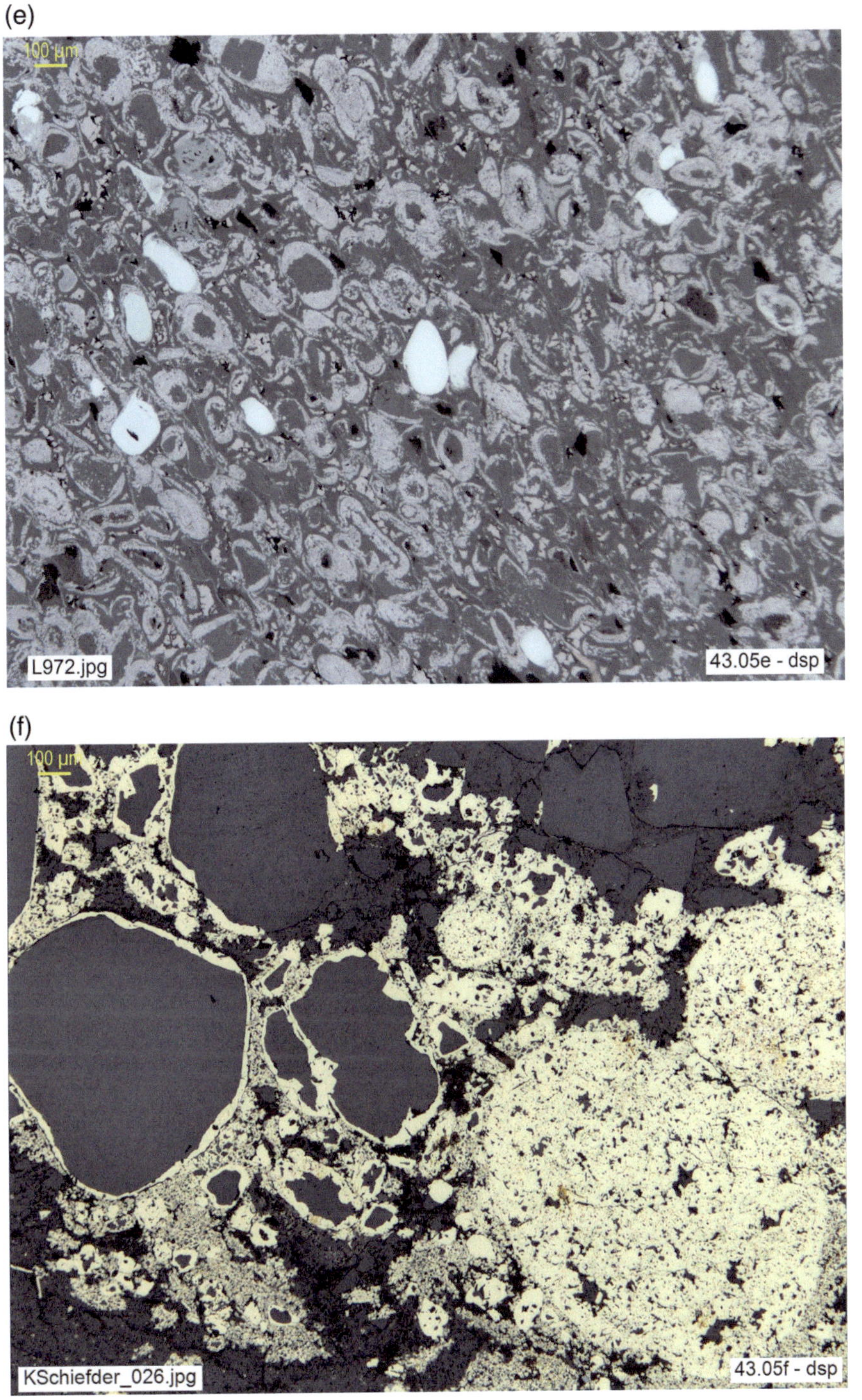

Fig. 1.43 (continued)

(g)

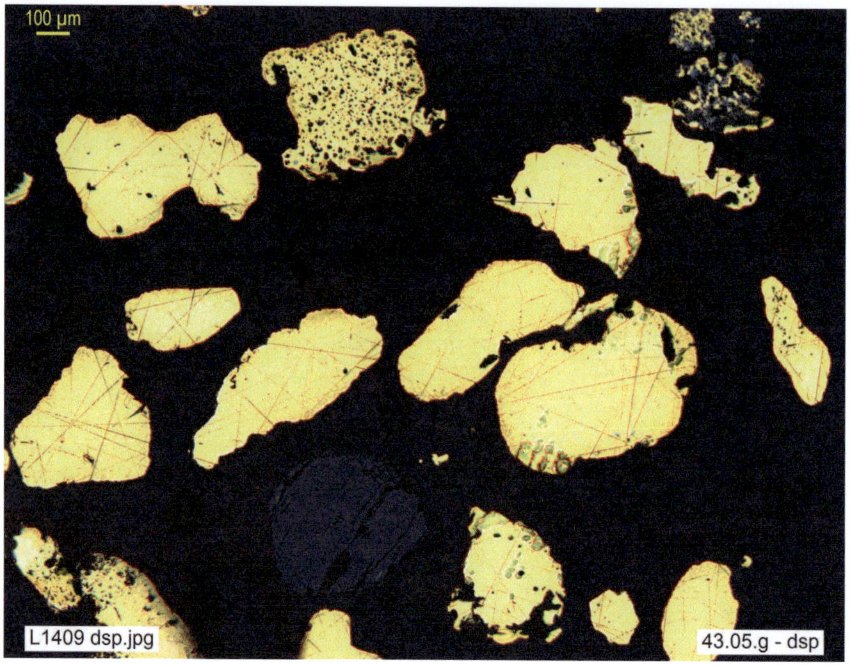

Fig. 1.43 (continued)

Modified textures result from the transformations that the rock undergoes after the deposition of the ore, tending to re-establish the equilibrium disturbed by changes in the environment. Usually, these are changes of pressure, temperature or chemistry, including the various types of brittle or ductile deformation. For clarity of exposition, **metamorphic** textures in general are treated with this group. This might seem contradictory in the case of *metamorphogenic deposits*, those essentially formed by metamorphic processes as opposed to those simply transformed by them (*metamorphosed deposits*). Indeed, it would seem more coherent, in principle, to consider the textures of metamorphogenic deposits as primary. However, in the most frequent deposits of this type, varied textures are observed, many the result of successive or cyclic processes that modify the preceding textural features, as in *orogenic gold concentrations*. In this case it is consistent to analyse the textural picture as a whole, including the various generations deposited and their successive modifications. Something similar could be said of *skarn* concentrations, in which various events involving alteration or metasomatism and recrystallization are frequent and in which, in addition, episodes of higher and lower temperature usually follow one another.

The first modification to be considered is the **re-equilibration**; that is, the spontaneous transformation of the ore to reach a new equilibrium by adaptation to changing conditions. For example, in the case of hypogene ores, during the cooling process the crystal lattice contracts and may have to adjust by expelling non-stoichiometric atoms or atoms of incompatible dimensions, which are segregated into new mineral phases. In supergene ores, by contrast, the increase of temperature (by diagenesis, burial or metamorphism) facilitates recrystallization of colloidal or cryptocrystalline aggregates; thus they minimise their chemical potential, increasing their grain size and reducing the specific surface, resulting in secondary crystalline aggregates. Some unstable or metastable minerals spontaneously re-equilibrate, transforming into a stable polymorph, as in the substitution of marcasite for pyrite; this process is known as *paramorphic transformation*.

Exsolution or *unmixing* occurs in a high-temperature solid solution that becomes unstable on cooling and decomposes into two or more mineral phases, intergrown in proportions determined by the original composition. The minority phase appears included in the main phase, forming corpuscles whose orientation is conditioned by the structure of the latter (Fig. 1.44a–g) and whose distribution sometimes reflects a primitive compositional zoning. The reality of this process is confirmed by experimental results. However, not all textures initially attributed to unmixing are due to this process, as shown by the frequent intergrowths between chalcopyrite and sphalerite.

At high temperature, chalcopyrite may contain a certain amount of Zn (up to 4.7% in massive sulfides, according to Huston et al., 1995, or up to 12.7 at% in solid solution in the *iss* phase at 800 °C, according to Kojima and Sugaki 1985). This is segregated by cooling into disseminated ZnS particles in chalcopyrite (typically star-shaped: the well-known *sphalerite stars*: *cf* Fig. 1.30b). When sphalerite is the dominant mineral, disseminated textures (known as *emulsion of chalcopyrite* in sphalerite: Fig. 1.44c) may also occur. These were traditionally interpreted as exsolutions, but more recently, with the support of experimental data, this explanation has been questioned in favour of two alternative mechanisms involving either co-precipitation or replacement.

Laboratory investigation of the Cu-Fe-Zn-S system shows that, under conditions similar to those common in hydrothermal deposits, the solubility of CuS in sphalerite is very low (<1 mol% at temperatures below 600 °C: Bowles et al. 2011), so that its exsolution could not explain the amount of disseminated chalcopyrite. According to the most frequently accepted opinion today, the dissemination of chalcopyrite is usually the result of a process of alteration or secondary replacement of sphalerite by chalcopyrite (*chalcopyrite disease* of sphalerite, very common in copper-rich massive sulfides: Fig. 1.44d). This is confirmed by textural evidence in general, yet in some cases *emulsion textures* with regular contours and smooth edges are observed, suggesting equilibrium conditions compatible with a process of co-precipitation or, in the case of Fig. 1.44 c, exsolution of chalcopyrite in sphalerite.

Colloidal deposits are prone to re-equilibration by increased crystallinity, resulting in a textural evolution from colomorphic textures, sometimes only visible as relicts or ghosts, to mossy, fibrous and microcrystalline textures (Fig. 1.44h–l).

(a)

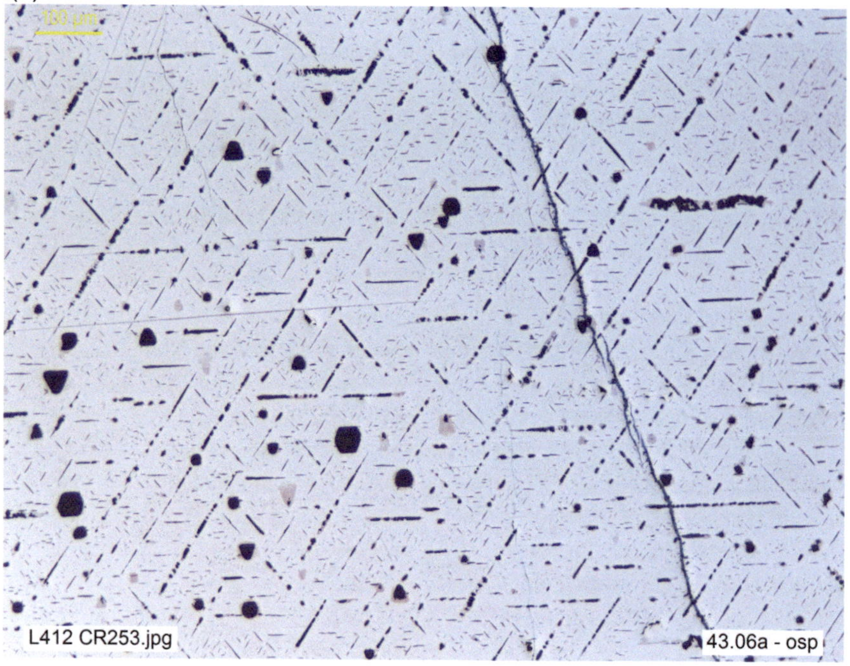
100 µm
L412 CR253.jpg
43.06a - osp

(b)

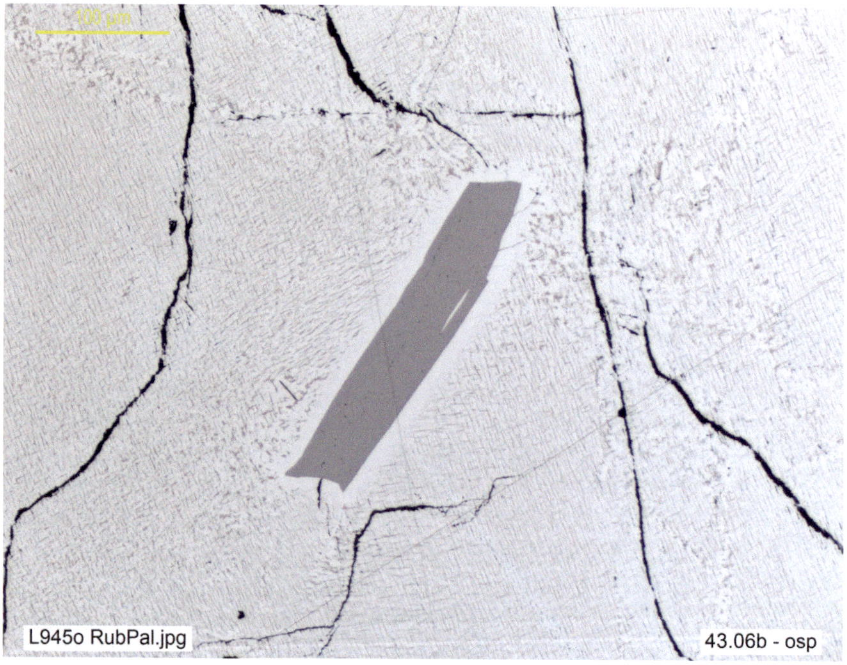
100 µm
L945o RubPal.jpg
43.06b - osp

◀ **Fig. 1.44** **a** Magnetite, *mt* (light gray), with exsolution of spinel, *spn* (dark gray), as minute cubic crystals, scattered or aligned following octahedral directions of mt (sometimes associated with unmixed ilmenite, *il*, pinkish gray), and as tiny oriented lamellae (also //111 of mt). Segregation of il in horizontal and much longer but scarcer lamellae. CR253, Lac Berta, Quebec, Canada (**osp**). **b** Exsolution of titaniferous phases in titanomagnetite: fine lattice of ulvöespinel (brown, oriented //100, with the typical cloth-like texture) and thick isolated crystal (//111) of ilmenite (purplish gray). The absence of ulvöespinel in the immediate surroundings of this large ilmenite crystal is due to their Ti depletion by its unmixing (an exsolution criterion). FM, Foskor Mine, Palabora, Rep. South Africa (**dsp**). **c** Sphalerite with exsolution of minute chalcopyrite particles, distributed at random or aligned, bearing no relation to microcracks or to the grain borders of the sphalerite grain. The aligned chalcopyrite particles mirror in part the sphalerite lattice, while their separation increases with their size (*compare to Fig. 1.44d*). VA1-480.70 massive sulfide ore, Masa Valverde, FPI, drill hole 3, Huelva, Spain (**dsp**). **d** Partial replacement of sphalerite (and pyrite) by chalcopyrite (*ccp disease*). The texture excludes unmixing because, *unlike* Fig. 1.44c: the contacts between both phases are irregular or jagged (reaction borders), the chalcopyrite distribution is irregular and is spatially related to the outer margins of sphalerite or to microcracks (i.e. the presence of ccp depends on external supply). VA3-668.80 massive sulfide *feeder*, Masa Valverde, FPI, drill hole 3, Huelva, Spain (**dsp**). **e** Native platinum (white, slightly yellowish) included in chromite (almost black) and with two clearly visible native osmium inclusions (bluish white, relief: one lenticular, center right, and one lamellar, top right). Additional unmixing of Os in very thin, inconspicuous lamellae, which are distinctly visible only with *i.c.*, *cf* Fig. 1.44f). CR 297, Folx Gulch, Goodnews Bay, Alaska (**osp**). **f** Same origin as Fig. 1.44e, with interference contrast, *i.c.* Relief makes the Os unmixing texture in Pt clearly visible, and shows the octahedral and cubic orientation of the Os lamellae, // (111) and (100) of Pt, resp. CR 297 (**o + p. w**). **g** Cu-Ni ore with early gersdorffite (white, hypidiomorphic) included and partially corroded by massive pyrrhotite (pinkish brown). Included in po, a sphalerite particle (gray) and successive exsolutions of chalcopyrite, ccp (yellow), and pentlandite, pn (white), can be seen. While ccp occurs in conspicuous lamellae or thicker lenticles, pn forms very thin cords or filiform inclusions, which can be easily overlooked and whose cross-cutting relationships show that they are younger than ccp. CR373, Frood Mine, Sudbury, Canada (**dsp**). **h** Ghost colomorphic texture in massive sulfides: incipient crystallization of pyrite gel, in which syn-sedimentary banding, folded by submarine slip, are still visible (*comp.* Figure 1.42c–e). VA3-514, Masa Valverde, FPI. Drill hole 3, Huelva, Spain (**dsp**). **i** Emulsion of native As and stibarsen, by immiscibility of both phases upon crystallization of primitive As-Sb gel, still recognizable as concentric banded ghost texture (*comp.* Figure 1.42g, h). CR 345, Engineer Gold Mine, Atlin District, BC, Canada (**osp**). **j–l** Mineralogical and textural evolution in Mn oxide ore: crystalline pyrolusite, secondary to manganite (gray, few relicts), **j**: partially replaced by radial fibrous psilomelane (gray botryoidal mass); **k**: progress of alteration, with appearance of mossy variety of psilomelane; **l**: incipient recrystallization of mossy psilomelane in radial fibrous textures. CR 289, Amalienhohe, Waldalgesheim, Bingerbrück, Germany (**osp**). **m.** *Bird's eye* decay texture of pyrrhotite, totally replaced by the so-called *intermediate product* (cryptocrystalline aggregate of pyrite, marcasite, etc.), with minor gold, arsenopyrite and quartz. CR-295, La Rinconada, Puno, Peru (**osp**). **n–p** (**osp**, **osp** and **o + p**, resp). Decomposition of *nickeliferous pyrrhotite* (pyrrhotite + pentlandite), included in unaltered chalcopyrite (**n**). The neoformed minerals are pyrite + marcasite (secondary to pyrrhotite) and violarite (from pentlandite). **o**, **p** shows the same process on granoblastic pyrrhotite, whose texture remains (ghost) despite advanced mineralogical transformation; the neoformed marcasite (//0001 of pyrrhotite) mirrors the corresponding orientation of each pyrrhotite grain. Pyrrhotite relicts, brownish-pink, show light green anisotropism, contrasting with the yellowish-white color and isotropy of pyrite and the greenish-white color, bluish-green anisotropism and lamellar habit of marcasite; violarite (pink, isotropic) is cryptocrystalline and appears darker. HD 6771, Horbach, Schwarzwald, Germany

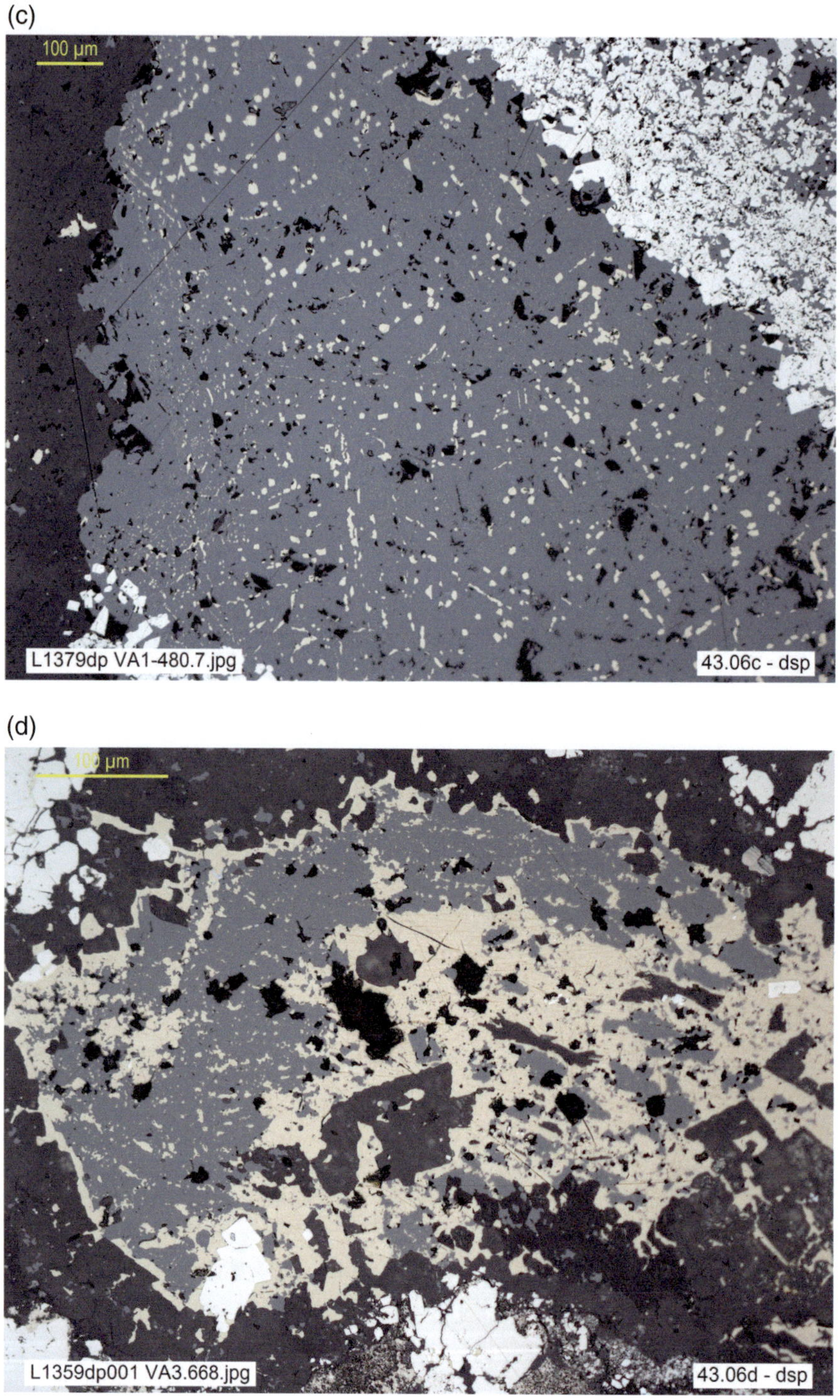

Fig. 1.44 (continued)

(e)

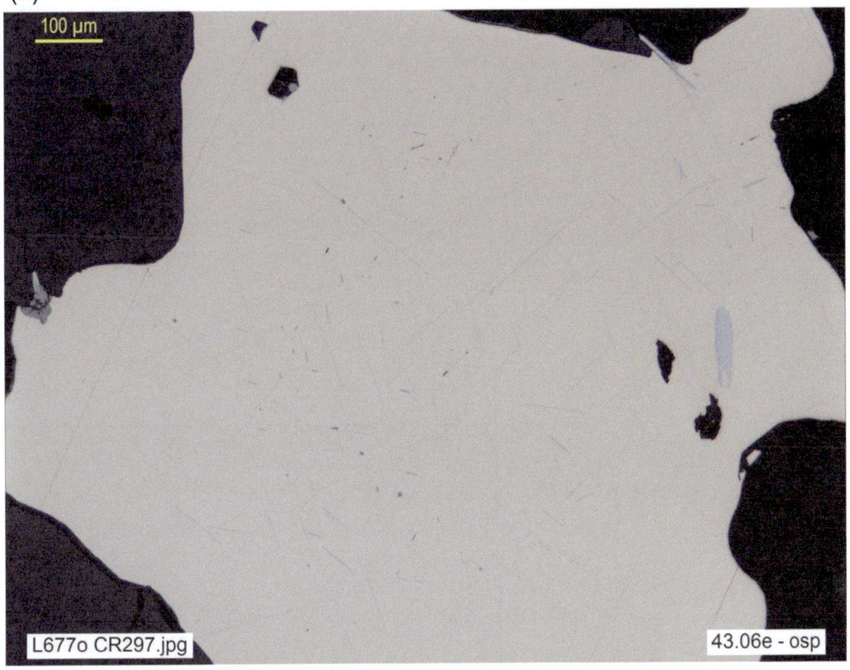

(f)

Fig. 1.44 (continued)

(g)

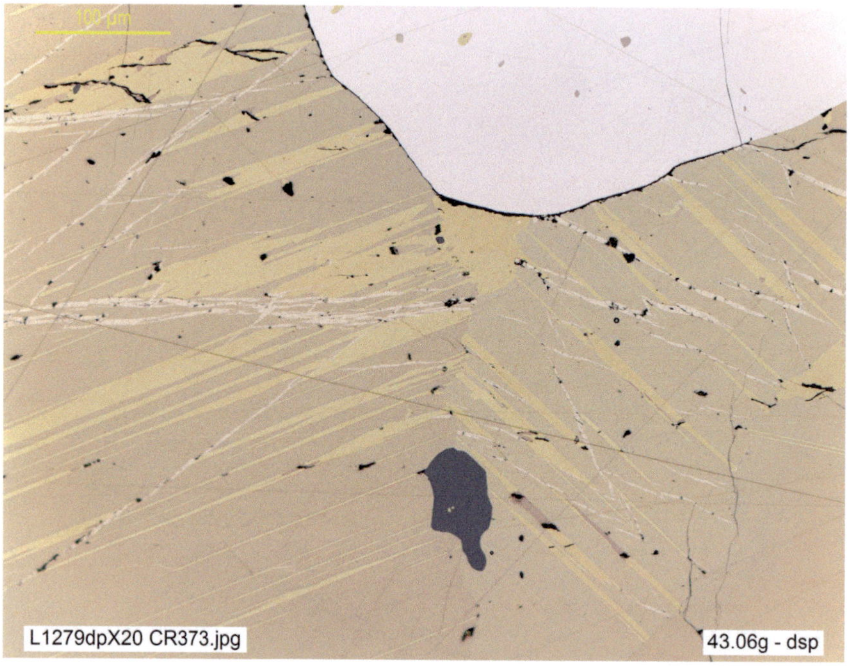

(h)

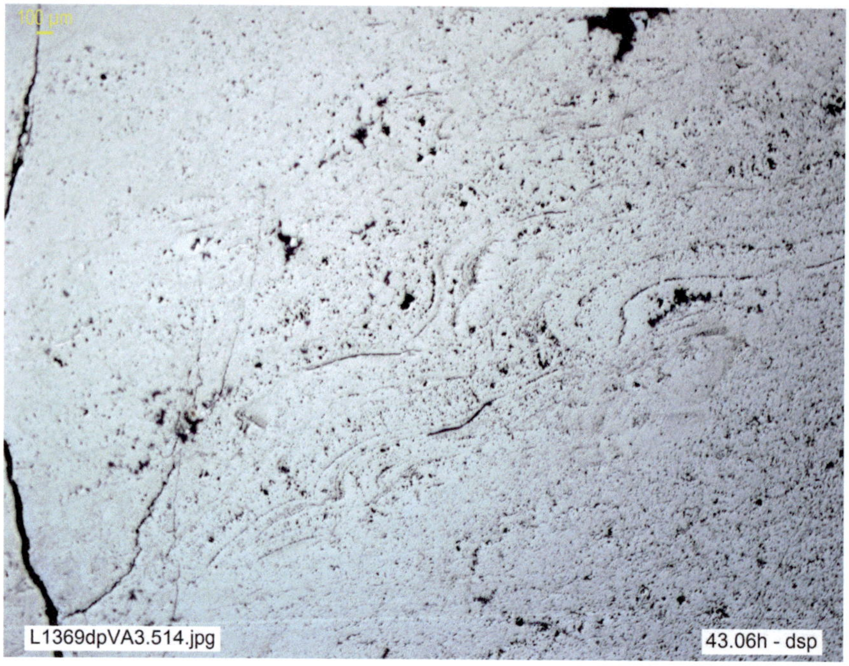

Fig. 1.44 (continued)

Fig. 1.44 (continued)

(k)

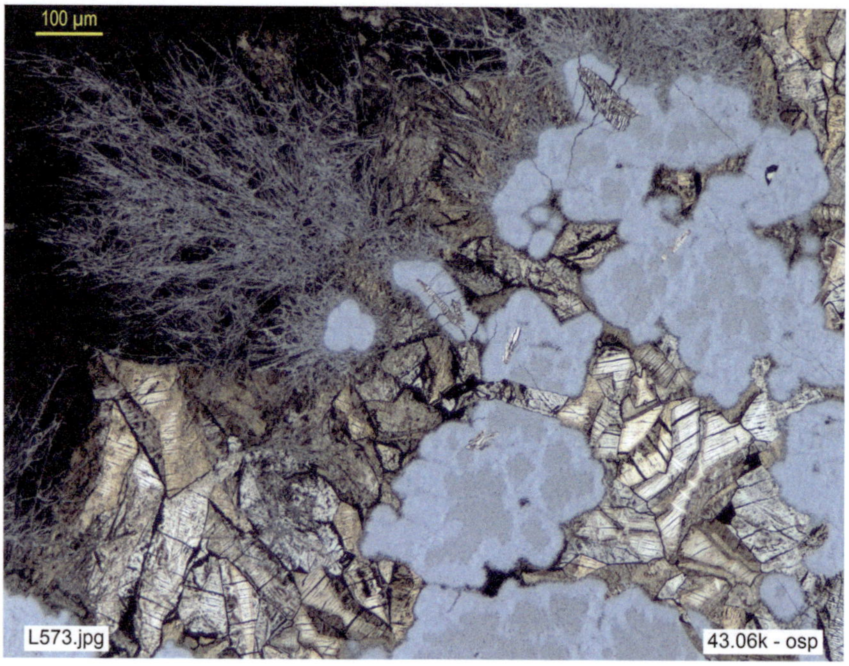

(l)

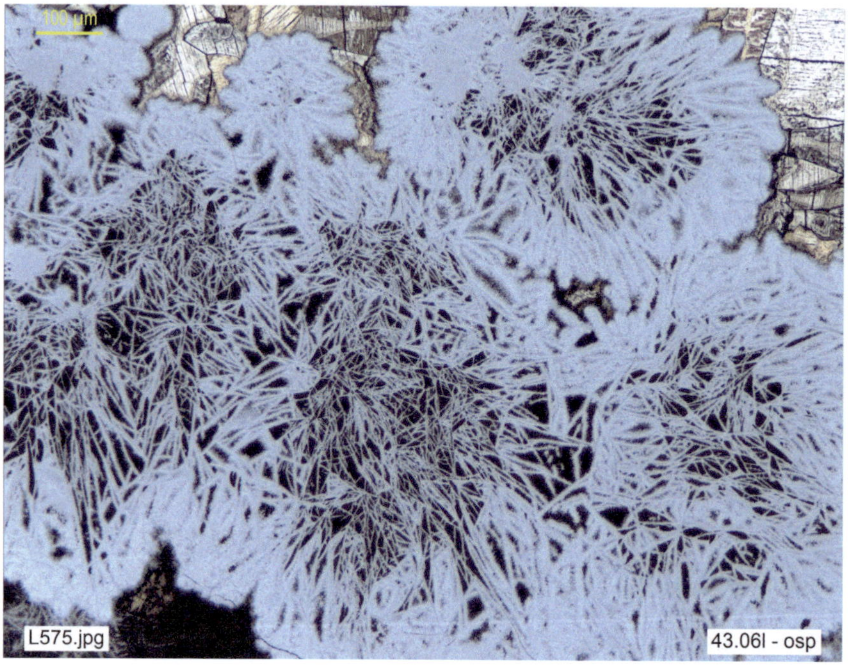

Fig. 1.44 (continued)

Decomposition or **decay** textures comprise intergrowths produced by the decomposition of a precursor mineral phase into several others by changes of temperature, pressure, f_{O2}, f_S, by reaction with other phases, by radioactivity, etc. (*cf.* Figure 1.44m–p). The existence of *eutectoids* has been experimentally established for some mineral phases. These points, determined on the composition-temperature stability diagrams, may be used as geological thermometers, if the simultaneous precipitation of the corresponding mineral phases, intergrown in equilibrium, is demonstrated by textural evidence. On cooling a high-temperature phase of eutectoid composition, it decomposes precisely at the eutectoid temperature, so that thereafter only the low-temperature phases are stable, whose simultaneous precipitation can be recognised by their typical finely intergrown, symplectitic textures (Table 1.2, *cf.* Figure 1.31a–c). The process is similar to that of crystallization of a eutectic, except that it occurs by transformation in the solid state and not by direct precipitation from solution.

(m)

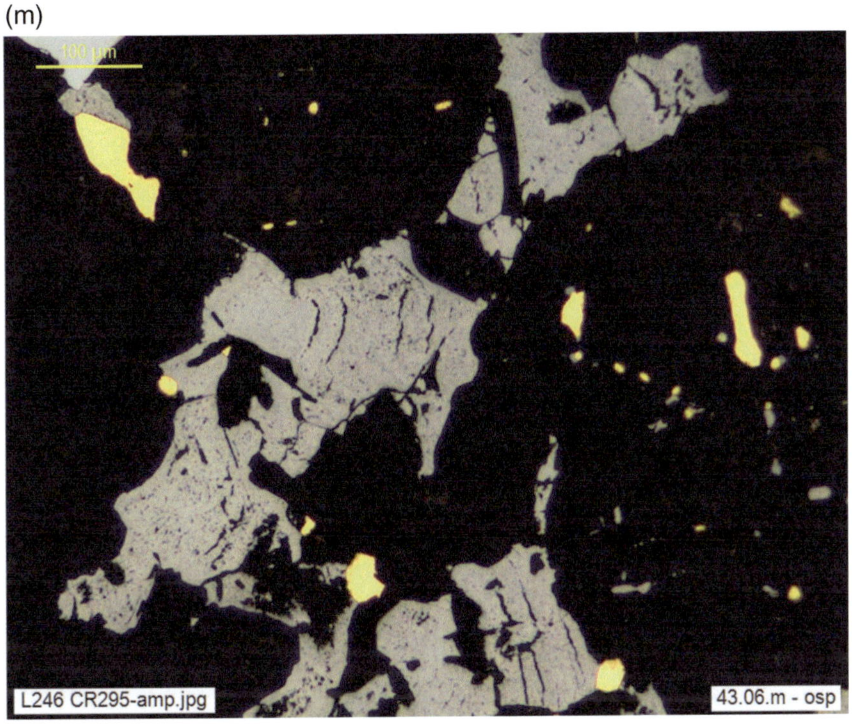

(n)

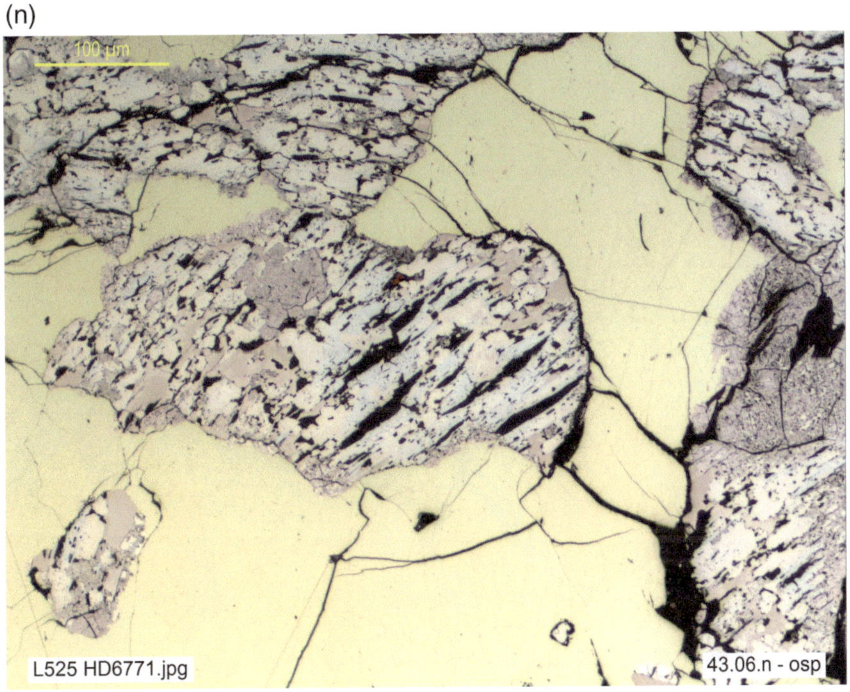

Fig. 1.44 (continued)

(o)

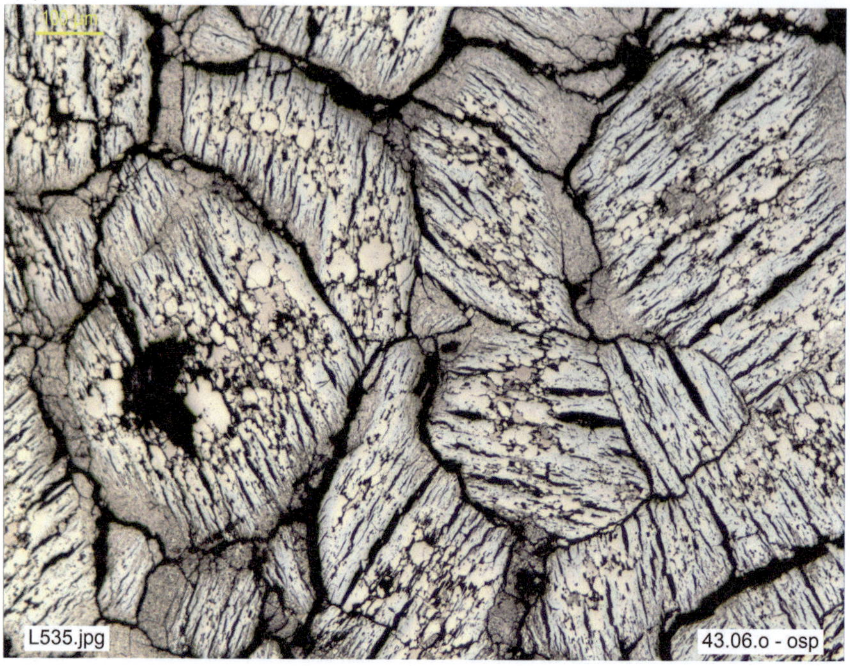

(p)

Fig. 1.44 (continued)

Metamorphism produces textural, mineralogical and, in the case of metasomatism, chemical changes in any rock. The *textures* that can be expected in metallic ores are equivalent to those found in common rocks, with the difference that the recrystallization processes in the former are more rapid and complete, so it may be more difficult to identify the starting textures. Thus, for example, *granoblastic* textures are frequent in massive sulfides (Fig. 1.45a, b), even with a very low metamorphic grade, as are large *porphyroblasts* in more advanced grades. Otherwise, the description of metamorphic processes—dynamic, thermal, regional, and so on—and of metamorphic textures (vbgr. Figure 1.45c–k) uses the same formal criteria for ore deposits as for petrology, which the interested reader will easily find in standard texts on metamorphic petrology (cf. Fettes and Desmons 2007).

Fig. 1.45 a, b (**dsp** and **d + p**, resp). Granoblastic aggregate of pyrrhotite, with interstitial sphalerite and chalcopyrite and idioblastic arsenopyrite. M170800.1, Sulitjelma, Norway. **c** Granoblastic franklinite ore (light gray), with willemite and silicates (gray). CR 364, Franklin Furnace, New Jersey, USA (**dsp**). **d** Massive granoblastic ilmenite (with fine emulsion of hematite, interstitial goethite and small inclusions of pyrrhotite). HD6764, Abu Galqua, Egypt (**dsp**). **e** Recrystallized chromitite cumulus texture with interstitial olivine and minute pentlandite inclusions. CR 355, Prootboom Mine, Bushveld, RS Africa (**dsp**). **f** Diablastic texture in hematite aggregate, with relics of magnetite in process of martitization. CR 370, Mina Recanzoni, Alidos, Mexico (**dsp**). **g** Idioblast of magnetite included in massive microcrystalline hematite. CR 371, BIF Minas Gerais, Brazil (**dsp**). **h** Recrystallized massive sulfide ore: granoblastic pyrrhotite, including pyrite idioblasts and interstitial chalcopyrite and sphalerite. M170800.1, Sulitjelma Mine, Norway (**dsp**). **i** Idio-poikiloblastic pyrite growth in deformed phyllite, with fine dissemination of framboidal pyrite and rutile marking the schistosity. Poikilitic inclusions in pyrite idioblasts preserve external orientation—parallel to the phyllite schistosity—indicative of their post-tectonic growth. VA14-377.80, FPI, Masa Valverde, drill hole 3, Huelva, Spain (**dsp**). **j** Pyrite poikilo-porphyroblast in metamorphosed massive sulfide ore (sphalerite, chalcopyrite, pyrrhotite). M170800.2, Lergruvbakken Mine, Roros, Norway, (**dsp**). **k** Dynamic metamorphism in massive sulfide ore: pyrite (brittle: microclasts) and chalcopyrite (ductile: matrix) breccia, with traces of covellite and sphalerite. Mpy001.r, Moeche, Coruña, Spain (**dsp**)

(c)

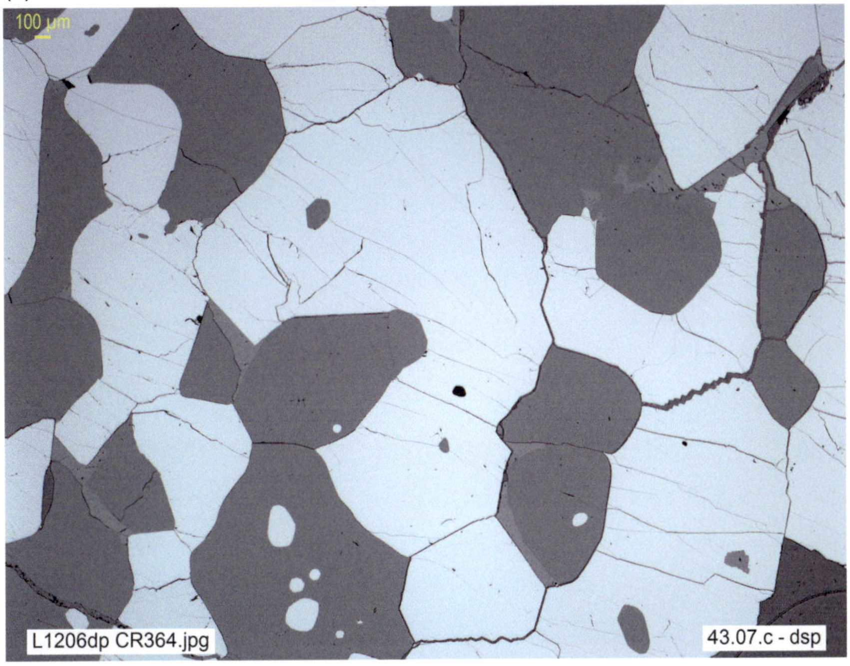

(d)

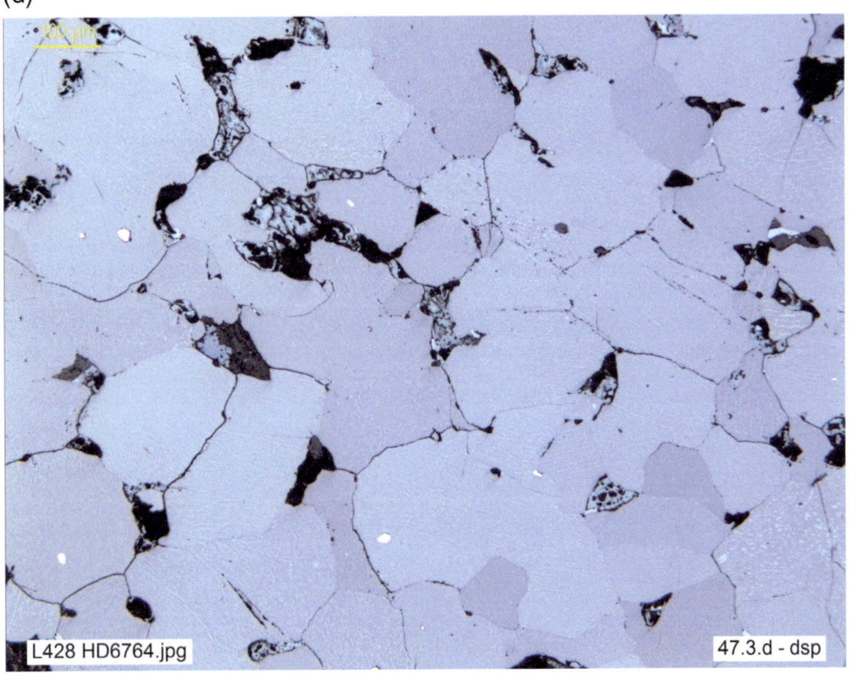

Fig. 1.45 (continued)

(e)

(f)

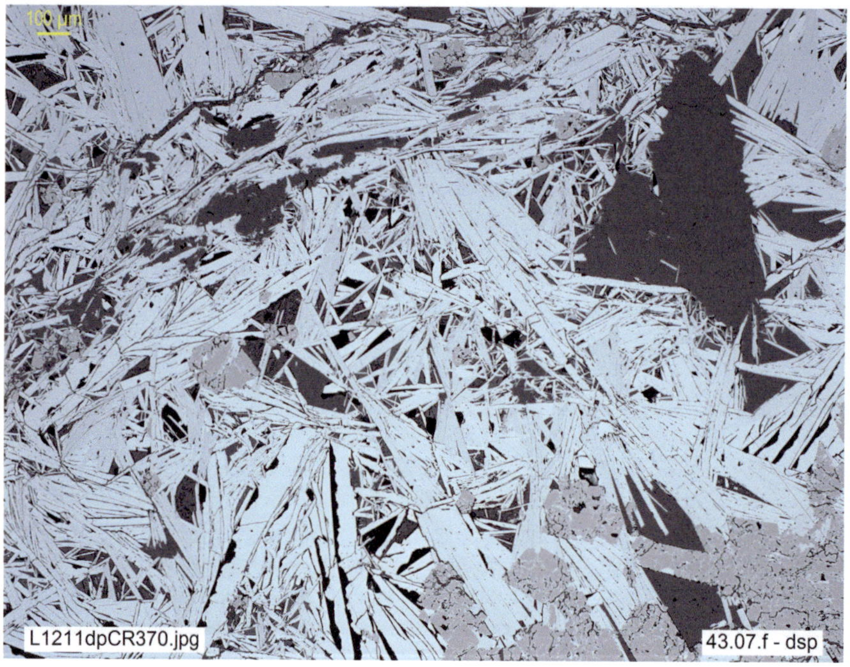

Fig. 1.45 (continued)

(g)

Fig. 1.45 (continued)

(i)

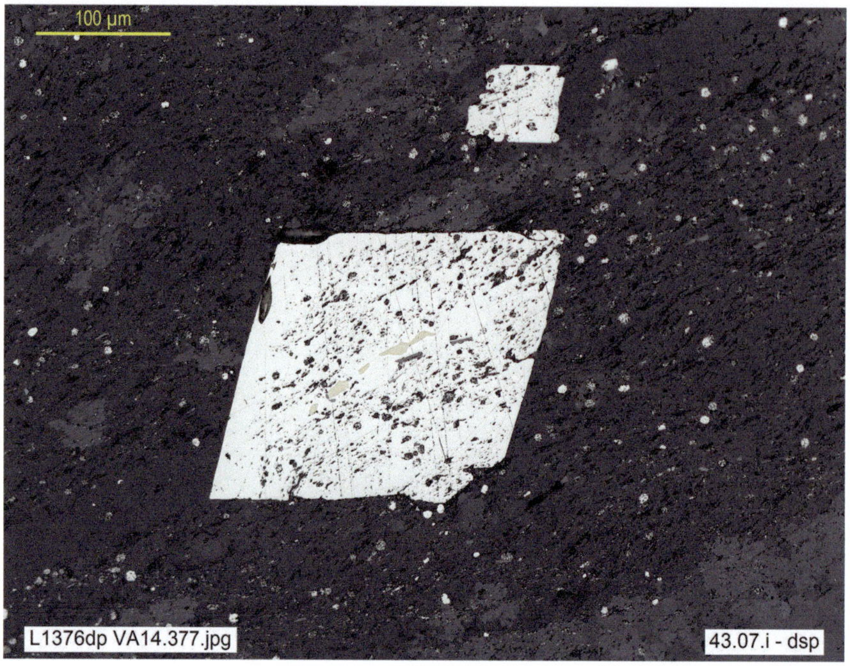

(j)

Fig. 1.45 (continued)

(k)

Fig. 1.45 (continued)

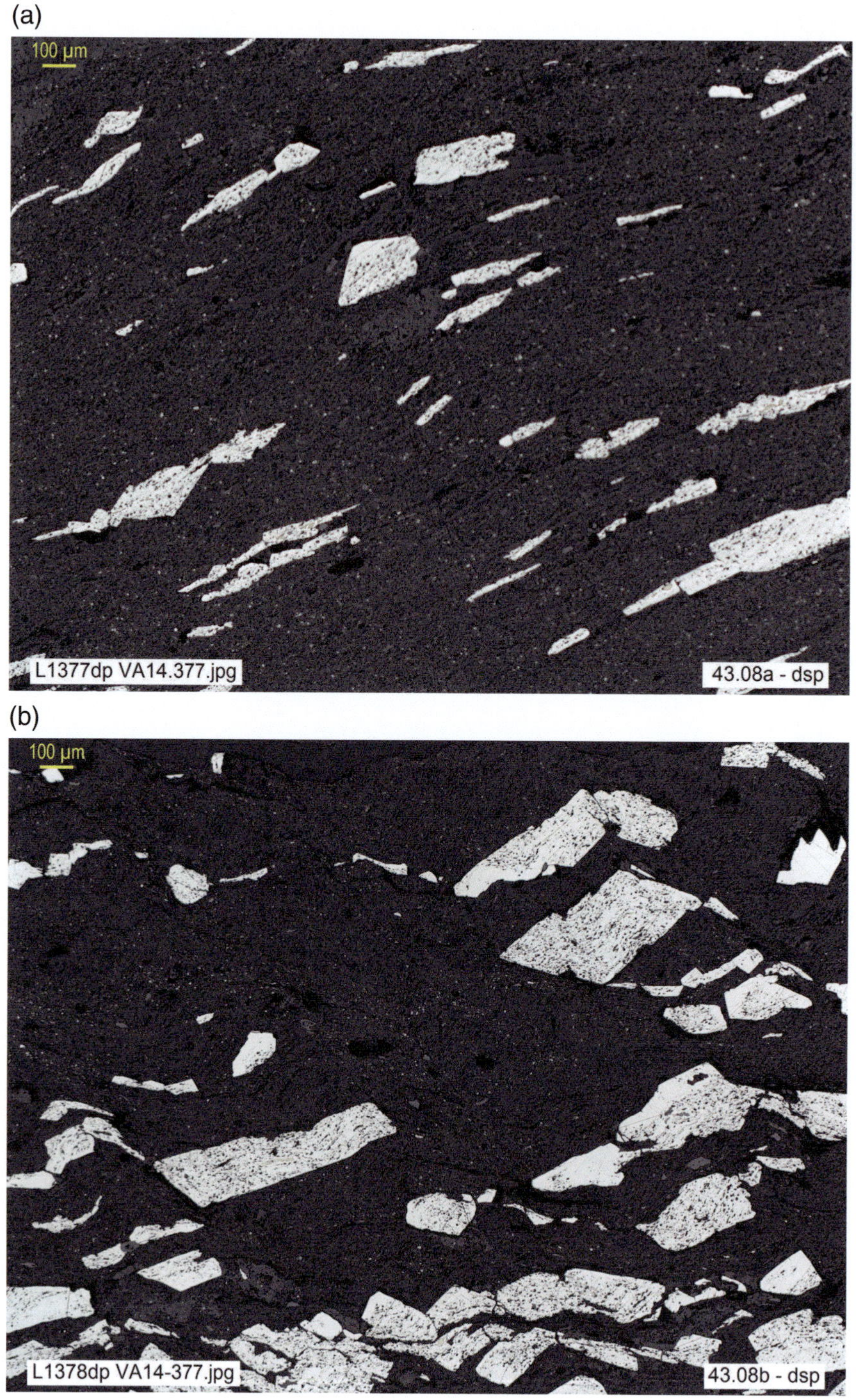

Fig. 1.46 **a** Fluid-assisted re-orientation of pyrite by pressure solution of phenoblasts, small-scale mobilization and re-precipitation in pyrite tails, resulting thin bands perpendicular to the maximum stress and elongated parallel to schistosity. VA14-377.80, FPI, Masa Valverde, drill hole 3, Huelva, Spain (**dsp**). **b** Same process as 47 a, followed by microfolding of the schistosity and solution transfer to post-tectonic pyrite phenoblasts, which include acicular rutile and phyllosilicates whose orientation mimics the folded schistosity of the phyllite. VA14-377.80, FPI, Masa Valverde, drill hole 3, Huelva, Spain (**dsp**)

Remobilization is the redistribution of the metal content of an ore deposit, usually facilitated by fluid influx and thermal activation linked to metamorphism. Although it is actually metamorphic mobilization in most cases (Fig. 1.46a, b), it can also occur in other environments, such as diagenetic or supergene. The associated textures are varied and combine evidence of corrosion or pressure solution in the source zones with that of transport and dissemination: infill of veinlets, microcracks, pores, breccia cementation, sealing of schistosity planes, intergranular films, etc. Frequently, there is also evidence of replacement of the host rock related to these veinlets (*cf.* additional comments under *pressure-solution:* Table 1.3 and Fig. 1.53). In some cases, such as those of massive sulfides, mobilization in a *ductile regime* is also postulated for the more plastic ores, such as galena or chalcopyrite, so that, due to the effect of differential pressure, these are embedded in open spaces and veinlets or cementing clasts of the more rigid ores, such as pyrite or arsenopyrite (*cf. Deformation:* Table 1.3 and Fig. 1.52).

Replacement, a fluid-assisted process of *substitution* in the solid state, atom by atom, of one mineral phase by another, is very common in ore deposits and provides a concrete image of what on a large scale is known as *metasomatism*. Here it is to be analysed only from the point of view of the textures observable under the microscope. Nevertheless, it is worth bearing in mind that there are also other scales or complementary sources of information, such as chemism, that should never be ignored. From a conceptual point of view, replacement textures are easily distinguished from infill (primary) textures by evidence of the disequilibrium features related to the substitution process (reaction rims, relicts, ghosts, jagged contacts, pseudomorphs, etc.). They also differ from the textures related to reequilibration described above since, in the latter, the transformation is internal, occurring inside the grain, while replacement comes from the outside (for example, from the edges towards the core, or advancing from veinlets towards the interior of the aggregates).

In fact, hydrothermal infill and replacement are often found in association, with one or the other as main process, depending on the location of the sample or the type of deposit. Therefore, the microscopist must get used to deciphering textural features that may appear ambiguous (as shown in Figs. 122.4a, b, vol. 1).

Reaction rims (Fig. 1.47a–e) are irregular and jagged grain contacts resulting from corrosion of the original grain (*primary mineral*, unstable in the new conditions), as the first step in the deposition of the phase that replaces it (*secondary or neoformed* mineral).

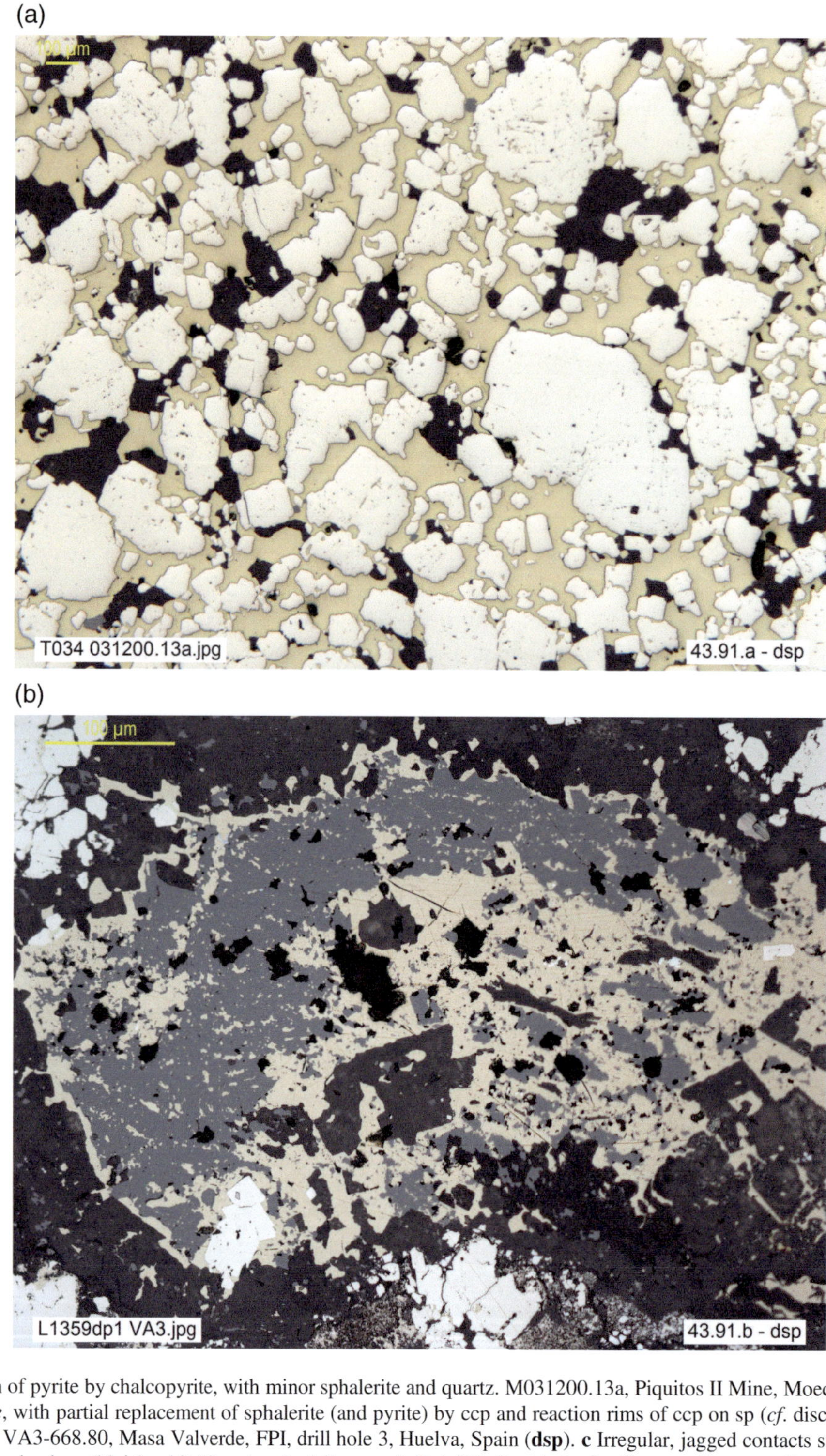

Fig. 1.47 **a** Corrosion of pyrite by chalcopyrite, with minor sphalerite and quartz. M031200.13a, Piquitos II Mine, Moeche, Coruña, Spain (**dsp**). **b** *Chalcopyrite disease*, with partial replacement of sphalerite (and pyrite) by ccp and reaction rims of ccp on sp (*cf.* discussion Fig. 1.44c and d). Massive sulfide *feeder* VA3-668.80, Masa Valverde, FPI, drill hole 3, Huelva, Spain (**dsp**). **c** Irregular, jagged contacts suggesting replacement of sphalerite (dark gray) and galena (bluish white) by stannite (olive gray). Rub483, Davidschacht, Saxony, Germany (**dsp**). **d** Reaction rim between ilmenite (pinkish brown) and massive titanomagnetite (light brownish gray), with corrosion of magnetite and neo-formation of spinel (dark gray) and ilmenite. CR253, L. Berta, Quebec, Canada (**dsp**). **e** Jagged contacts of quartz gangue and ore, suggesting corrosion of quartz by bismuthinite (gray, with relict skutterudite, bright gray, as a discontinuous and thin rim on bm), followed by native Bi (white, scratched), which corrodes all other ores, defining the crystallization sequence for the hydrothermal infill: quartz-bismuthinite-skutterudite-Bi. C-15, Loma La Pizarra, Villanueva de Córdoba, Spain (**dsp**)

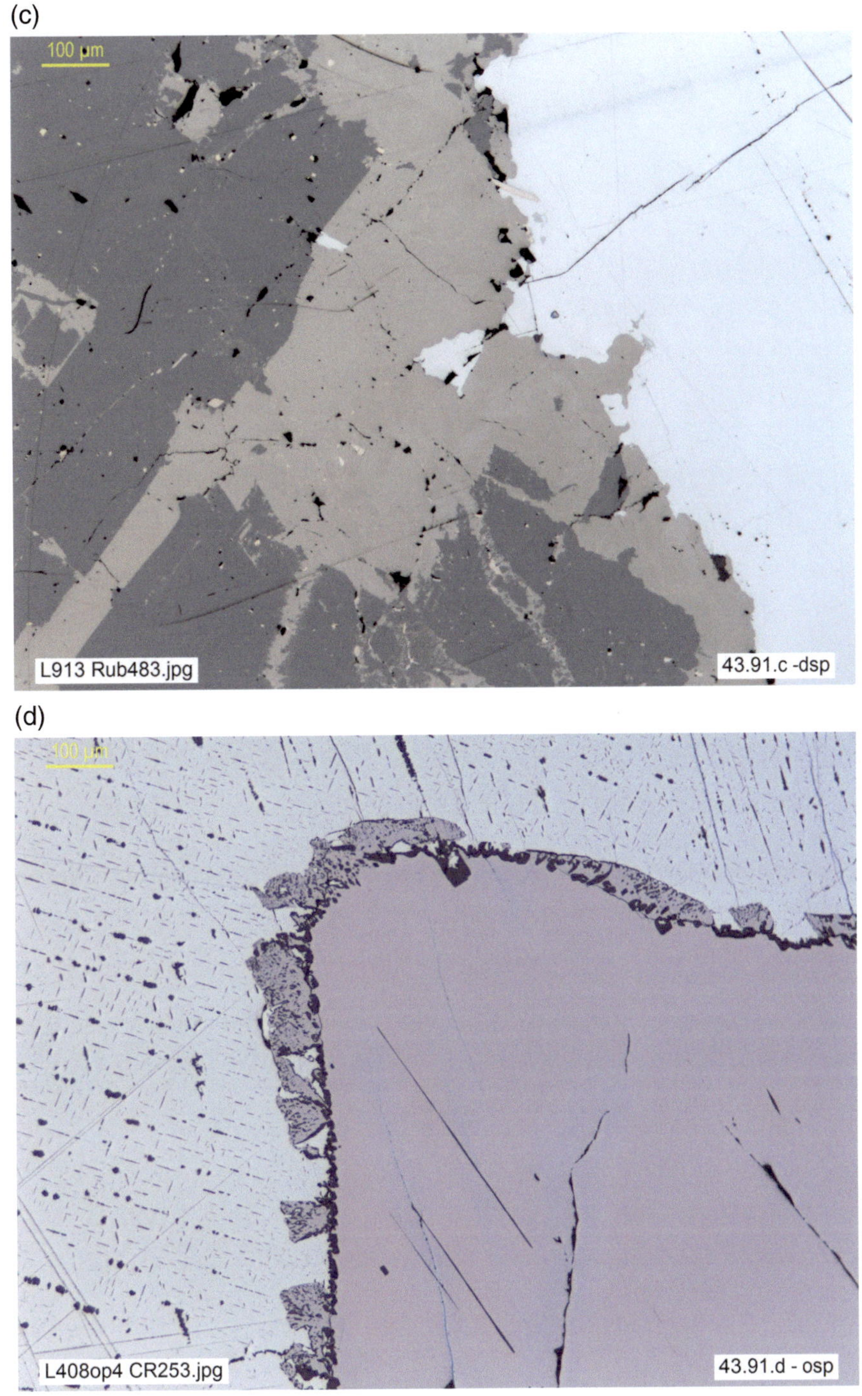

Fig. 1.47 (continued)

(e)

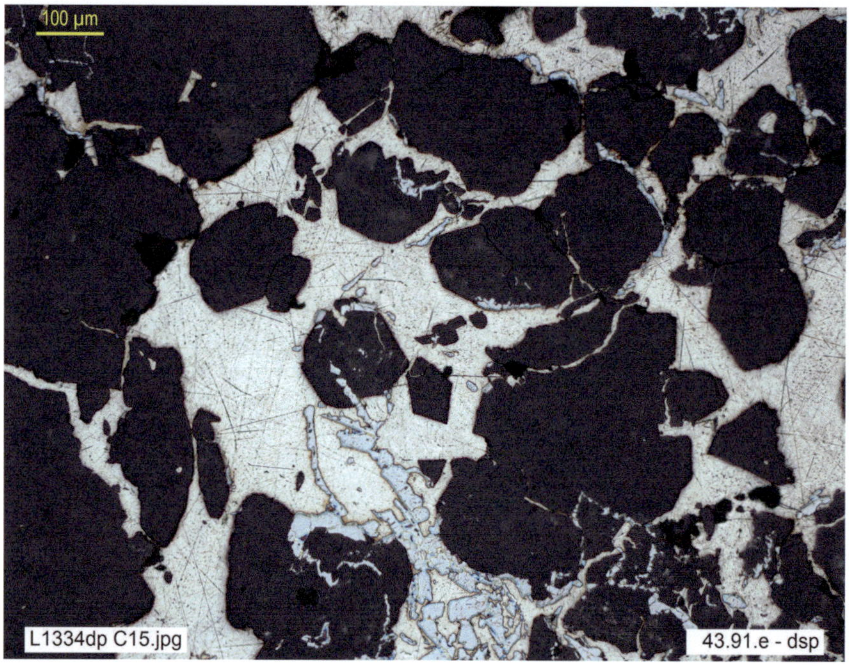

Fig. 1.47 (continued)

Often the replacement is incomplete and, at first, is limited to the immediate surroundings of microcracks or pores. Alternatively it progresses from the edges, forming a narrow rim of replacement (*corona or reaction rim* texture). At a more advanced stage, only a few corroded remains of the primary mineral are preserved as **relict** inclusions in the neoformed mineral, or relict traces of textures (Fig. 1.48a–d), or both (cf. Fig. cc8) may be left.

(a)

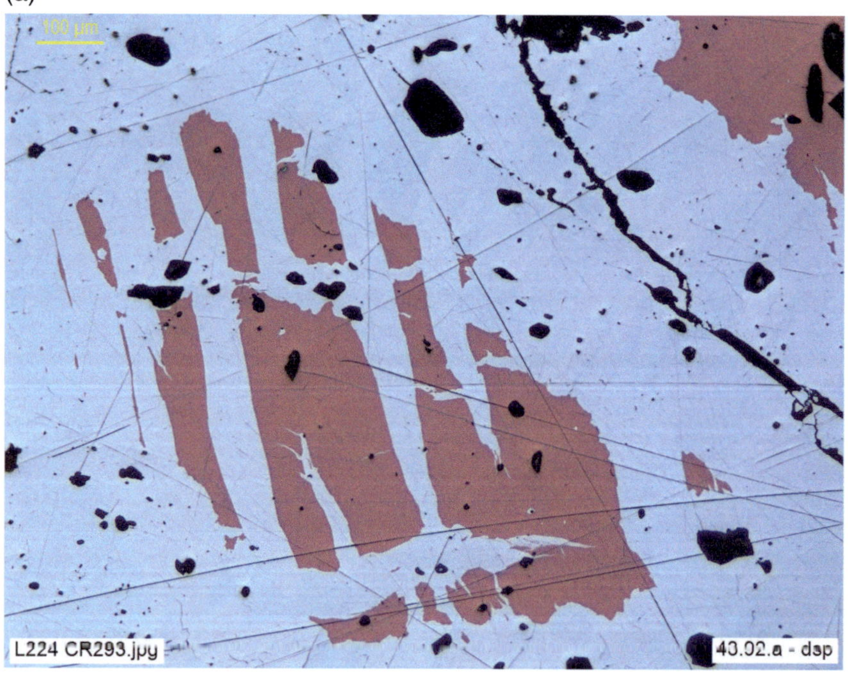

Fig. 1.48 **a** Relicts of bornite in chalcocite, Copperbelt. CR 293, Kipushi, Shaba, DR Congo (**osp**). **b** Massive chalcopyrite with hematite relicts. M031200.30 g, Maruxa Mine, Moeche, Coruña, Spain (**osp**). **c** Relict manganite (scarce, ragged remnants, dark gray) included in pyrolusite, *prl* (yellowish-white to gray, prismatic), itself partially replaced by psilomelane (gray, botryoidal: upper left), which also forms a *corona* rim on prl (right). CR 289, Amalienhohe Mine, Waldalgesheim Bingerbrück, Germany (**osp**). **d** Relict texture (micro-folding by syn-sedimentary sliding) in massive sulfides (meta-pyrite). FPI, VA3-514.00, Masa Valverde, Spain, drill hole 3 (**dsp**)

(b)

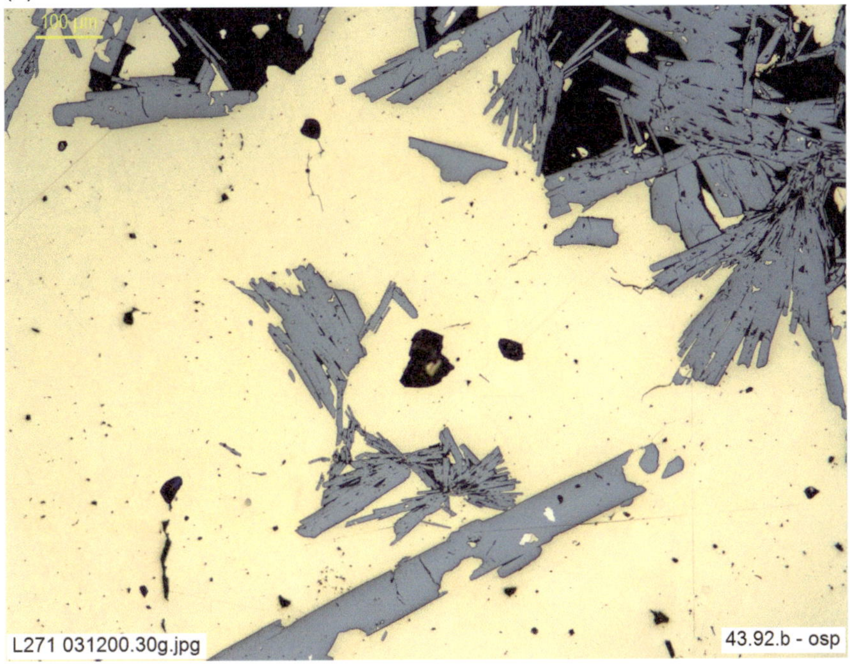

(c)

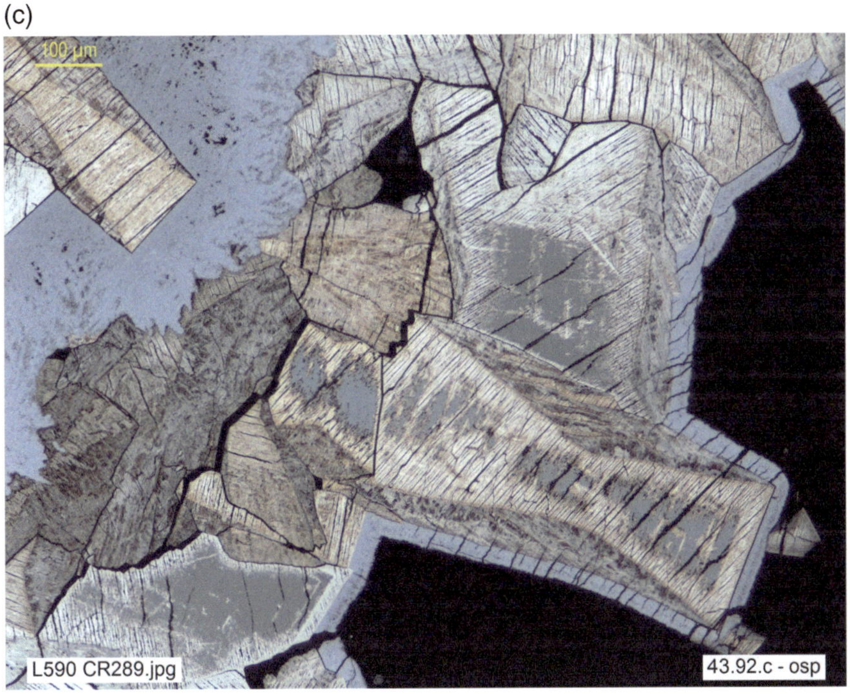

Fig. 1.48 (continued)

(d)

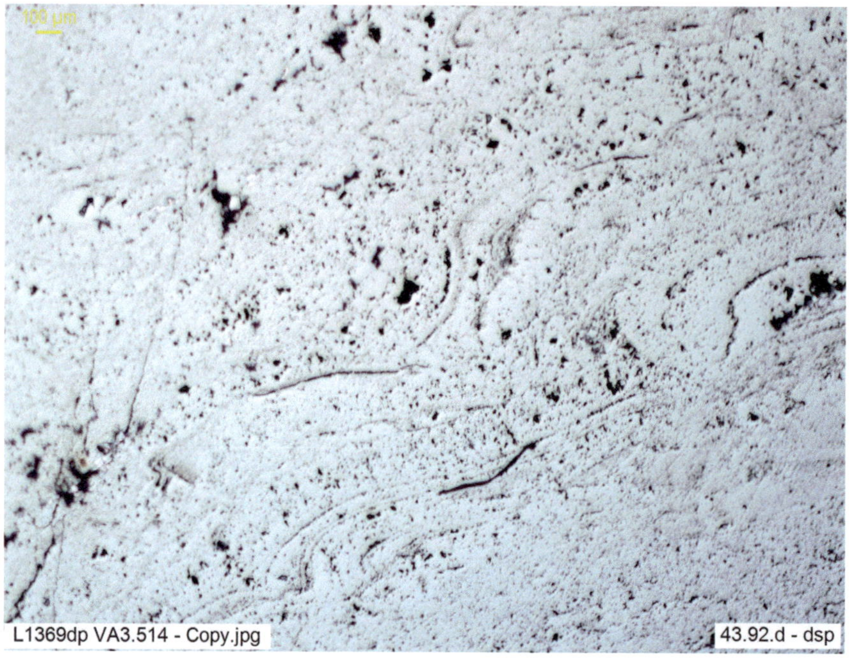

Fig. 1.48 (continued)

At other times the replacement is so advanced that it affects practically the whole primary mineral, yet it is still recognizable by some preserved feature (generally geometrical or morphological, such as cleavage, traces of twinning or growth zoning, crystalline contours, etc.). In this case the textures are known as **ghost** textures (Fig. 1.49a–d). The final result is a **pseudomorph** when the outline or external shape of the primary mineral, although totally replaced by the neoformed mineral, is preserved (*cf.* Figure 1.10a). Authors generally refer to these textures (relict, ghost, pseudomorphic) as *replica textures*, because they reproduce recognizable features of the primary mineral. This is useful as a general concept, but for precision the specific terms are preferred.

(a)

(b)

Fig. 1.49 **a** Pyrite ghosts (preserved as *skeletal* or *atoll-like* vestiges), replaced by secondary bornite (pinkish-brown) and chalcocite (gray). Rub26, Tsumeb, Namibia (**dsp**). **b** Brecciated podiform chromitite alteration: formation of Fe-enriched *coronas* or rims (ferritchromite, magnetite) on chromite microclasts and in microcracks. In addition, silicate and ilmenite neoformation (*cf* Fig. 1.49**c**, **d**). TP110607.21, Tapo Mine, Tarma, Dept° Junin, Peru (**osp**). **c** Advanced alteration process of chromite, chr (*comp.* Fig. 1.49b), almost completely transformed into a more iron-rich, gray variety (intermediate reflectance), including sparse darker relicts of primary chromite (centre left) and neoformed silicates. These (serpentine, chlorite: black) are oriented defining a chr-ghost, as they show the directions of octahedral cleavage (111) inherited from chr. Magnetite (white) and ferritchromite (very light gray) occur in microcracks. Sample TP140607.30, Tapo Mine, Tarma, Dept° Junin, Peru (**osp**). **d** Magnetite ghost, comprising tabular ilmenite (gray-brownish-pinkish) oriented following octahedral magnetite directions, of which some traces (light gray, center left) and sparse relicts remain between the ilmenite sheets. Sample 110806.22, Tapo, Tarma, Peru (**dsp**)

(c)

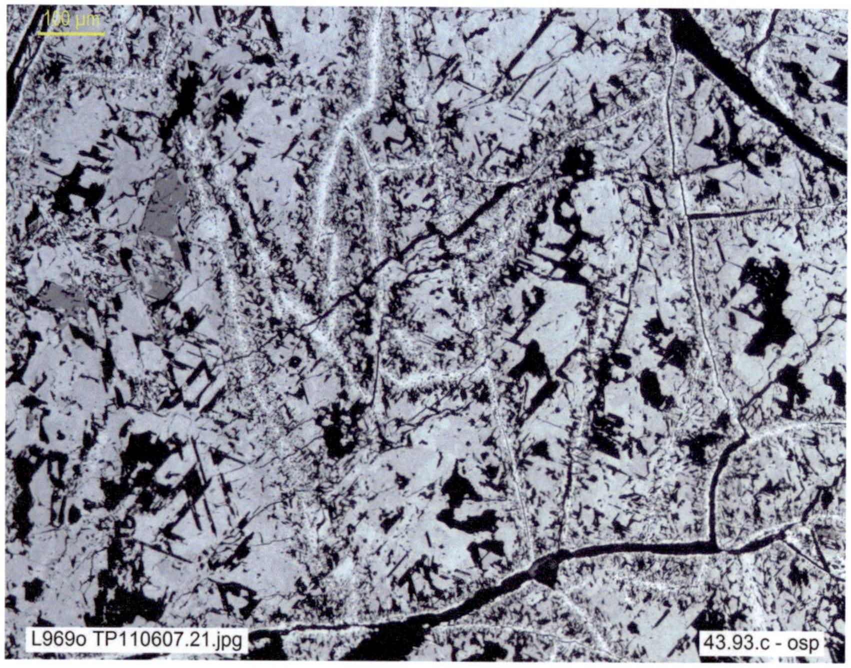

(d)

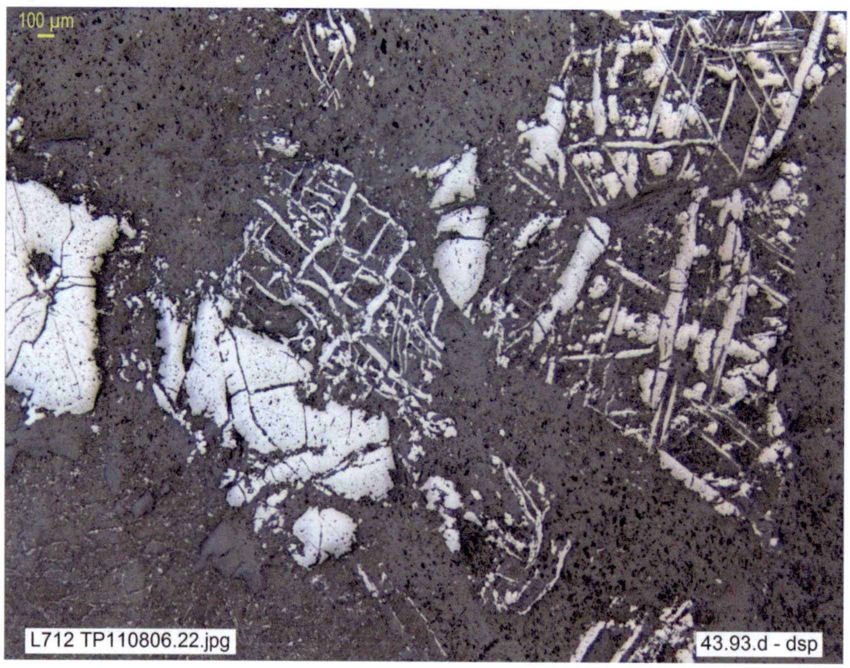

Fig. 1.49 (continued)

An important group of replacement textures comprises those caused by **supergene** processes, which alter the primary paragenesis to replace it by stable associations under surface conditions. They are recognized by the disequilibrium features just discussed and by the mineralogy. Nearest to the surface, the oxide *blanket* appears. The percolation of meteoric waters creates an oxidising environment (**oxidation zone**), where metals are leached out[4] and sulfides are replaced by oxides, carbonates, sulphates or silicates (goethite, lepidocrocite, hematite, malachite, azurite, jarosite, chrysocolla, clays, opal, etc.: Fig. 1.50d–h).

These secondary minerals are seldom seen distinctly (individualized) under the microscope, because of their low crystallinity, as with limonites in *iron caps* or *gossan* deposits. Even so, their textural study can provide valuable information for exploration:

- First, it allows us to distinguish between *autochthonous gossan* and *transported gossan*. The first type is formed in situ, replacing a mass of sulfides: it therefore identifies the position of the primary deposit at depth. The second is re-deposited; that is, it has undergone transport, so its position is no longer directly related to the primary deposit.
- Second, inspection of the ghost textures inherited by the autochthonous limonite (*massive*, *cellular*, *boxwork*, etc.: Fig. 1.50f, g), can provide information about the nature of the primary ore and guide strategic decisions about the ore's potential at depth (for more information see the monograph by Blanchard, 1968, a classic that still deserves attention, and the work of Blain and Andrew, 1977.

[4] Nevertheless, precious metals (Au, Ag, EGP) are sometimes enriched in the oxidation blanket and in laterites.

(a)

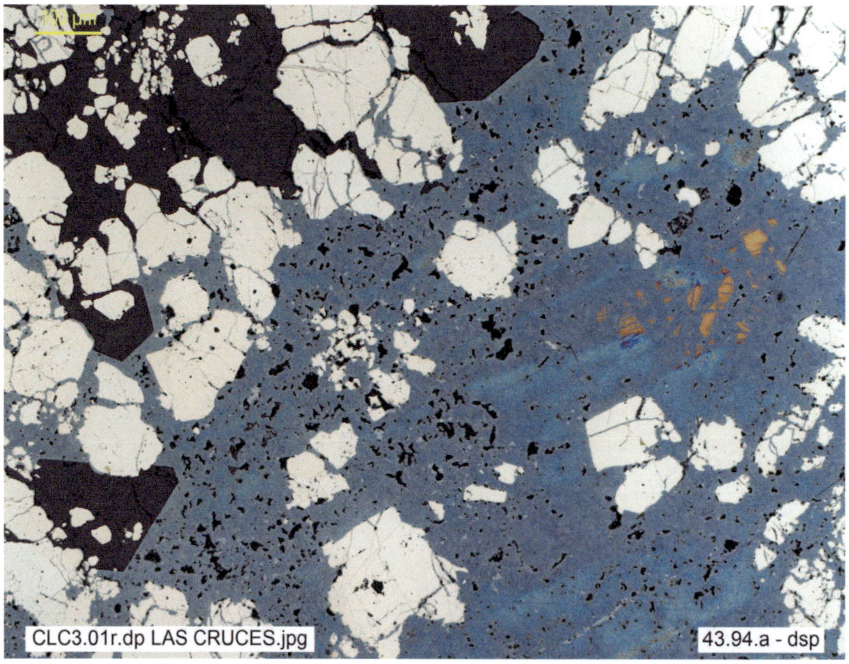

(b)

Fig. 1.50 **a** Supergene secondary enrichment in massive sulfides: intergrown digenite, chalcocite and covellite, replacing chalcopyrite (scarce brownish- yellow relicts left, darkened by tarnishing) and, to a lesser extent, also pyrite. CLC3.01r, Las Cruces Mine, Seville, Spain (**dsp**). **b**, **c** (**dsp** and **d + p**, resp). Evolutionary redox, from sulfide stability (*chalcocite*, cc, bluish-white) to oxidation conditions (*cuprite*, cup: gray, red RI, internal reflections; *malachite*, mch: bireflective, gray, green RI; *limonite*, lim: gray, orange-brown RI, sparse). Two generations of chalcocite, separated by a narrow *corona* of cup, define the sequence: cc1 (oval-shaped, center right)-cup (thin coating on cc1)-cc2 (final sulfide event), suggesting a short-lived oxidative event (between cc1 and cc2), before the final oxide-carbonate episode. CR-376, Tsumeb, Namibia. **d** Oxidation of native bismuth (white) to bismite (bismuth ochre, gray) in goethite-impregnated carbonate gangue. CR 372, Cobalt, Ontario, Canada (**osp**). **e** Zoned supergene alteration of galena (white) to anglesite and cerussite (both gray, cerussite is lighter and bireflective). CR 347, Webb's Consols Mine, New England, Australia (**dsp**). **f**, **g** (osp and **o + p**, resp). Detail of limonitic *boxwork* (goethite and probable cryptocrystalline cuprite, gray, red internal reflections), in PGE ore: disseminated sperrylite (white, isotropic). CR-386, Tweefontein Pt Mine, Bushveld, South Africa. **h** Brecciated and totally limonitized arsenopyrite-quartz vein, with pseudomorphic replacement of arsenopyrite microclasts by limonite. CR-316, Magros Mine, Beariz, Orense, Spain (**dsp**)

(c)

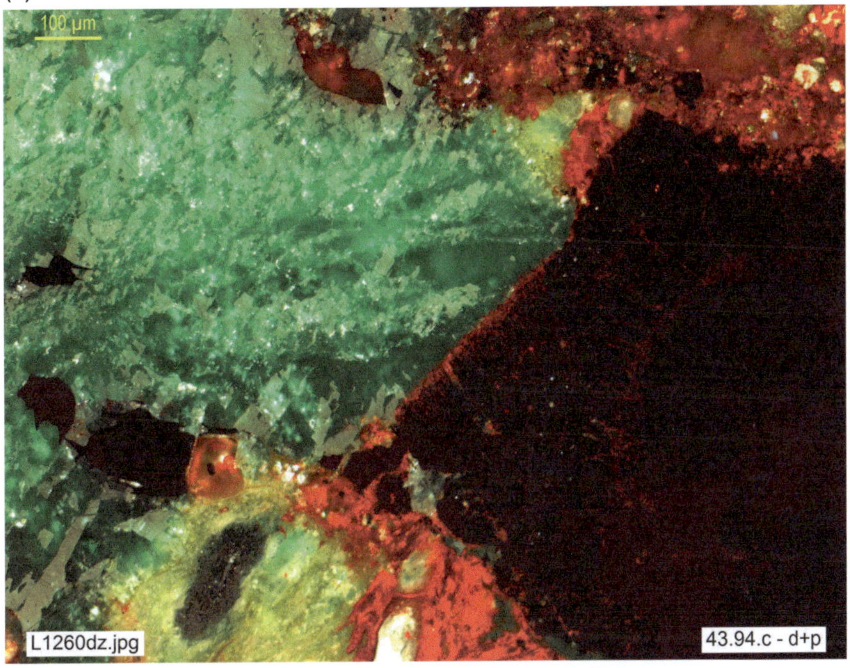

(d)

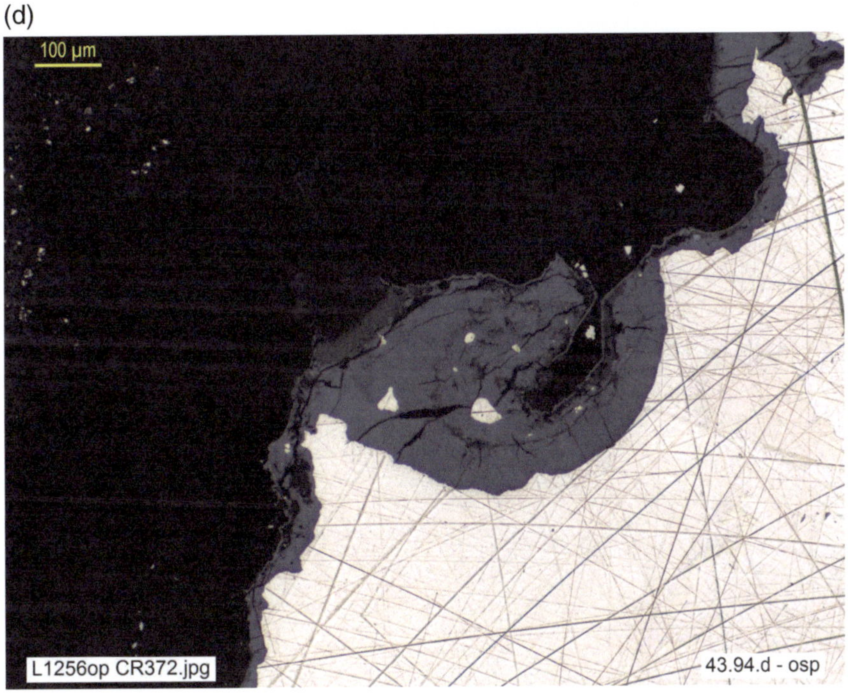

Fig. 1.50 (continued)

(e)

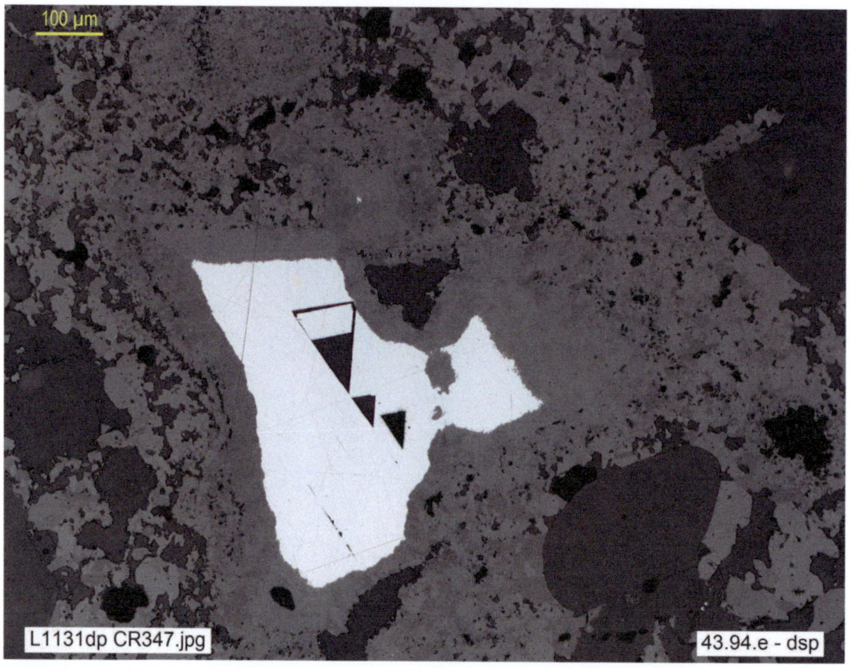

(f)

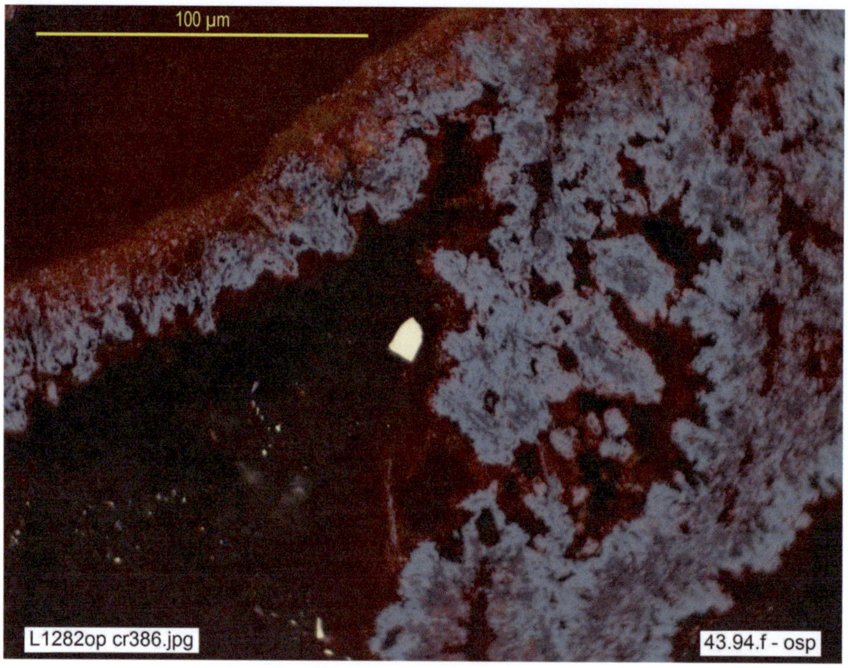

Fig. 1.50 (continued)

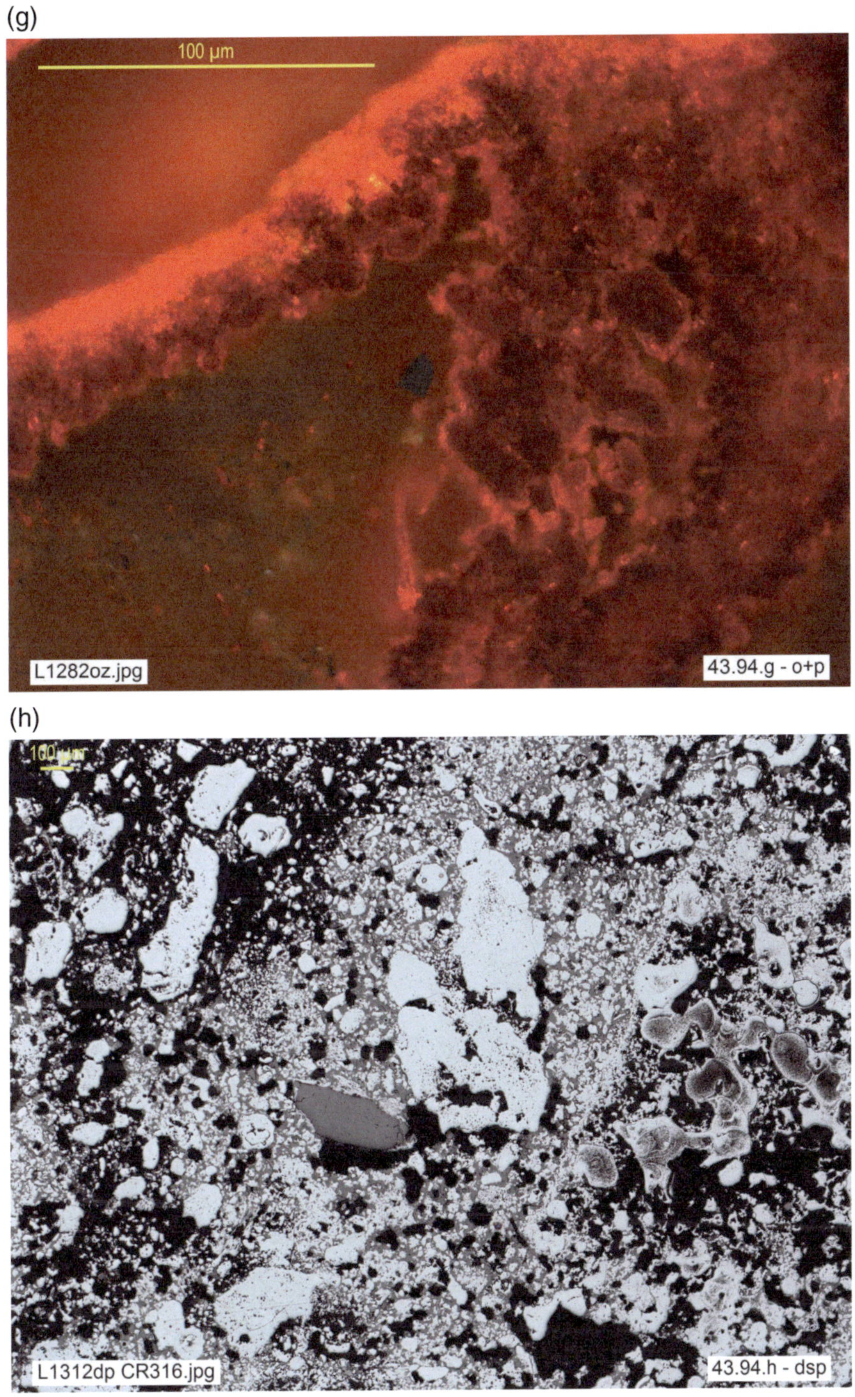

Fig. 1.50 (continued)

At a deeper level, below the water table (or *cementation zone*), reducing conditions prevail and the metals leached in the oxidation zone are no longer soluble, producing a selective precipitation of ***secondary sulfides*** that can greatly increase the metal grade of the primary ore (*supergene enrichment blanket*). For example, in Las Cruces, Seville, mine grades locally exceeding 8*wght%* Cu have been obtained in the cementation zone, as a result of the supergene replacement of primary sulfides, such as pyrite (0% Cu) and chalcopyrite (34% Cu), by secondary sulfides that are much richer, such as calcocite (79% Cu), covellite (66.5% Cu), digenite (76.5% Cu) or djurleite (78% Cu). The textures are, as in the oxidation zone, of disequilibrium by replacement, but in this case the neoformed minerals are sulfides. Frequent, among others, are relicts and ghosts of pyrite or chalcopyrite replaced by chalcocite or digenite, or *coronas* or microcracks of secondary copper sulfides on cores or aggregates of primary sulfides (Fig. 1.50a–c).

Infiltrative textures result from the precipitation of ores from solutions that infiltrate or percolate an orebody or *host rock*; in general, they can be described as **impregnation** textures, usually porous. This is a very general concept that includes hypogene and supergene processes and various precipitation mechanisms, such as pore and vein filling (Fig. 1.51a, b, g, h) or selective replacement by dissolution of the host and precipitation of neoformed minerals (Fig. 1.51c–f). The term *infiltration* has been applied by Korzinskii (1970) to designate a type of metasomatism, the most effective alternative to *diffusion* from a stationary fluid, but it has other meanings as well. It has been applied to processes such as those described in the supergene alteration zones; however, even though the textural evidence may be similar, the concept *infiltrative* refers in this text to processes —usually surficial/sedimentary (*descendant*)—not identified by other more specific terms, for instance infiltrative concentrations of Fe and Mn oxides, karstic, lateritic (Fig. 1.51g–h), and so on.

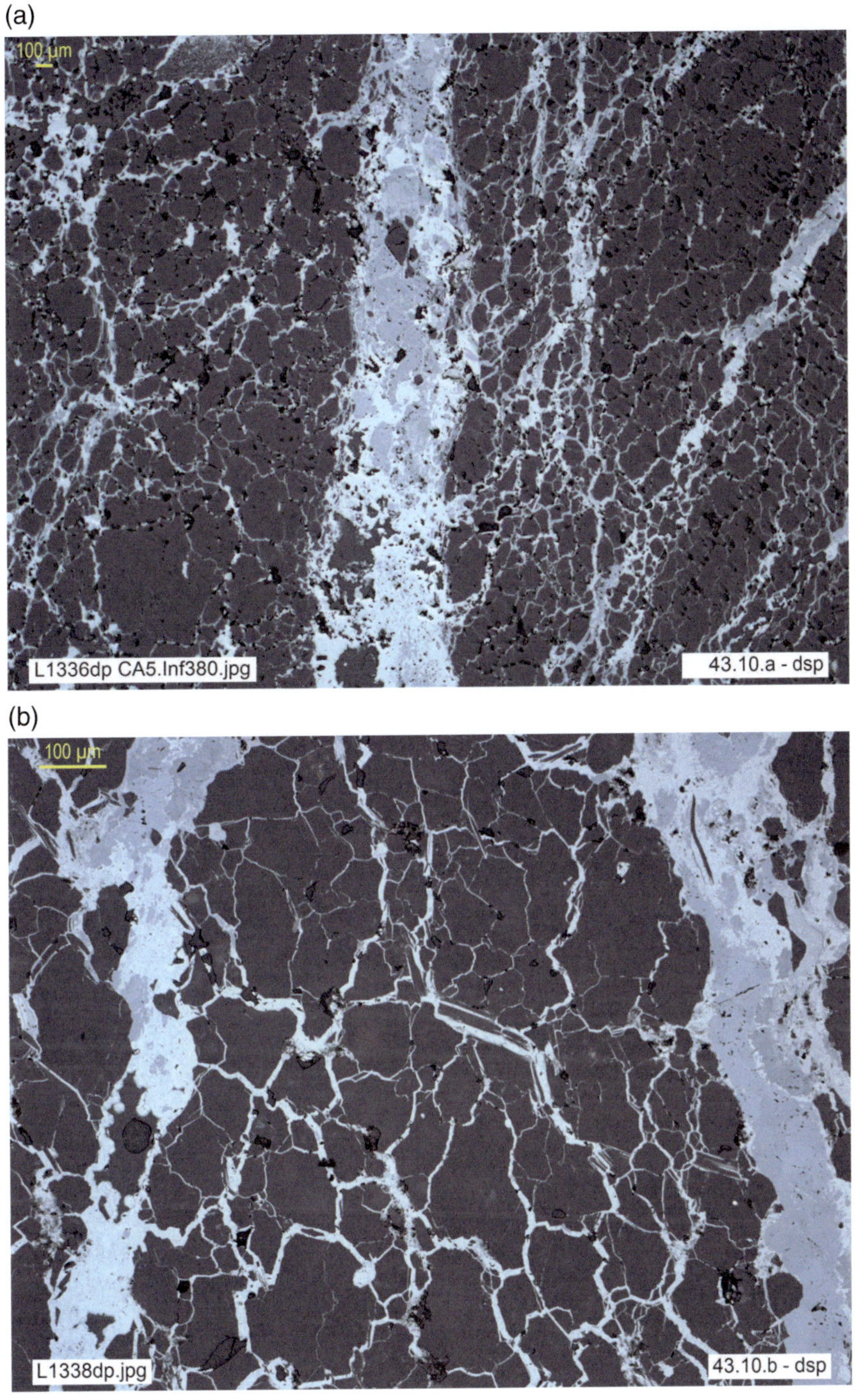

Fig. 1.51 **a** Supergene infiltration in oxidation zone on shattered quartz vein, with microcracks cemented by limonite-goethite. MCA-5-Inf380, Cantal, Lomo de Bas, Murcia, Spain (**dsp**). **b** Limonite-goethite infiltrative texture: detail of Fig. 1.51a (**dsp**). **c** Infiltrative-metasomatic process in *Rotliegendes* sandstones, in contact with *Kupferschiefer*: interstitial pyrite impregnation, with preferential replacement of porous zones (matrix and RF, rock fragments, mudstones). Kupferschiefer MRub010211, Spremberg, Brandenburg, Germany (**dsp**). **d** Detail of the central RF grain of Fig. 52c: porous and reactive (shaly) rock fragment, in whose interior pyrite substitution progresses more rapidly (**dsp**). **e** Advanced impregnation and total replacement of RF by pyrite (same rock as Fig. c, **dsp**). **f** Lithological and mineralogical control of infiltration along a mudstone-sandstone contact. It is very scarce in the impermeable layer (clayey mudstone, bottom), while in the reactive sandstone layer replacement by pyrite is complete on RF, but weak and limited to a very thin reaction rim on quartz sands, and partial in the porous matrix, whose sedimentary features, as the load cast (bottom right), are replicated by pyrite (same rock as Fig. c, **dsp**). **g** Infiltrative gold concentration in lateritic crust, composed of limonite (bluish gray), clay minerals and organic matter (dark gray): gold infiltrates in microcracks and pores. LAT4.13, Minvoul Greeenstone Belt, Gabon (**dsp**). **h** Microcracks in lateritic crust sealed by infiltrative gold and subsequently bent, dislocated and folded as the sediment compacted. LAT4.13, Greeenstone Belt de Minvoul, Gabon (**dsp**)

(c)

(d)

Fig. 1.51 (continued)

(e)

(f)

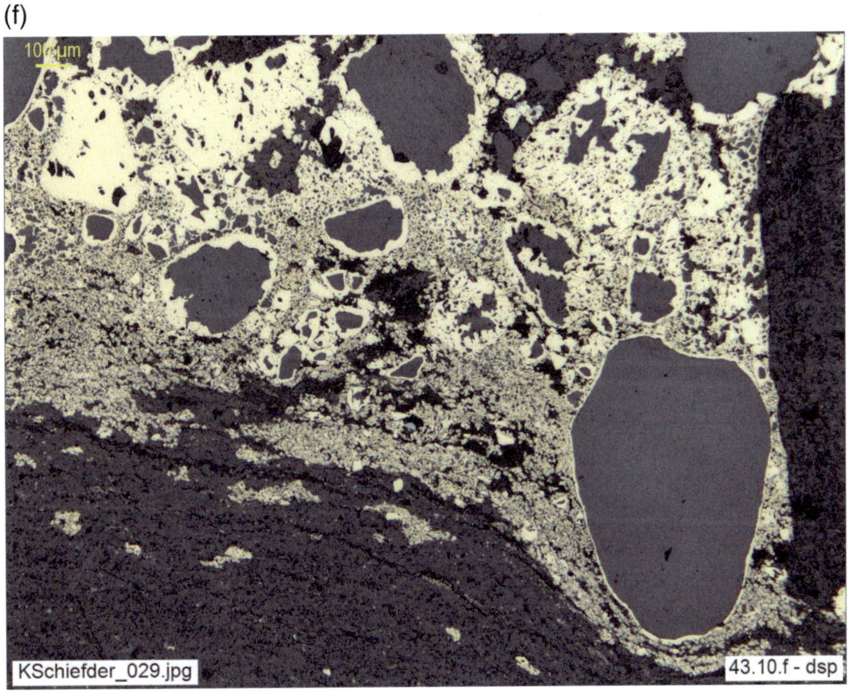

Fig. 1.51 (continued)

(g)

(h)

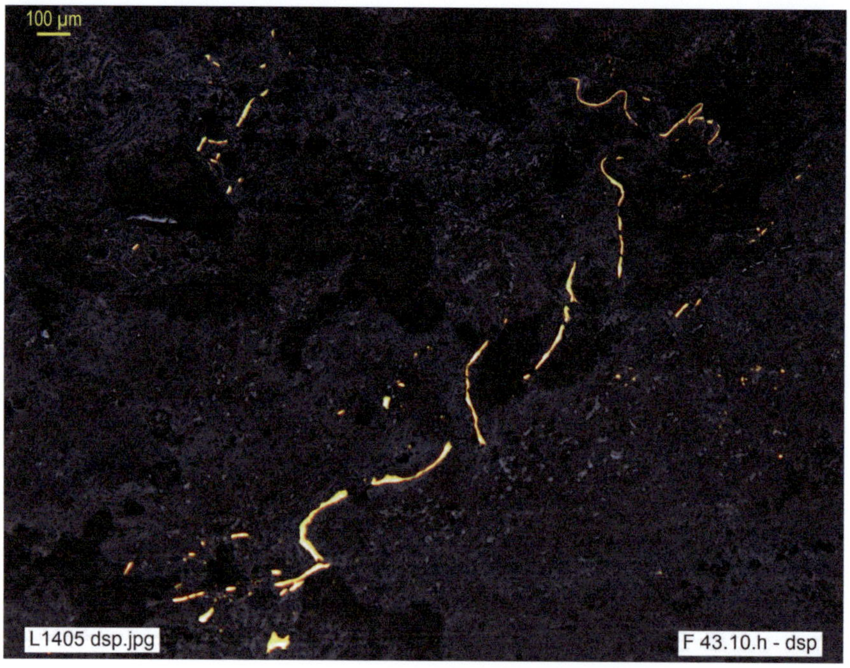

Fig. 1.51 (continued)

Deformation textures, as in common rocks, allow us to distinguish the intensity and type of deformation. They are classified by the same criteria. In the *brittle regime,* deformation-related schistosity does not develop and the textures are characterized by the absence of orientation and the presence of angular *clasts of* tectonic origin (Fig. 1.52a–e): this is the domain of **breccias** and **cataclasites.** In the *ductile regime*, by contrast, a plastic type of deformation develops oriented textures, characterized by a schistosity linked to deformation (Fig. 1.52f–i): this is the domain of **mylonites**. The intensity, estimated by the proportion of crushed matrix in the total of deformed rock, ranges from *breccia* type (<10%) through *protocataclasites/ protomylonites* (10% to 50%) and *cataclasites/mylonites* (50% to 90%), to *ultracataclasites/ultramilonites* (>90%).

In metallic ores are some peculiarities worthy of attention. In the first place, *recrystallization* is easier and more frequent than in silicates. Recrystallization is recognizable in various forms of *blastesis* (*cf.* Figure 1.45b, g–i, etc.) and can overprint previous deformations. Also frequent in some species (molybdenite, stibnite, graphite, etc.) are *internal deformations* by various mechanisms that allow them to accommodate differential stresses, such as pressure twins, buckling, translation, and so on. On the other hand, there are great differences in the rheological behaviour of the various metallic ores (for example, galena is much more ductile than pyrite and can show a plastic behaviour that contrasts with the rigidity of pyrite or sphalerite in adjacent grains: Fig. 1.52f); but in addition such behaviour can vary markedly, as do pressure and temperature (pyrite switches to ductile behaviour under catazonal conditions, as shown experimentally by various authors, also by comparison of various massive sulfide domains in the Iberian Peninsula, *cf.* Castroviejo et al. 2011). Therefore, the interpretation of deformation textures requires a combination of microscopic observation with field data and comparison with experimental results.

(a)

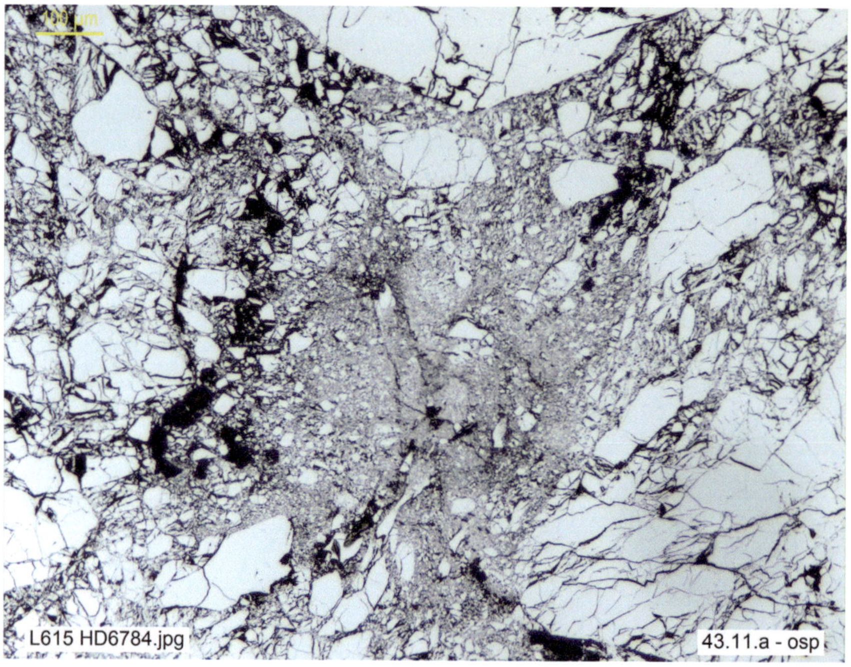

(b)

Fig. 1.52 **a** Skutterudite microbreccia. HD 6784, Schneeberg, Saxony, Germany (**osp**). **b** Brecciation of massive sulfide ore: cataclastic metapyritite cemented by chalcopyrite. Brittle behaviour of pyrite and ductile behaviour of chalcopyrite. FPI 291,100, Herrerias Mine, Huelva, Spain (**dsp**). **c** Brecciation of podiform chromitite (gray) and related Fe mobilization towards the periphery (ferritchromite and magnetite, light gray and white, resp.). TP 110607.21, Tapo Mine, Tarma, Junín, Peru (**osp**). **d** Native silver-cemented löllingite microbreccia.CR 384, Cobalt, Ontario, Canada (**dsp**). **e** Incipient microbrecciation of linnaeite ore (crackle microbreccia, transitional to the mosaic or puzzle types: penetrative fragmentation, with no or scarce displacement of clasts). CR 340, Müsen, Siegerland, Germany (**osp**). **f** Ultramylonitic Pb–Zn ore: banded, ductile galena matrix hosting microclasts of brittle minerals (quartz, sphalerite and pyrite), which also appear crushed forming fine powdery bands. CR-272, Cierco Mine, Lérida, Spain (**dsp**). **g** Protomylonitic berthierite and stibnite vein infill: brittle behavior of the former (harder, predominantly in clasts), unlike the more plastic stibnite, present in partially recrystallized microcrystalline bands. TF Pn 12, Almuradiel, Ciudad Real, Spain (**dsp**). **h** Tectonic banding of hydrothermal sulfide ore (sphalerite, chalcopyrite, with minor pyrite and galena): sheared lenticular bands of chalcopyrite-sphalerite-galena; microclasts of pyrite, quartz and carbonates. SV.1RGb, Rio Gibranzos, Caceres, Spain (**dsp**). **i** Ultramylonitic layered massive sulfide ore, partially recrystallized; tectonic banding of pyrite and quartz-carbonate gangue, transposition of the original syn-sedimentary banding. VA3-502.75, FPI, Masa Valverde, Huelva, Spain (**dsp**)

(c)

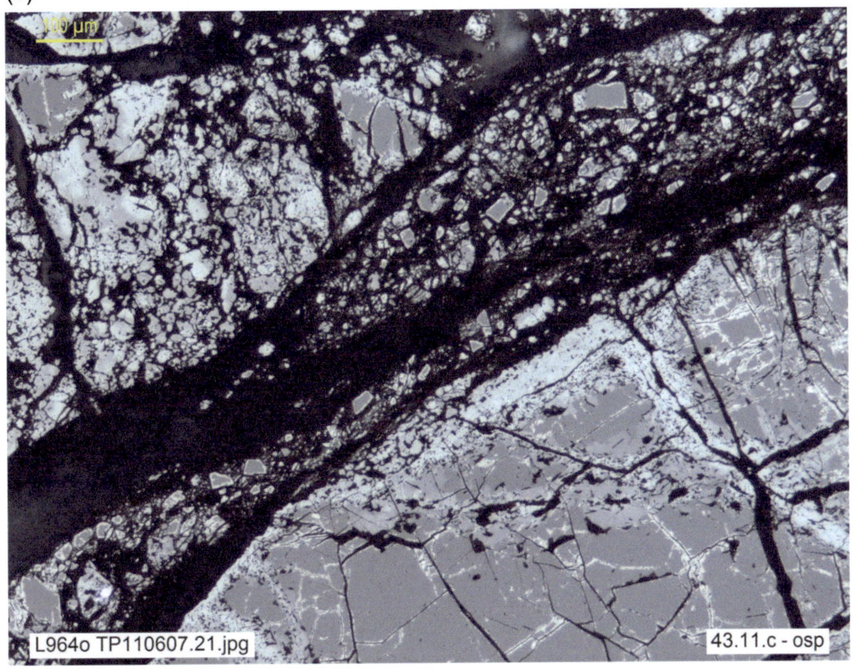

(d)

Fig. 1.52 (continued)

(e)

(f)

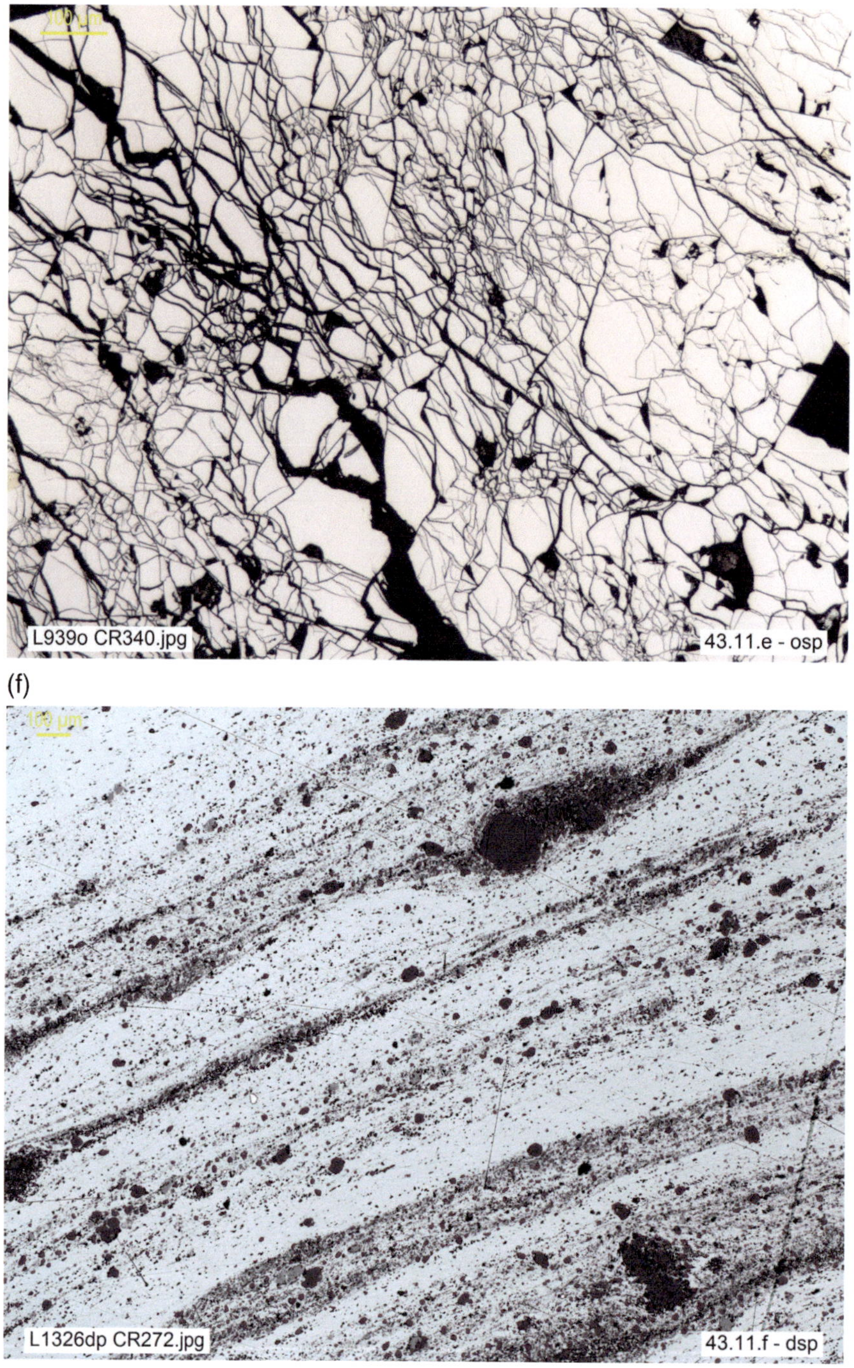

Fig. 1.52 (continued)

(g)

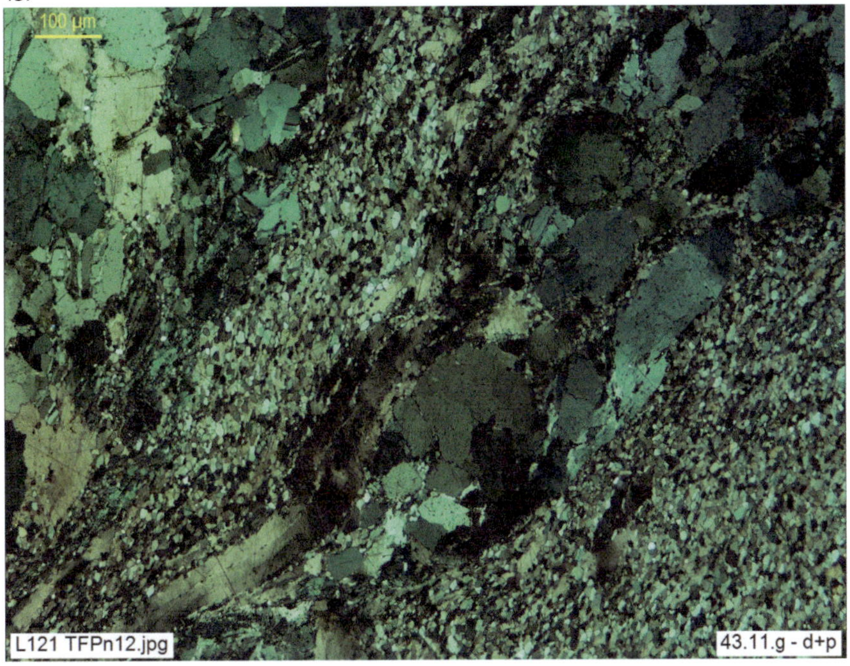

(h)

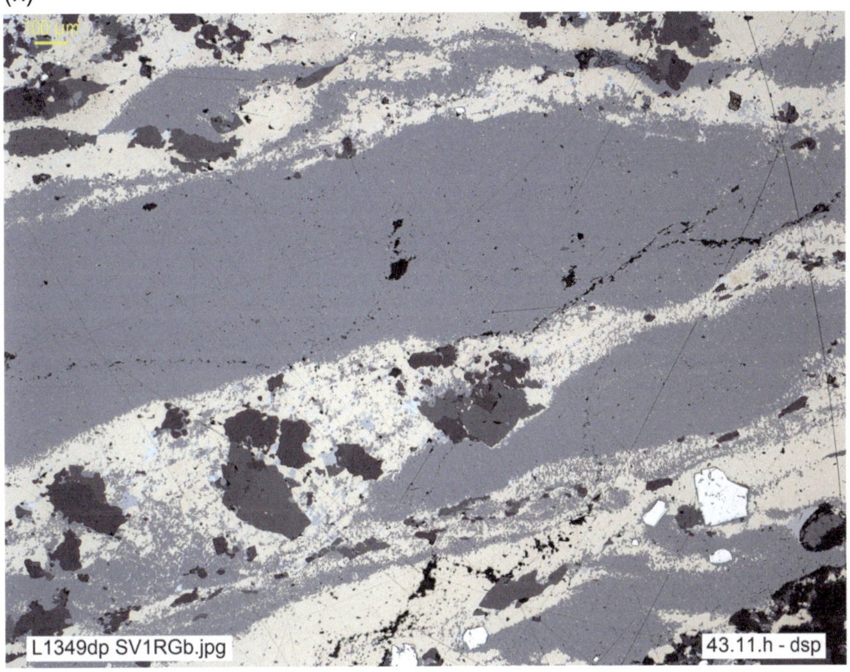

Fig. 1.52 (continued)

(i)

Fig. 1.52 (continued)

Under certain conditions of pressure, temperature and deformation intensity (usually in epizonal or *greenschist facies* conditions of regional metamorphism and with deformations of moderate intensity), **fluid-assisted strain** mechanisms may predominate. These imply a combination of dissolution and re-precipitation, controlled by differential stress and orientation. Ore dissolution is enhanced in the points of each grain that are subjected to the highest pressure (**pressure solution**), in a reaction to restore the equilibrium disturbed by the deformation (Fig. 1.53a, b, c).

(a)

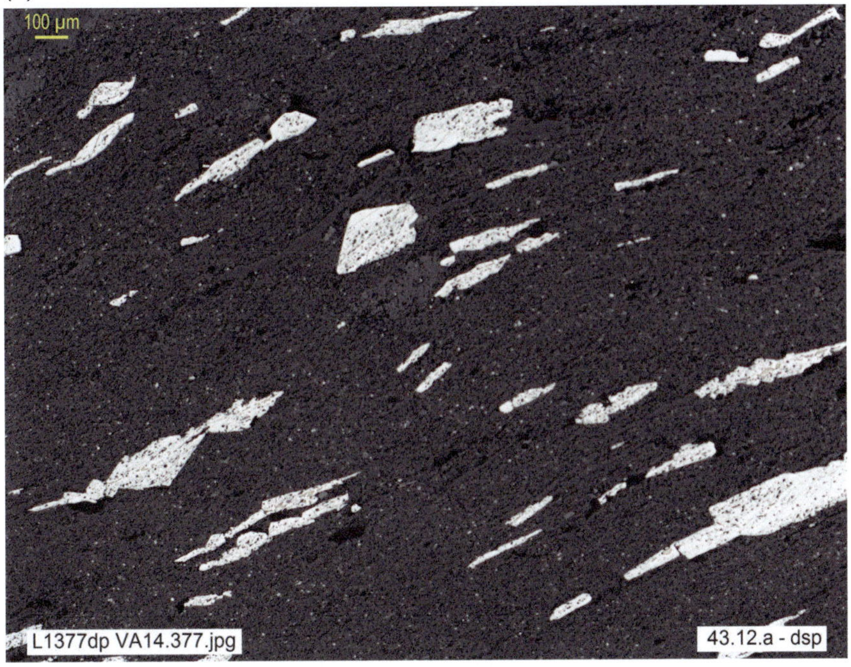

Fig. 1.53 **a** Pressure solution in disseminated pyrite idioblasts, reduced to elongated relicts perpendicular to maximum pressure (// schistosity). FPI: VA14-377.80, Masa Valverde, drill hole 3, Huelva, Spain (**dsp**). **b** Incipient dissolution by pressure in massive sulfides (pyrite, chalcopyrite, sphalerite): selective dissolution of pyrite grains, which are shortened in the compression direction (NE-SW). 031200.10, Piquitos II Mine, Moeche, Coruña, Spain (**dsp**). **c** Detail of the pressure solution process in pyrite (including minor chalcopyrite and sphalerite): the contacts between pyrite grains, initially cubic, acquire concave and lobed or indented shapes at the points where the grain is dissolved as response to direct pressure transmitted by the neighboring grain. M031200.10, Piquitos II Mine, Moeche, Coruña, Spain (**osp**). **d**, **e** (**dsp** and **d + p.w**, resp.). Solution transfer of pyrite to be re-deposited in points of favorable orientation on relicts of the same mineral, which are encrusted and welded by the neo-formed py. The resulting py idioblast shows an irregular, complex internal structure, appreciable by contrasts of color and polishing hardness or enhanced by *interference contrast* (with Wollaston prism, **e**). M031200.17, Piquitos II Mine, Moeche, Coruña, Spain. **f**, **g** (**dsp** and **d + p.w**, resp). Complex zoning in pyrite idioblast, due to solution transfer, resulting in successive generations affected by dissolution and re-deposition, as shows the *interference contrast* image (**g**). M031200.17, Piquitos II Mine, Moeche, Coruña, Spain. **h**, **i** (**dsp** and **LT + p**, resp). Relict of pyrite phenoblast, with complex structure (including gangue and remnants of older py) and elongated outline, resulting from solution transfer. Pressure shadows or tails of chalcopyrite, chlorite and quartz (visible in **i**, *transmitted light*). 031200.17, Piquitos II Mine, Moeche, Coruña, Spain

(b)

(c)

Fig. 1.53 (continued)

The dissolved components can then be evacuated in solution, yet are mostly redeposited in the surroundings, at the nearest point of lowest pressure. Precipitation actually occurs on the two opposite sides of rigid grains or microclasts, the sides facing the direction of relative extension (minimum stress). As a result, a tail of oriented, sheltered crystallization (**pressure shadow**: Fig. 1.53h, i) appears following the dilational direction, thus increasing the length of the grains perpendicular to the maximum stress.

Solution transfer occurs when the dissolved material is mobilized and transported to areas where the lower pressure makes precipitation possible, either on pre-existing crystalline cores or grains, or as entirely neoformed minerals (Fig. 1.53d–g). Mobilization can cover very variable distances, from micro- to mega-scale. Usually the process may be recognized under the microscope, for instance by the presence of relicts of the primary grains preserved from dissolution, or by the evidence of orientation-controlled deposition, for instance pressure shadows or sheltered crystallization, in the neoformed phases. Nevertheless, if precipitation goes on after the stress relaxes or changes (e.g. by rotation of the stress ellipsoid), new crystals or aggregates may develop that cut the schistosity or engulf it.

Overall and from a geometrical point of view, fluid-assisted strain generally results in an elongation or extension of the grains (and, in general, of the ore bodies), which may be very pronounced in the direction transverse to the main stress, as well as a thinning in a direction parallel to the main stress (Fig. 1.53a, h).

(d)

(e)

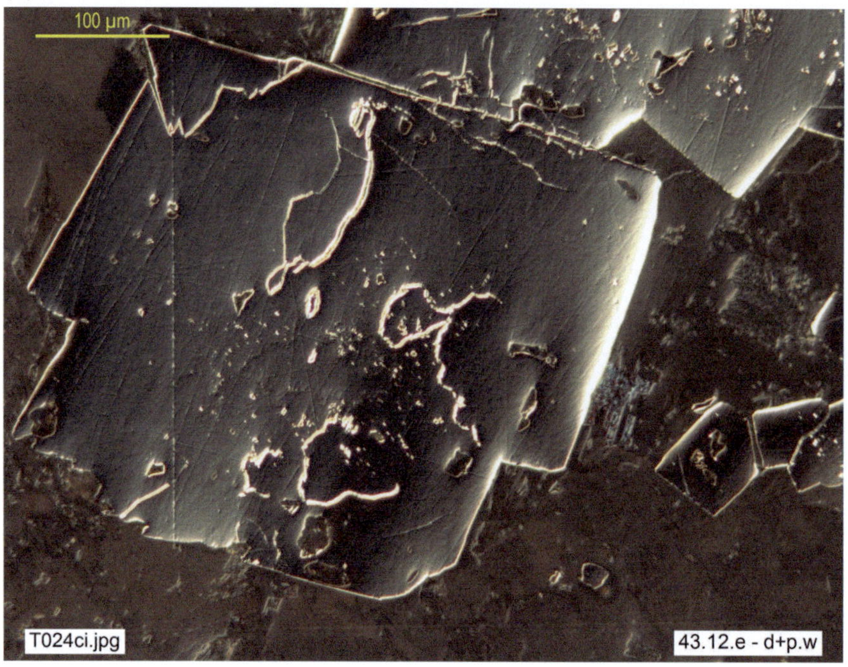

Fig. 1.53 (continued)

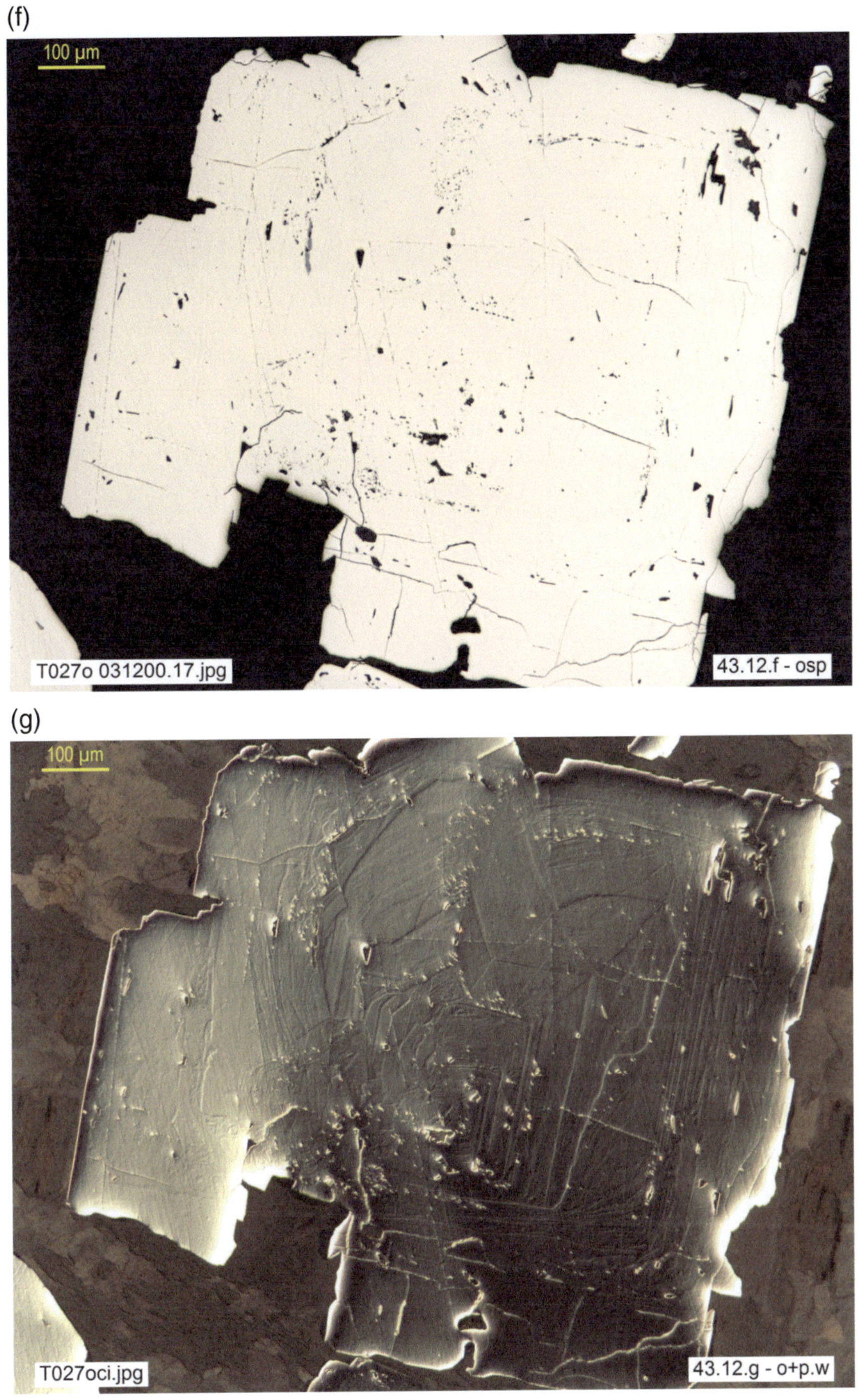

Fig. 1.53 (continued)

(h)

(i)

Fig. 1.53 (continued)

1.2.3 Geometallurgical Classification

Mineralogical characterization is crucial to industrial processing of mineral resources, and textural information is a key element of this characterization since it determines the possibility of physical separation of ore and gangue, the degree of grinding or comminution required, the accessibility to mineral phases of the reagents, and so on. This fits the general concept of geometallurgy, as the set of processes of extraction and processing of minerals of economic interest, taking into account the relevant geological information. In a more precise sense, *geometallurgy* refers to an holistic approach to mine planning that incorporates and relates metallurgical and geological information to *the geological block models that can be used for all aspects of mine planning* (Becker et al. 2016). In this geological information, *process mineralogy,* or the mineralogical characterization of the ore, plays a central role as predictive support for the design of plant processes. This concept, now increasingly accepted, extends the traditional concepts of Minerallurgy, ore processing, ore dressing or ore concentration, incorporating into the technical data as a key tool mineralogical-textural information and its numerical quantification.

Geometallurgical applications require at least three qualities from textural classification:

- It must be objective; that is, based on observable facts: it must be a purely descriptive classification
- It must apply the technical criteria most appropriate to the problems of ore processing; that is, the textural types must reflect the morphological properties or parameters that determine the behaviour of the particles/aggregates in the grinding and concentration processes
- It must be simple and clear; that is, easily manageable for practical purposes and unambiguously applicable by anyone.

Since the beginning of ore microscopy, metallurgical applications have been in strong demand. Attempts to systematize textural information for this purpose date back at least to the 1920s (e.g. Head 1928; Head et al. 1932). In the course of the twentieth century, various contributions were published (e.g. Bastin 1960; Edwards 1965; Rehwald, Ramdohr et al., in Freund 1966), among which Ramdohr (1980), who discusses many aspects of application, stands out. Of the various historical proposals for textural classification, consistent with the above-three requirements for geometallurgical application, those by Gaudin (1939), Amstutz (1962) and Craig and Vaughan (1994) should be mentioned. By virtue of its simplicity, the most practical one turns out to be the first, while the types in the other two coincide with it or are assimilable into it (Pérez-Barnuevo et al. 2013).

Table 1.4 Basic geometallurgical classification (mill particles)

Class	Type	Characterization
Liberated	Figure 1.55 Monomineralic	After shape and grain size (Table 1.1)
Middling	Figure 1.56 Simple contact	Straight and smooth or gently curved
	Figure 1.57 Irregular contact	Jagged, lobed, jigsaw, interpenetrated
	Figure 1.58 Stockwork	Network of cross-cutting veins and dissemination
	Figure 1.59 Corona	Thin layer of ore 2, rimming core of ore 1
	Figure 1.60 Emulsion	Minor ore in scattered tiny inclusions

Table 1.4 summarizes the basic types of textures present in *liberated* (monomineralic) grains and in *middlings* (grains of mixed composition, bi- or polymineralic) that have passed a milling circuit (Fig. 1.54a, b), classifying the mixed grains into five types of intergrowths comprising Gaudin's four types (1939), redefining his *Type II* (vein) as stockwork and adding a fifth type for irregular contacts not assimilable into Gaudin's (1939) intergrowths. This fifth type (Fig. 1.54b) tends to be scarcer or even disappear as the milling process progresses from grinding to finer grain sizes, since it is visible mainly in coarse fragments and comminution eliminates them.

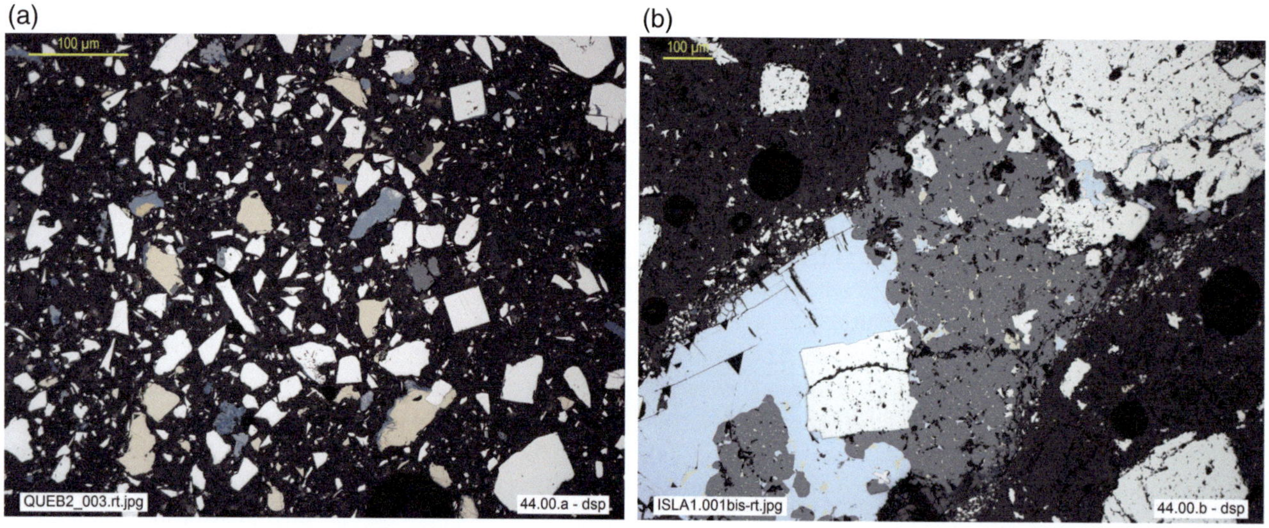

Fig. 1.54 **a** Appearance of a massive sulfide milling concentrate: liberated grains (monomineralic), mainly pyrite (**py**), chalcopyrite (**ccp**), sphalerite (**sp**), covellite (**cv**), chalcocite (**cc**), digenite (**dg**) and quartz (**q**), coexist with binary middlings (ccp-dg, ccp-cv, cv-dg, sp-cv…) of varied textures and scarce ternary middlings (ccp-py-cv). QUEB2.003, Riotinto, Huelva, Spain (**dsp**). **b** Quaternary middling (sp-gn-py-ccp) with irregular contacts and sp-ccp emulsion texture. Sample Isla1.001bs-rt, Riotinto, Huelva, Spain (**dsp**)

Figures 1.55, 1.56, 1.57, 1.58, 1.59 and 1.60 illustrate these six ideal types with images of actual ores. **Liberated** grains are defined as grains composed of a single mineral, either ore or gangue, separated from the other components in the raw aggregate or by natural processes, such as grain sorting in detrital sediments (Fig. 1.55a, b). Liberated grains can, in turn, be characterized by their shape and size, applying the criteria in Table 1.1. Mixed, non-liberated grains, composed of two or more components, are defined as **middlings** and are characterized below.

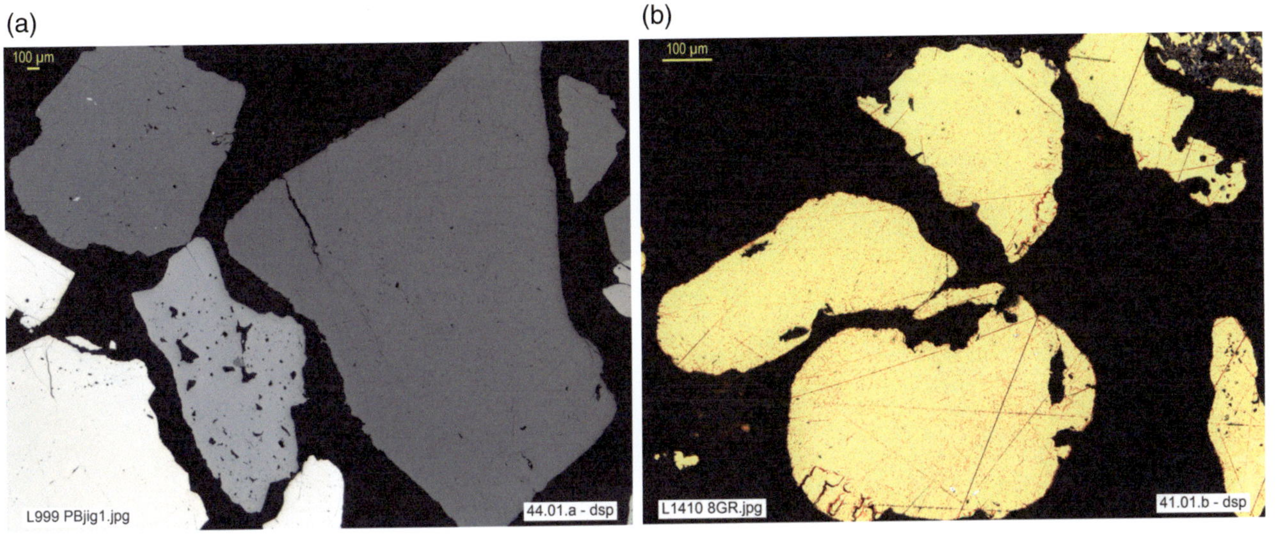

Fig. 1.55 **a** Liberated grains (monomineralic) of hübnerite (gray, main component), pyrite, sphalerite and tetrahedrite in milling concentrate. JIG-1, Pasto Bueno Mine, Ancash, Peru (**dsp**). **b** Liberated grains (nuggets) of gold in pan concentrate from stream sediments. 8-GR, Minvoul, Gabon (**dsp**)

Middlings can be *binary* (aggregate of two components), *ternary* (3), *quaternary* (4), and so on, depending on their composition. By the morphology of the intergrowth, they are defined according to the contact between the components, applying the criteria exposed for the description of mineral aggregates (cf. Table 1.2, Figs. 1.27, 1.28, 1.29, 1.30, 1.31, 1.32 and 1.34). A contact is called **simple** if it is rectilinear or gently curved; that is, susceptible to separation in later stages of the process (Fig. 1.56a); Fig. 1.27a, b.

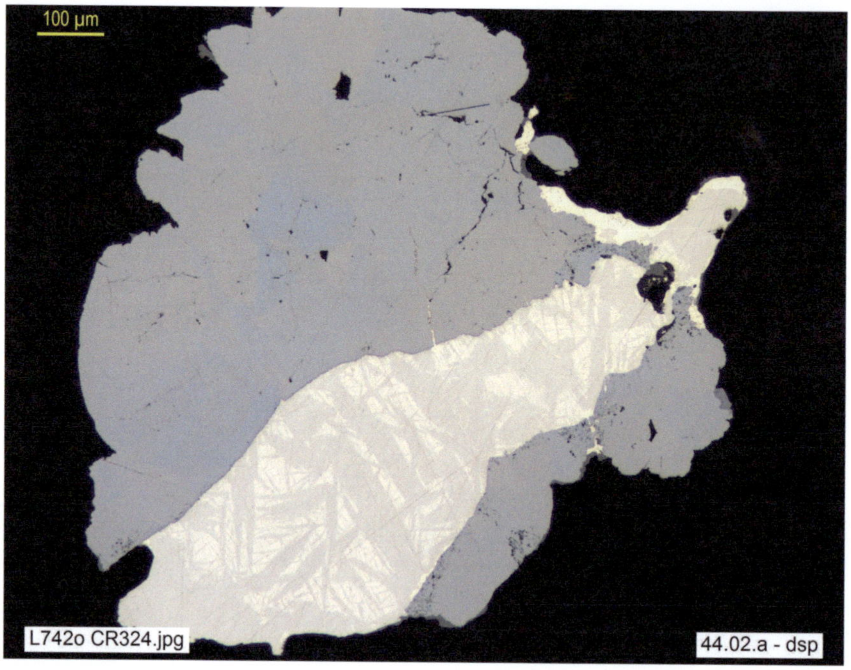

Fig. 1.56 Simple contact in löllingite-allargentum binary middling; the latter contains abundant exsolved silver amalgam (eugenite, the brightest phase). CR 324, Bouismas Mine, Bou Azzer, Morocco (**dsp**)

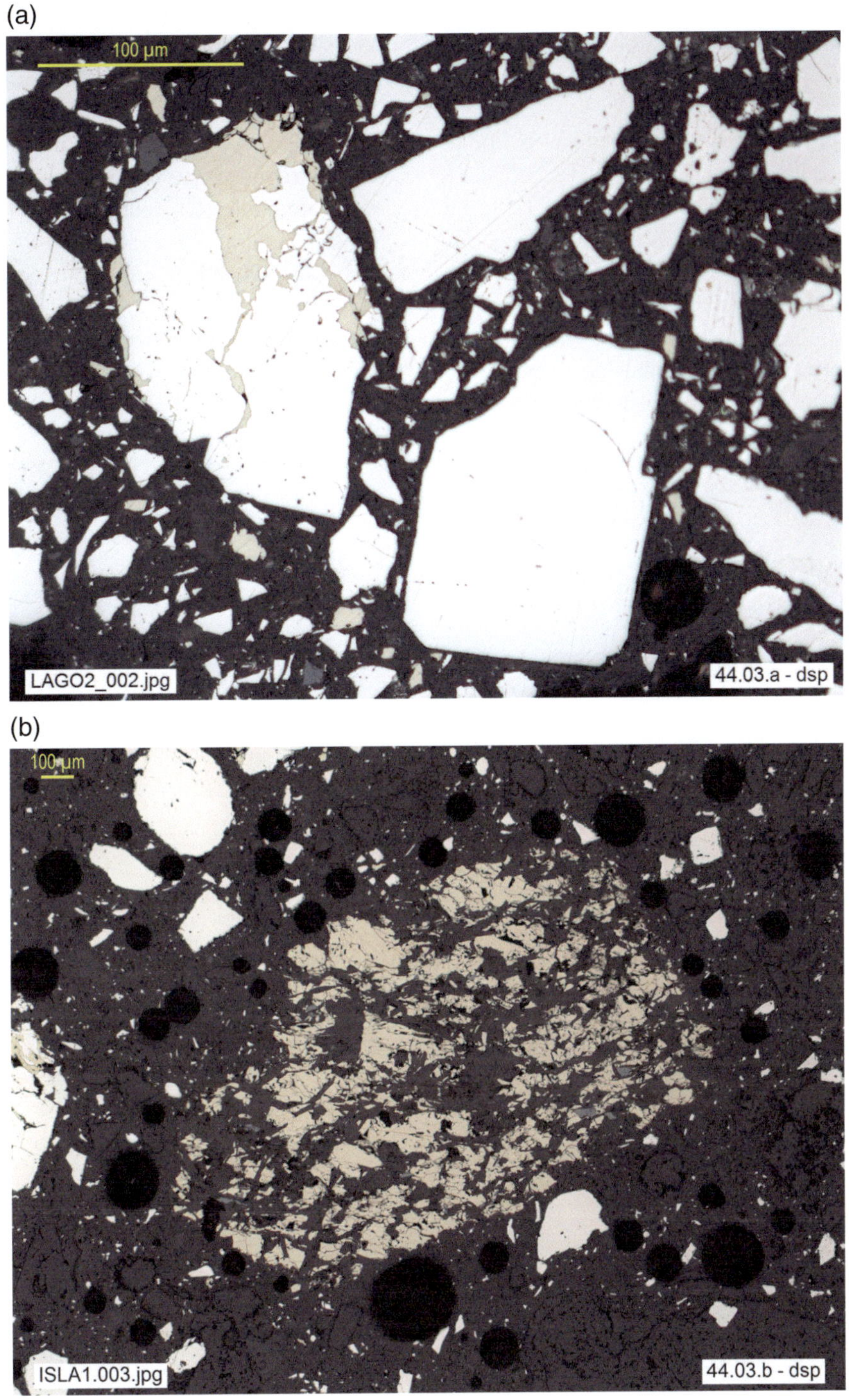

Fig. 1.57 **a** Irregular contact in pyrite-chalcopyrite binary middling. Sparse sphalerite and chalcopyrite and frequent liberated pyrite. LAGO2 concentrate, Riotinto, Huelva, Spain (**dsp**). **b** Middling of chalcopyrite and gangue (center, surrounded by black pores) with irregular contacts. Massive sulfide concentrate, sample ISLA1, Riotinto, Huelva, Spain (**dsp**)

The remaining contacts, broadly defined as *complex* in Table 1.2, from a geometallurgical point of view (Table 1.4) are characterized as:

- **Corona**, when the particle is surrounded, as a rim or crown, by a phase occupying the periphery (Fig. 1.59a, b); it is one of the variants of the *concentric* type of Table 1.2 (Fig. 1.29a). It could be described as a particular *encrustation* (Fig. 1.35a, b).
- **Emulsion**, when one or more components occur in minute inclusions disseminated in the main phase (Fig. 1.60a, b); equivalent to the emulsion type considered so far yet broader and free from genetic implications (it could comprise disseminations, relicts, impregnations and other fine intergrowths, for instance).
- **Stockwork**, when the particle comprises a network of cross-cutting veinlets and minute scattered corpuscles (Fig. 1.58a, b); it mimics at the microscopic scale the homonymous texture observed at outcrop scale, in stockwork ore bodies (as defined in the text corresponding to Table 1.2, Fig. 1.34a).
- **Irregular**, when the contacts are complex (lobed, jagged, interpenetrated, etc., *cf.* Figure 1.28a–c) and erratic, lacking the regularity implicit in the previous definitions (Fig. 1.57a, b).

(a)

(b)

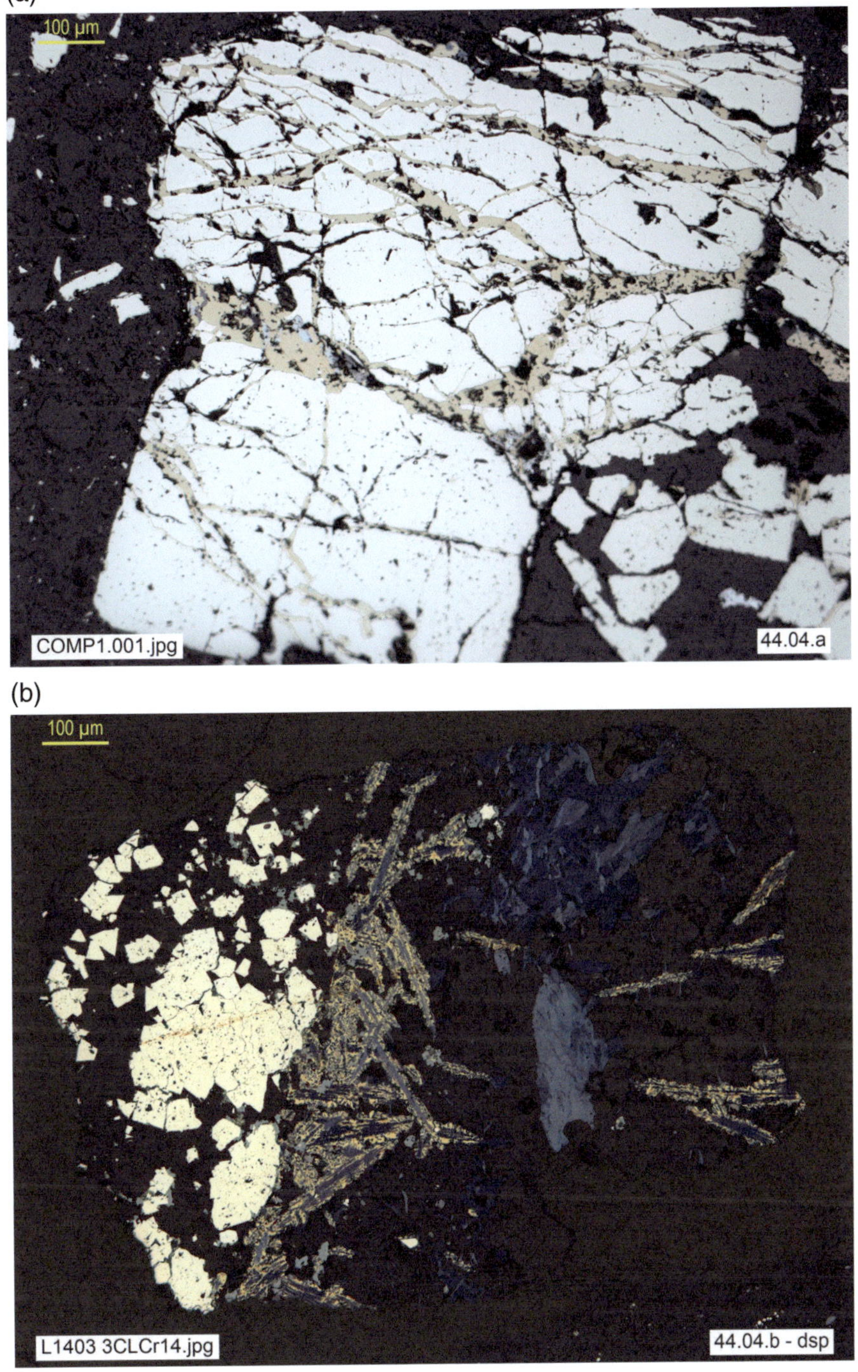

Fig. 1.58 **a** Pyrite-chalcopyrite microstockwork, with traces of sphalerite, galena and bismuthinite associated with ccp. Concentrate COMP1.001, Riotinto, Huelva, Spain (**dsp**). **b** Complex texture, stockwork-like, in senary middling comprising gangue (very dark)-pyrite-covellite-chalcopyrite-chalcocite-digenite. 3CLC.R14-AMCO, Las Cruces, Seville, Spain (**dsp**)

(a)

Fig. 1.59 **a** Senary middling enclosed by *corona* of acanthite (gray, major component), and comprising a pyrite core, chalcopyrite, scarce galena and proustite; lath-shaped stephanite inclusions in acanthite highlighted by their immediate and inevitable light etching (darkening and abundant black dots). CR 331, Fresnillo, Zacatecas, Mexico (**dsp**). **b** Corona of galena (light gray) on a core of native antimony (white, bright, top), dyscrasite (brownish white, in the center), pyrargyrite (bluish gray) and antimony ochre (*cervantite*, dark gray). CR 398, Hiendelaencina, Guadalajara, Spain (**dsp**)

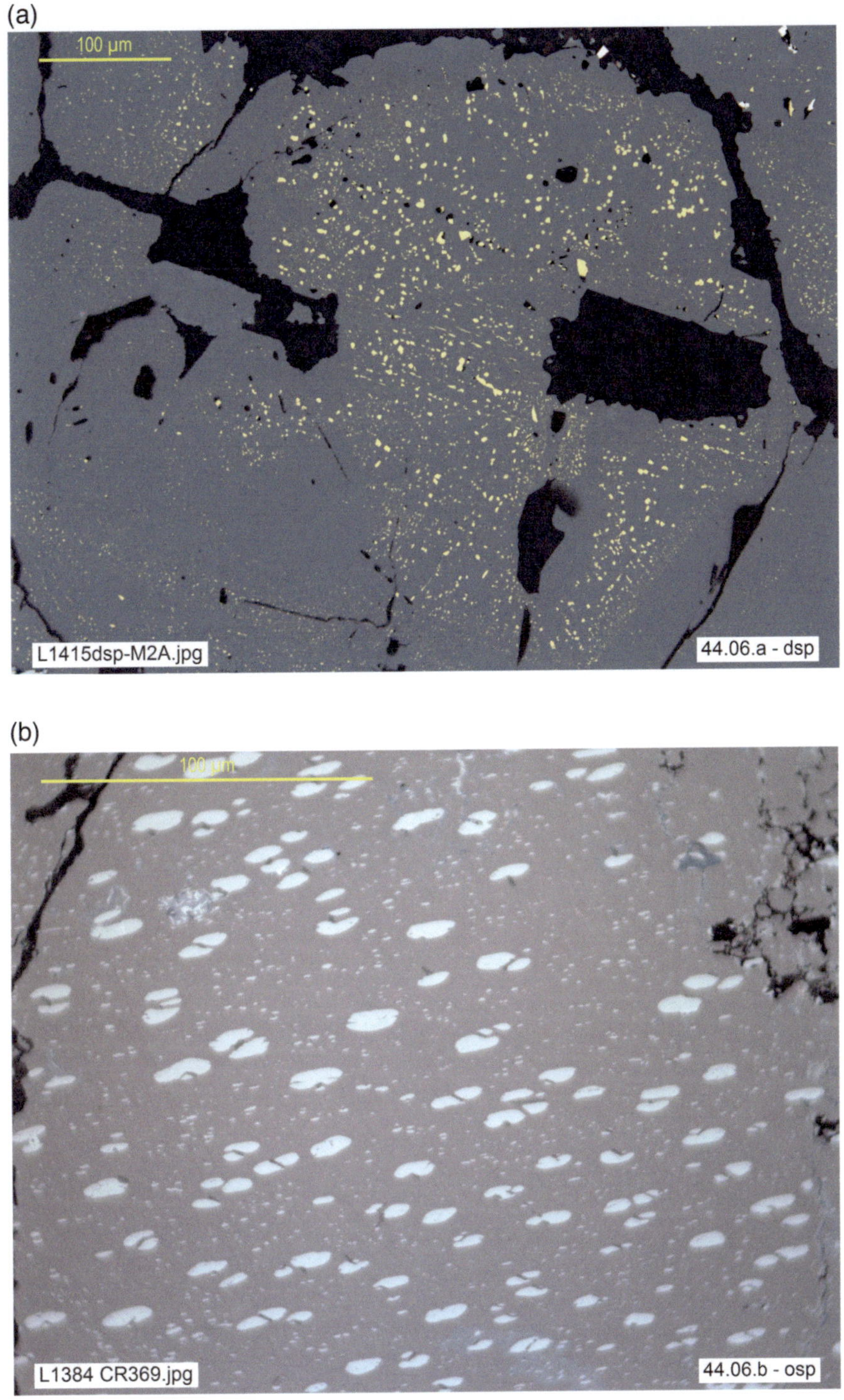

Fig. 1.60 **a** Fine emulsion of chalcopyrite in sphalerite, marking the reticular planes and faint zoning of sp. M.2A, Bienaventurada Mine, Peru (**dsp**). **b** Ilmenite (pinkish brown, main ore) with very fine emulsion of hematite (white, bright, bimodal grain size distribution: $\approx$ 10 to 15 µm and $\leq$ 3 µm) and ilmenorutile; the latter occurs in dark gray, tiny, lenticular corpuscles ($\approx$ 3–5 µm: barely visible despite immersion and high magnification), associated to the coarser generation of hematite. CR369, Ljosland, Iveland, Norway (**osp**)

Nevertheless, the variety found in industrial processes challenges any classification and demands an approach that provides the necessary flexibility and understanding in the application of these characterizations. A brief demonstration of their practical use for geometallurgical characterization applied to actual processing problems is comprised in the following section (cf. 1.3.4). Firstly, and as a snapshot of the method, a few examples are briefly discussed to show applied textural analysis. This is intended to facilitate the control of the process and to assess the foreseeable behaviour of the ore in plant from a qualitative point of view. Then it addresses the need to complete this information with quantitative data, supported by automated methods that provide an acceptable performance for the industry, such as the AMCO system (cf. § Part II), based on optical microscopy and currently in use at the Applied Microscopy and Image Analysis Laboratory of the Madrid School of Mines, LMA (ETSI Minas y Energía, Universidad Politécnica de Madrid).

1.3 Interpretive Keys: How to Define and Interpret Textures

The correct interpretation of ore textures is a complex task, requiring a combination of accurate observation and rigorous logical analysis. As a guideline and without pretending to be exhaustive, some recommendations on the *observation methodology* and some basic *interpretive keys* are proposed, whose application is illustrated by simple examples of the most common natural processes (§ 1.3.1 to § 1.3.3).

Readers may try their own analysis, applying the principles discussed so far to the textures considered in the previous chapters, and may then, as self-evaluation, critically compare the conclusions. To facilitate this, in each of the cases discussed below, some illustrative photomicrographs from the descriptive section (Part II, vol. 1) are cited together with the interpretation criteria.

Finally (§ 1.3.4), some examples of textural analysis are very briefly discussed from the point of view of *process engineering*. The aim is to see the use of this information, facing real industrial problems with a *geometallurgical* approach. This is done firstly through a qualitative outlook, to be followed up by the compilation and use of quantitative data (cf. chapter on automation: *Part II*). Well aware that the amplitude and the variety of this important field of knowledge exceed by far the scope of the present text, interested readers are referred to the abundant literature on the subject of Geometallurgy and Process Mineralogy (e.g. the updated presentation of current techniques in the monograph by Becker et al. 2016).

Before proceeding to take forward this task, a brief reflection about the preliminary requirements for any sound interpretation of microtextures seems timely.

First and foremost, for a reliable interpretation the observer must be aware of the problems related to the **scale of observation**:

- Especially, the natural hierarchy of the processes generating mineral concentrations must be recognized, but also their complexity. For example, a hydrothermal filling process may be followed by a deformation episode or by a re-equilibration of the ore during cooling and, finally, by supergene secondary enrichment: each event may leave its textural imprint, but not all have the same importance or act on the same scale. It is the observer's mission to make a contextualized and realistic analysis, indispensable for a correct interpretation.
- On the other hand, there are data that are perceptible at one scale and not at another. For example, a zoned, intergrowth or finely banded geometry at the microscopic scale may be imperceptible in a hand sample. But also, a coarsely banded texture, clearly visible at the outcrop scale, may appear massive at the microscopic scale, if the polished section comprises a single band.
- Consequently, microscopic observations should never be understood in isolation, but must be complemented by information at other scales, especially field and mine data. A texture correctly interpreted as hydrothermal infill under the microscope may actually correspond to a veinlet in a stockwork body visible in the field, while a sample from the infill of a large vein may appear under the microscope as a massive, undifferentiated texture. To correctly establish a paragenetic sequence, microscopic data are not enough: it is essential to take into account the field relations, in particular the crosscutting relations that define the different generations of veins. In a word: it is a matter of correctly applying the basic principles of geological methodology.

Secondly, it must be taken into account that the ore is only a part of the real rock that is worked as orebody, or extracted as ROM (Run-Of-Mine ore). Actually, ore deposits are usually made up mostly of common **gangue** minerals (quartz, silicates, carbonates, etc.), whose textures can also be very significant and, in any case, should never be ignored. A correct interpretation has to account simultaneously for the observations on ore and gangue and the relationships between them. It is therefore highly recommended to get into the habit of comparing, systematically and as a matter of principle, observations with transmitted and reflected light.[5] And, in practice, the need for complete mineralogical information for industrial applications

[5] The use of polished thin sections may be very useful for this purpose.

(Geometallurgy) can make this recommendation an unavoidable requirement.

Another source of problems is the confusion between **observational** data and interpretation. The former must be objective and unquestionable, regardless of the observer. Interpretations sometimes have, inevitably, a certain subjective character, but a correct textural description remains unchanged, regardless of how one interprets it, and is always a very valuable source of information even if ideas change. Hence the need to approach the textural study, at first, with *rigorously descriptive criteria*. Even so, these criteria can vary according to the point of view, depending on the objective of the study: paragenetic sequence, grain size analysis, intergrowth characterization, liberation grade, modal analysis, process control in industrial plants, etc. Textures are often complex but, anyway, the description must clearly reflect the *most relevant features*.

Starting from a description that satisfies these requirements, it is time to discuss the textures observed under the microscope, aiming at a full understanding of the information they may convey; that is, at their interpretation. Although it is not within the scope of this text, it is worth recalling contributions such as the classic work by Stanton (1972), which offers a solid discussion—chapters 8–10—of the mechanisms that control crystal growth, as well as their interactions and the resulting forms, which will define the textures (since then, much work has been done on the subject, cf. e.g. Barton 1991; Craig 1990a, b; the abundant references cited by Bowles et al. 2011; etc.).

As an introduction to the **genetic interpretation**, the microscopic textures resulting from a few typical processes of mineral genesis will be discussed. Beforehand, it is important to distinguish **primary** textures from **modified** textures. The latter are very frequent, given the marked tendency of metallic ores to rebalance, but even in these cases a careful observation often allows to discover preserved traces of primary features (*relicts*) sufficient for a correct interpretation.

To illustrate the practical application of interpretive criteria, two important processes, **hydrothermal filling** and **metasomatic alteration/replacement**, characterized by primary and modified textures, respectively, are compared. **Colloidal** textures and their evolution are also briefly characterised. The microscopist's interpretation must always be consistent with observational data at all scales and, most particularly, at the outcrop/hand sample (*mesoscopic*) scale.

1.3.1 Cavity or Vein Infill

Hydrothermal vein infill can be recognized by some of the following criteria at the mesoscopic scale:

- As far as the geometrical relationship to the host rock is concerned, the hydrothermal infill minerals are younger than the host rock and are hosted in the latter in empty spaces or cavities, such as geodes, vacuoles and, above all, seams or veins; that is, cross-cutting structures defined by sharp contacts (open fractures). The vein edges are clear-cut and would ideally fit if the movement was reversed by closing the open fracture. Brecciation is common, as are angular clasts of the host rock, which could also fit if recomposed like pieces of a jigsaw puzzle.
- The internal structure of the vein is independent of that of the host rock. Deposition in parallel bands from the edges is frequent, sometimes resulting in a symmetrical geometry with respect to the centre, corresponding to the deposition of successive generations. Geode alignment is sometimes observed in the central band (incomplete filling in the later stages).

The mineral content of the vein is dictated by the hydrothermal fluid and often differs from that of the host rock.

There is usually an increase in crystallinity and grain size while moving away from the edge or salband towards the centre of the vein (progressively slower cooling). Occasionally features of rapid cooling are found, such as *chilled* margins or narrow bands along the borders of the vein with very fine grain size, in contrast to megacrystals in the centre (*vbgr.* pegmatites).

On a microscopic scale, hydrothermal filling processes typically produce:

- Smooth or straight contacts between grains (suggestive of equilibrium): Fig. 1.40a–c.
- Sequential crystallization features, with idiomorphic development of early crystals, while later minerals are more often allotriomorphic and interstitial or occur cementing microcracks: Fig. 1.41c–e.
- *Cockscomb* or *palisade* textures: Fig. 1.40c.
- Encrustations, often banded: Fig. 1.35a–c and Fig. 1.40j.
- Occasional colomorphic textures, usually in marginal zones of the deposit or in posthumous geodes; also in situations of marked disequilibrium, even at relatively high temperatures, with extreme oversaturation and very rapid (*flash*) precipitation: Fig. 1.42a, b, f.
- *Corona* textures (or *cockade* textures, cementing breccias, with sequential deposition around clasts, usually of host rock): Fig. 1.29a.
- Often evidence at various scales of successive cycles of fracturing and hydrothermal discharge, whose relative chronology can be established by cross-cutting evidence and infill relationships: Fig. 1.41g, h.

1.3.2 Alteration and Replacement

Metasomatic replacement processes have often been the subject of controversy and cannot always be strictly separated from other processes. Thus, for example, it is not uncommon for hydrothermal filling to be accompanied by fluid re-equilibration processes, involving reactions with the host rock or with early minerals and partial replacement of these, evidenced by their corrosion. On the other hand, metasomatism as a form of hydrothermal alteration propagates most easily through a network of fractures and microcracks, in which small scale evidence of infilling can be found.

All this makes terminology sometimes confusing, for example when terms such as *veins* (actually, in this particular case, tabular bodies or *mantos*, usually stratabound and without evidence of infill) or *clasts* (in reality, undigested remnants of the host rock) are used in this context. In the following, to avoid misunderstandings, these common but improper or debatable terms are written in quotation marks (“**veins**”, “**clasts**”).

In general, metasomatic replacement processes are recognised at the mesoscopic scale by criteria such as:

- Diffuse, transitional contacts, frequently associated with dissemination of the ore in the host rock, in decreasing abundance as one moves away from the ore body.
- Topomineral deposition; that is, selective precipitation at the points where the replaced material had the appropriate composition or higher chemical reactivity.
- The “veins” or mineralized structures can have any geometry, independent of the host, but they tend to mimic the more reactive lithologies, without strictly following mechanical fracturing patterns.
- The edges on either side of the “vein” do not fit together, since they are defined by independent reaction fronts following their own paths.
- The internal structure does not have to be symmetrical.
- The “clasts” or fragments of the host rock usually have curvilinear and irregular contours, which cannot be fitted together like pieces of a jigsaw puzzle; however, their orientation and internal structure may be in continuity with those of the host rock (absence of movement, as opposed to clasts in true veins).
- Preservation of relict structures inherited from the host rock (stratification, banding, folds, schistosity, etc.).

On a microscopic scale, substitution processes can be recognized by

- Jagged, irregular grain-to-grain contacts, reaction rims: Figs. 1.47c, 1.44d, 1.28a.
- Geometrical relations, between neighbouring grains, that usually allow to distinguish primary or replaced minerals from secondary or neoformed minerals. The former show mostly concave contacts; also, oriented replacement (that is, conditioned by their own lattice or structure: favourable crystallographic directions, twins or cleavage planes). The latter show convex contacts, with lobes or indentations penetrating the replaced mineral, usually favouring suitable directions or zones of weakness of the host (e.g. aggregates of scarce crystallinity or enhanced porosity, micro-fractured zones). Figs. 1.41a, 1.49c, d, 1.43f.
- Disequilibrium mineral assemblages, evidenced by superposition of parageneses that are not stable under the same conditions or by the coexistence of phases suggesting disequilibrium, such as copper oxides and sulfides: Fig. 1.50a–c.
- Presence of pseudomorphs: Fig. 1.10a–d.
- Remnants of the primary mineral, preserved as relicts or ghosts: Figs. 1.48b, 1.44o–p, 1.31g.
- Development of bands or alignments of neoformed minerals, usually idiomorphic, which overprint pre-existing features or structures and mimic them or grow over them: Figs. 1.45i, 1.33a, b, 1.51c–f.

1.3.3 Colloidal Textures

The interpretation of colloidal textures has often been controversial, because they frequently combine primary and modified features and because some of them can be generated by different processes. In principle, as primary textures, they fill open spaces, cavities or fractures, which explains their presence among the hydrothermal filling textures. However, their scarce or null crystallinity makes them unstable, so that they generally show a marked tendency to spontaneous crystallization, resulting modified textures known as *metacolloids*.[6]

The conditions of formation are another debatable issue. Colloidal textures were traditionally attributed to precipitation of a gel at low or very low temperature, but are nowadays interpreted rather as the result of strong supersaturation (typically due to sudden disequilibrium, for

[6] According to Ramdohr (1980), natural gel materials are usually metacolloidal because all colloidal precipitation products are metastable. Therefore, metacolloidal textures should be considered primary when they acquire their crystalline character immediately after the precipitación. However, given the features mentioned above, they will be treated here as modified when crystallization is clearly visible, maintaining the original criterion of Grigoriev (1928).

instance boiling or fracturing and loss of pressure in an epithermal context), which causes massive, very rapid (*flash*) precipitation, even at relatively high temperatures. This does not invalidate, for other cases, the traditional interpretation.

How can colloidal textures be recognized? Some basic criteria, linked to their genesis, help to recognize *primary* colloidal textures:

- Spheroidal contours, explainable by the effect of surface tension in liquid phases (botryoidal, spherulitic, reniform, etc.): Fig. 1.42a, b.
- Low or null crystallinity, explainable (i) by the low temperature and its effects on the kinetics of deposition; or (ii), at high temperature, by strong supersaturation, cause of flash deposition: immediate, fast and massive precipitation, leaving no time for crystal growth: Fig. 1.42c–h.
- Color banding (diffusion bands or Liesegang rings), often showing concentric structures. It is interpreted, with experimental support, as the result of diffusion of an electrolyte in a gel: Figs. 1.10b, 1.44i, 1.25a.

Some common *modifications* are:

- Shrinkage cracks, due to dehydration of the original deposit, as seen in laboratory gels and, sporadically, in some mineral deposits.
- Post-depositional restructuring, which usually results in a peculiar secondary crystallization, with development of prismatic or columnar crystals on the aphanitic mass, often in radial disposition and perpendicular to the spheroidal outlines, or even crossing several colored bands, without interrupting them. These textures cannot be explained by primary filling: they imply re-equilibration (crystallization) of a pre-existing amorphous or cryptocrystalline deposit.

1.3.4 Geometallurgical Applications

The development of *process mineralogy* or the application of mineralogical information to the control and improvement of plant processes (concentration of ores) has been continuous since the beginning of the last century (*vbgr.* Schwartz 1923). And, in the last decades, it has grown up exponentially. Furthermore, it has expanded to the exploration and evaluation of mineral resources and, importantly, to predict and control the environmental impact of mining and metallurgical activities. Given the nature and the objectives of the present text, it is out of its scope to address the multiplicity of problems and techniques of this broad field of knowledge, but there is a very abundant literature on the subject and the interested reader can find basic information in specific works, such as Jones (1987), Cabri and Vaughan (1998), Petruk (2000) or Becker et al. (2016), without forgetting classics such as Gaudin (1939). However, the text would be incomplete without, at least, a brief review of the industrial applications of ore microscopy, among which geometallurgical issues and, more specifically, process mineralogy problems occupy a prominent place.

First thing, the adequacy of mineralogical information will be briefly examined. A rapid characterization of some selected ores should easily introduce to the implications for ore processing. The tool to use is mineralogical characterization, involving description and discussion of their textures, presented in photomicrographs (Fig. 1.61a–k). In order to deepen the understanding, preference has been given to debatable textures, suitable to discuss and sharpen concepts, rather than the more obvious (and easier) textures, such as those in Figs. 1.55, 1.56, 1.57, 1.58, 1.59 and 1.60, timely explained (cf. *Geometallurgical classification*: § 1.2.3) and therefore not needing further comment.

To conclude, the present *automation* trends in reflected light microscopy of ores are summarily addressed. These trends compel the mineralogist to increase steadily the performance and accuracy in the collection of information for mineralogical characterization and its applications. For this purpose, optical microscopy must provide a performance comparable to SEM-supported methods. And it does, as shown in the next chapter (cf. Chap. 2, this volume).

To start discussing and applying textural interpretation, a few examples of the practical significance for ore processing of the textures observed under the microscope are summarily analysed, from a qualitative point of view.

Scheelite ore: Fig. 1.61a shows a concentrate from a skarn (scheelite) mine. It is composed essentially of particles of the first type in Table 1.4 (liberated) and it can be seen that scheelite is the dominant species and its *liberation* is *practically total*, but despite this the concentrate is relatively poor, as there has not yet been an effective separation of the gangue (silicates, carbonates, rutile and some steel chips from the mill), made difficult in part by the abundance of fines in the ore, which should not be lost.

(a)

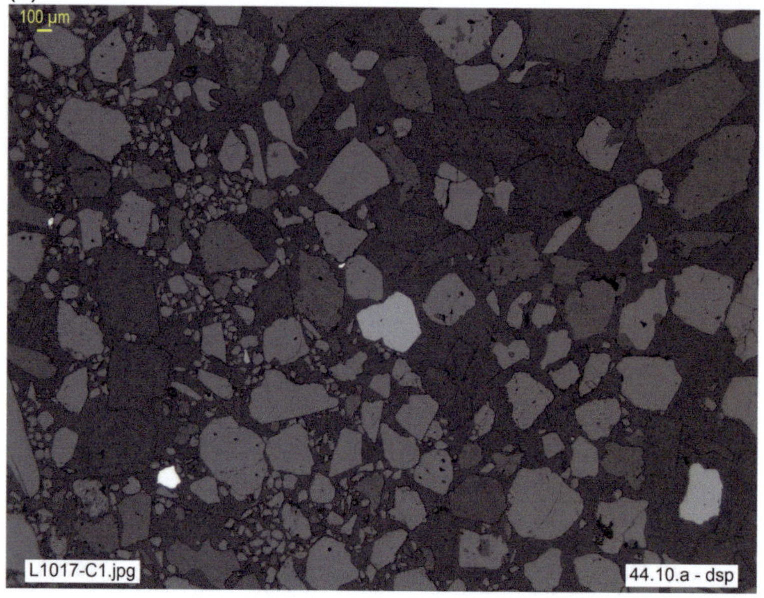

(b)

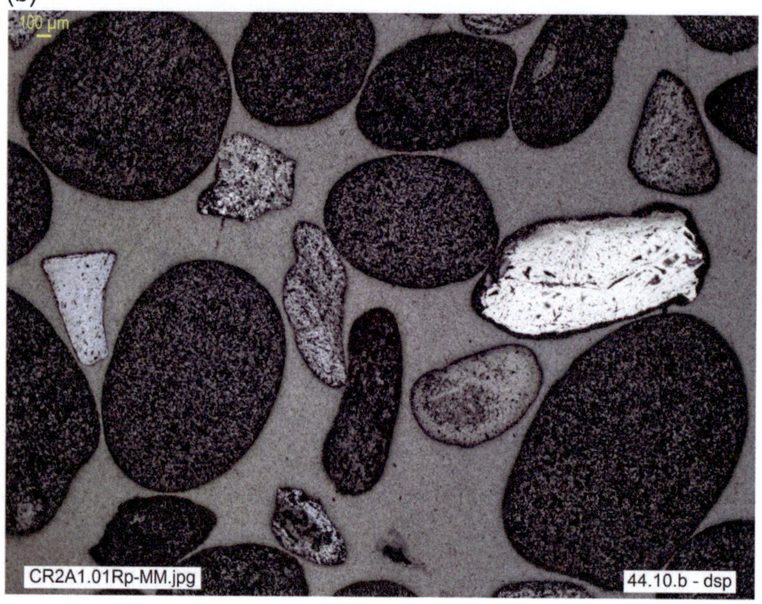

Fig. 1.61 **a** Concentrate of scheelite (light gray) and gangue (quartz, pyroxene, epidote and carbonate, different shades of gray, all darker than scheelite). Minor components (clearly more reflective than sch) are: rutile (grayish-white, two clearly visible crystals) and a few minute, accidental steel chips (white, bright). C1, Los Santos Mine, Cáceres, Spain (**dsp**). **b** Detrital concentrate of nodular monazite (gray, hosting dark silicate inclusions), with a few limonite and manganese oxide nodules or sands (varying shades of light to bright gray). CR2A1, Matamulas, Ciudad Real, Spain (**dsp**). **c** Typical monazite nodule (ore of Fig. b): monazite (light gray) with fine interstitial inclusions of silicates (darker) and traces of limonite (lighter). 251114M1, Matamulas, Ciudad Real, Spain (**dsp**). **d** Abundant silicate inclusions (dark: biotite, muscovite, quartz) and rutile crystallite (bright, in the center) in monazite nodule. CR1B3, Matamulas, Ciudad Real, Spain (**dsp**). **e** Aggregate of chalcocite and digenite (partly framboidal or corona-like) with traces of pyrite; scarce interstitial chalcedony gangue gives consistency to the aggregate. QUEB1, head concentrate, Riotinto, Huelva, Spain (**dsp**). **f** Aggregate of miargyrite (bluish gray), with brecciated and corroded relicts of arsenopyrite (white) and carbonate gangue (black). CR398, Hiendelaencina, Guadalajara, Spain (**dsp**). **g** Middling comprising mainly intergrown digenite (bluish gray) and chalcocite (faintly brownish gray), which include residual pyrite (white) and chalcopyrite (yellow, dark tarnishing), remnants of supergene alteration; traces of bornite and gangue. QUEB1, Riotinto, Huelva, Spain (**dsp**). **h** Sphalerite with inclusions of pyrite, sparse galena and pyrrhotite + chalcopyrite forming a coarse emulsion, in recrystallized, metamorphic massive sulfide ore. 170,800.2, Roros, Norway (**dsp**). **i** Middling of crystalline pyrite and interstitial chalcopyrite, with scarce sphalerite. SAL1.008-rt, Riotinto, Huelva, Spain (**dsp**). **j** Pyrite-chalcopyrite middling. Pyrite is the main component, while chalcopyrite occurs intergrown with both types of pyrite: crystalline and melnikovitic (or submicroscopic, porous). Minor gangue. SAL1, Riotinto, Huelva, Spain (**dsp**). **k** Middling composed of finely intergrown pyrite, chalcopyrite, tetrahedrite, sphalerite and galena; although they are not exsolutions, the resulting geometry could be assimilated to the "emulsion" type (sensu lato). 3CLC R18-AMCO, Cobre Las Cruces Mine, Seville, Spain (**dsp**)

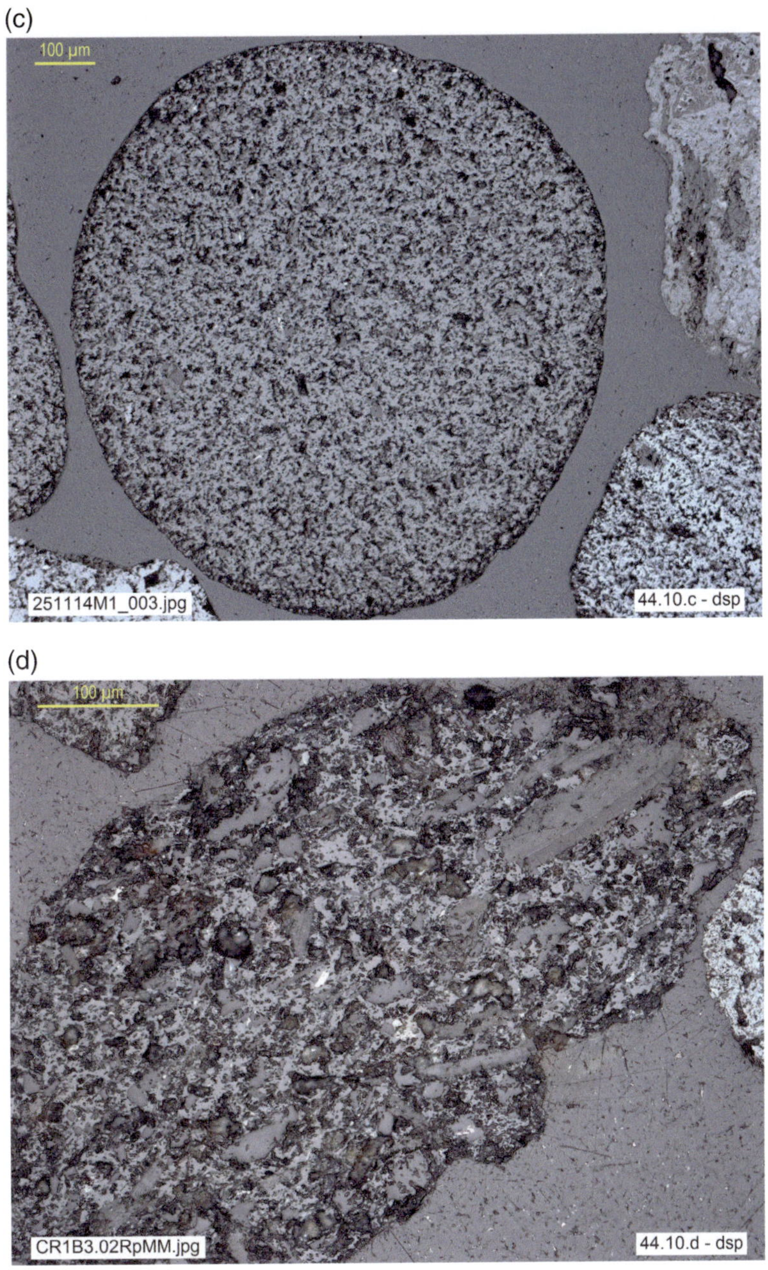

Fig. 1.61 (continued)

REE (Rare Earth Elements) ore: Figs. 1.61b to d correspond to a **detrital rare earth ore** (*monazite-Nd*). As shown in Fig. 1.61b, sedimentary attrition and grain sorting have anticipated, on their own, a remarkable upgrade of the ore. Favoured by the high density of monazite ($\sim$5, i.e. 4.6–5.4 g/cm^3), nature granted an efficient gravimetric concentration process: despite the presence of some grains of iron and manganese oxides, the concentrates are rich in monazite. And most monazite occurs in nodules already liberated before entering the plant.

However, as shown in Fig. 1.61c, d, the monazite nodules are not totally pure, since they contain fine inclusions of disseminated silicates, in a texture that could be described as "emulsion" (sensu lato, since it does not at all correspond, formally, to a true exsolution) following the aforementioned typology of Gaudin (1939). Although most of the nodules are of the type shown in Fig. 1.61c, that is with 90% or more monazite content, there are also some poorer nodules, that is with a lower monazite grade (Fig. 1.61d), although they are comparatively scarce. In both cases, the textures observed in the silicate inclusions and their fine grain sizes preclude the possibility of an economic separation of these inclusions by physical methods, because milling would need to be too fine: leaching of the REE (extraction by chemical means) should be a better option.

VMS (Volcanic-hosted Massive Sulfide) ore: Fig. 1.61e corresponds to a grain composed of digenite, calcocite and gangue, in a concentrate from the supergene enrichment zone of massive sulfide ores from the Iberian Pyrite Belt. Theoretically, it should be considered as middling, since it is composed of three mineral species. However, for practical purposes, it should rather be classified as liberated. This offers a better fit to the information needs of the plant engineer, since in a case like this:

- First, it does not seem convenient for the plant process to try and separate digenite and calcocite, both sulfides rich in copper and which are going to undergo the same metallurgical treatment.
- Next, its qualification as liberated or middling depends on the amount of gangue (chalcedony).
- Normally, depending on the nature of the ore, a threshold is set, below which the gangue is disregarded and the grain is considered as liberated. This threshold can vary greatly from one ore to another and must be agreed with the plant engineer (at the very least, it must be stated in the report), taking into account the foreseeable treatment or concentration process.
- In this case (<10% gangue) it seems advisable to consider it as liberated (*s.l.*), although strictly speaking it would be a ternary middling.

(e)

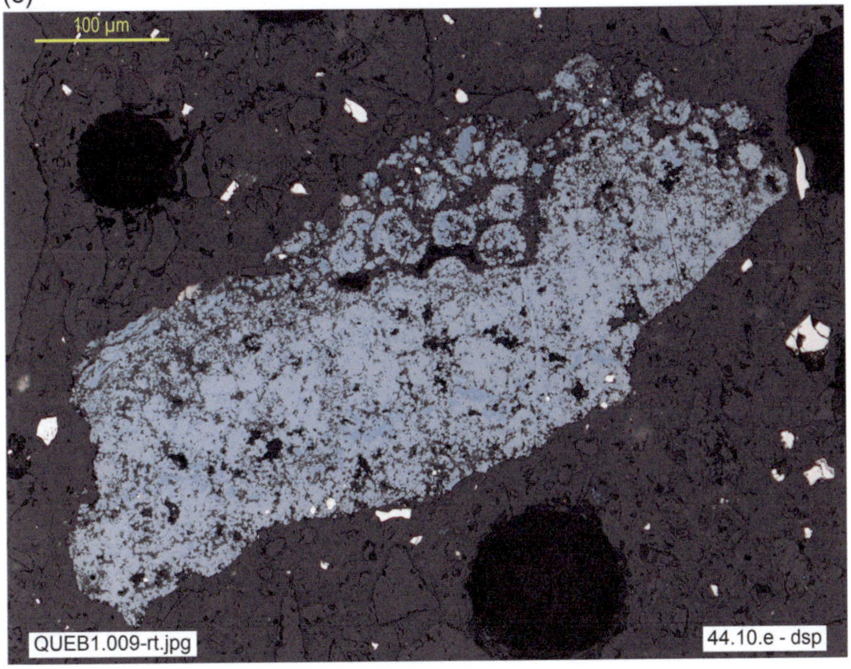

(f)

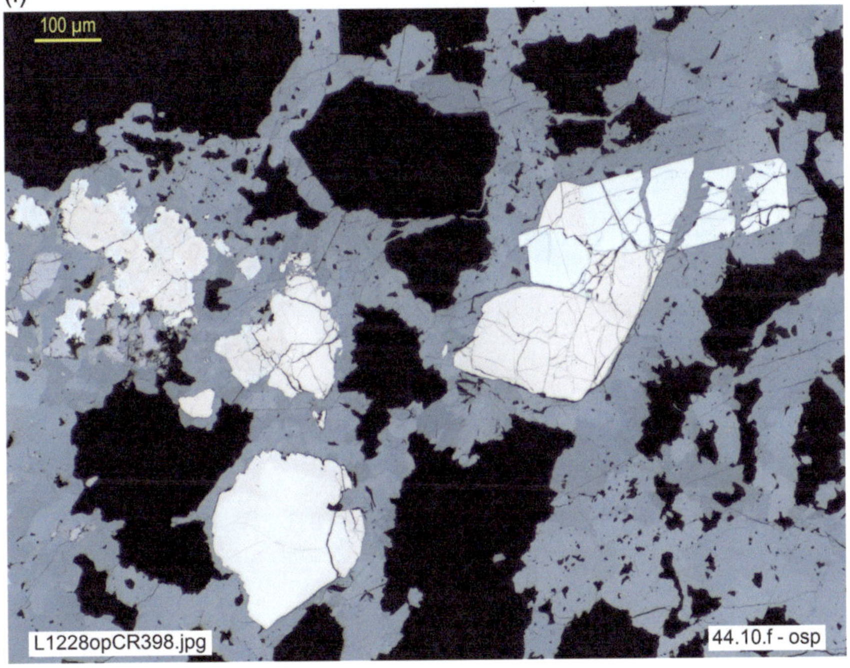

Fig. 1.61 (continued)

Silver ore: Turning to an argentiferous ore, Fig. 1.61f shows a ternary middling, composed of miargyrite, arsenopyrite and gangue. In terms of morphology and taking into account, in detail, their jagged or lobed contours, the contacts between the phases could be described as irregular contacts.

However, considering the overall appearance of the contacts (generally smooth and rounded) and the low hardness of the ore, a second thought suggests to consider another option. Given the features observed, it seems reasonable to predict that, in a subsequent milling process to eliminate the undesirable components (arsenopyrite and gangue), fragmentation will occur preferentially following the contacts between phases, thus facilitating liberation as in the case of simple contacts, so it would be more advisable to classify it mostly as type 44.02 (simple contact).

(g)

(h)

Fig. 1.61 (continued)

Supergene-enriched *vs.* metamorphosed VMS ores: Fig. 1.61g, h correspond to VMS massive sulfides in two different geological contexts:

1. Transition zone from the primary ore to the zone of secondary sulfide enrichment (*cementation* zone) in the Iberian Pyritic Belt:
   Figure 1.61g shows a middling composed essentially of digenite, chalcocite, pyrite and chalcopyrite, with sparse bornite and gangue: theoretically, a senary middling with complex internal structure (e.g. irregular chalcopyrite borders). In reality, for practical purposes, the interpretation is easier: since it is not necessary to separate the copper sulfides, which together make up the ore, and given the scarcity of the gangue, it should rather be understood, in practical terms, as a binary middling (copper sulfides + pyrite), with simple contacts (those of pyrite).
2. Primary ore from the Norwegian metallogenic province, characterized by a strong metamorphic imprint:
   Figure 1.61h shows a metamorphosed ore whose recrystallization has partly overprinted the original texture (likely emulsion) of the protolith. However, it can still be described as a (coarse) emulsion texture of pyrrhotite and chalcopyrite in sphalerite, forming part of a senary middling of pyrite, sphalerite, pyrrhotite, chalcopyrite, gangue and galena.

(i)

(j)

Fig. 1.61 (continued)

(k)

Fig. 1.61 (continued)

Complex sulfides of the Iberian Pyrite Belt: Massive sulfide ores can present difficult concentration problems, conditioned precisely by their textures, as occurs in the Iberian Pyrite Belt with the so-called *complex sulfides*. The *complex sulfides* are polymetallic ores (Zn, Cu, Pb, Ag, Au, etc.), finely intergrown and occasionally with traces of deleterious components (carrying As, Hg, Se, Sb, etc.), whose environmental impact must be kept under control, complying with strict regulations.

These issues affect mainly the primary sulfides. Actually, secondary ores from the supergene enrichment blanket (or cementation zone) comprise chalcocite, digenite, djurleite, covellite, and can be extraordinarily rich in copper. The same supergene process causing Cu enrichment (including the previous oxidation) can cause deleterious minerals to disappear, for instance, in the case of arsenopyrite, to be oxidized and replaced by scorodite.

Figure 1.61e, g illustrates different moments of this process in the secondary sulfide zone:

- In progress (residual pyrite and chalcopyrite) in Fig. 1.61g.
- Complete, with total replacement of the primary ore by copper-rich secondary sulfides, in Fig. 1.61e.
- Finally, an extreme example of ore upgrading through massive replacement by secondary sulfides (digenite, djurleite, chalcocite), with total disappearance of the ore, can be seen in Figure cc8 (*cf. chalcocite description*).

Figures 1.61i–k correspond to primary *complex sulfide ores*:

- The first image (Fig. 1.61i) shows a binary middling—with practically no gangue and scarce sphalerite—of type *irregular contact*, and illustrates the difficulty of liberating chalcopyrite by milling without producing a considerable amount of fines, while in the second (Fig. 1.61j) this difficulty is extreme: the contacts are also irregular, but additionally chalcopyrite shows such fine-sized intergrowths with *melnicovite* pyrite—with grain size sometimes below the resolving power of the microscope—that this texture, from the practical point of view of process engineering, would rather fit type *emulsion, s.l.*
- Fig. 1.61k, by contrast, illustrates a very favourable case: although the intergrowths remain irregular and are sometimes very fine-sized, the polymetallic ore is sufficiently rich in copper, zinc and lead to make feasible their direct extraction by leaching (hydrometallurgy), thus making unnecessary fine grinding to pre-concentrate the ore.

1.4 Concluding Statement: Automated Ore Analysis

To conclude, it is important to recall that current industrial processes require continuous information, if possible in *real time* and supporting with *quantitative data* each of its stages, for their control and optimization. The traditional support techniques, based on the point counter coupled to the ore microscope and handled by a human operator, have demonstrated their historical effectiveness, but the current demands—analysis of millions of points with immediate mathematical data processing, performed as close as possible to *real time*—exceed the human capacity for direct observation and measurement, therefore requiring *automation*.

Automated textural classification, coupled with mineralogical recognition by computer vision (Pirard 2004, 2016; Berrezueta and Castroviejo 2007; Castroviejo et al. 2009, 2010; Catalina 2018; Catalina and Castroviejo, 2017), is already a realistic alternative. To start, Gaudin's (1939) classification provides a suitable reference framework for the mathematical definition and automated analysis of textures of industrial interest, today possible according to the methodology developed by Pérez Barnuevo (2014).

This provides access to **complete and automated mineralogical-textural information**, applicable to ore processing and based on optical microscopy. An example of this is the AMCO system (cf. § Chap. 2, vol. 2). This offers a much more affordable alternative (both in price and infrastructure requirements) than electron microscopy-based systems, apart from comparative advantages in some cases, such as the direct and reliable characterization of iron oxide ores.

References

Amstutz GC (1962) How microscopy can increase recovery in your milling circuit. Mining World 24:19–23

Barton BP (1991) Ore textures: problems and opportunities. Mineral Mag 55:303–315

Bastin ES (1960) Interpretation of ore textures. Waverly Press Inc., Baltimore

Becker M, Wightman EM, Evans CL (eds) (2016) Processs mineralogy. SMI-JKMRC Monograft Series in Mining and Mineral Processing No. 6. Univ. Queensland, Australia

Berrezueta E, Castroviejo R (2007) Reconocimiento automatizado de menas metálicas mediante análisis digital de imagen: un apoyo al proceso mineralúrgico. I: ensayo metodológico. Rev. Metalurgia 43 (4):294–309

Best MG (1982) Igneous and metamorphic petrology. Freeman & Co, New York

Blain CF, Andrew RL (1977) Sulphide weathering and the evaluation of gossans in mineral exploration. Minerals Sci Eng 2:119–1150

Blanchard R (1968) Interpretation of leached outcrops. Bull Nevada Bur Mines, Reno, NEV, USA, p 66

Bowie SHU, Taylor K (1958) A system of ore mineral identification. Min Mag London 99:265–277

Bowles JFW, Howie RA, Vaughan DJ, Zussman J (2011) Rock-forming minerals. Non-silicates: oxides, hydroxides and sulphides, vol 5A, 2nd edn. Geological Society, London.

Cabri LJ, Vaughan DJ (eds) (1998) Modern approaches to ore and environmental mineralogy. COM/IMA Short Course Series, vol 27, Ottawa, Ontario

Castroviejo R, Brea C, Pérez-Barnuevo L, Catalina JC, Segundo F, Bernhardt HJ, Pirard E (2009) Using computer vision for microscopic identification of ores with reflected light: preliminary results. In: Williams et al (eds) Proc. 10th biennial SGA Meet, vol 2. Townsville, Australia, pp 682–684. ISSN: 9780980558685

Castroviejo R, Catalina JC, Bernhardt HJ, Pirard E, Segundo F, Brea C, Pérez Barnuevo L (2010) A fully automated system for multispectral ore microscopy. In: IMA2010 (20th General Meet. Internat. Mineralogical Association). Acta Mineral. Petrogr. Abstr. Ser. (HU ISSN 0324-6523/HU ISSN 1589-4835), Szeged, Budapest, p. 281. Proc. CD_ROM, Session EG53_Geometallurgy and Process Mineralogy, Code 4NF15.

Castroviejo R, Quesada C, Soler M (2011) Post-depositional tectonic modifications of VMS deposits in Iberia and its economic significance. Mineralium Deposita 46:615–637. ISSN: 0026-4598

Catalina JC (2018) Desarrollo de un Sistema Multiespectral para Microscopía de Menas. Ph.D. thesis, ETS Ingenieros de Minas y Energía, Universidad Politécnica de Madrid. https://doi.org/10.20868/UPM.thesis.52452; enlace: http://oa.upm.es/52452/

Catalina JC, Castroviejo R (2017) Microscopía de reflectancia multiespectral: Aplicación al reconocimiento automatizado de menas metálicas. Rev Metalurgia Madrid 53(4 (Oct–Dec.)):e107. https://doi.org/10.3989/revmetalm.107. CSIC / CENIM (ISSN-L: 0034-8570)

Craig JR (1990a) Textures of the ore minerals. In Jambor JL, Vaughan DJ (eds) Advanced microscopic studies of ore minerals, chap 9. MAC short course, vol 17, pp 213261.

Craig JR (1990b) Ore textures and paragenetic studies—some modern case histories and sources of comparative data. In: Jambor JL, Vaughan DJ (eds) Advanced microscopic studies of ore minerals, chap 10. MAC short course, 17, pp 263–317

Craig JR (2001) Ore-mineral textures and the tales they tell. Can Mineral 39:937–956

Craig J, Vaughan D (1994) Ore microscopy and ore petrography, 2nd edn. Wiley, New York

Edwards AB (1965) Textures of the ore minerals and their significance. Brown, Prior, Anderson Pty. Ltd., Australia

Fettes D, Desmons J (2007) Metamorphic rocks: a classification and glossary of terms (recommendations of the IUGS/SCMR). Cambridge University Press, Cambridge

Freund H (ed) (1966) Applied ore microscopy. Theory and technique. Collier-Macmillan, New York

Gaudin AM (1939) Principles of mineral dressing. McGraw-Hill, New York

Grigoriev J (1928) Structures of mineral intergrowths in ores. Mem Russ Min Soc 57:11–56

Head RE (1928) Microscopic study of ore: aid in Cu leaching. Eng Min J 126:13–15

Head RE et al (1932) Statistical microscopic examination of mill products of the copper queen concentrates of the Phelps Dodge Corporation, Bisbee, Arizona. US Bur. Mines Tech.

Huston DL, Sie SH, Suter GF, Cooke DR, Both RA (1995) Trace elements in sulfide minerals from Eastern Australia volcanic-hosted massive sulfide deposits: Part I. Proton microprobe analyses of pyrite, chalcopyrite and sphalerite. Part II. Selenium Levels in Pyrite. Econ Geol 90:1167–1196

Jones M (1987) Applied mineralogy: a quantitative approach. Graham and Trotman, London. ISBN 978-0-86010-510-7

Kojima S, Sugaki A (1985) Phase relations in the Cu-Fe-Zn-S system between 800 and 500 °C. Min J 12:15–28

Korzhinskii DS (1970) Theory of metasomatic zoning. Clarendon Press, Oxford

Park CF, MacDiarmid RA (1981) Yacimientos Minerales. Omega, Barcelona

Pérez-Barnuevo L (2014) Caracterización automatizada de texturas de menas mediante análisis digital de imagen para su aplicación geometalúrgica. Ph.D. thesis. ETSI Minas y Energía, Universidad Politécnica de Madrid

Pérez-Barnuevo L, Pirard E, Castroviejo R (2013) Automated characterisation of intergrowth textures in mineral particles. A case study. Minerals Eng. 52:136–142. ISSN 0892-6875

Petruk W (2000) Applied mineralogy in the mining industry. Elsevier, Amsterdam

Pirard E (2004) Multispectral imaging of ore minerals in optical microscopy. Mineral Mag 68(2):323–333

Pirard E. (2016) Optical microscopy. In: Becker M, Wightman EM, Evans CL (eds) Process mineralogy, Chap 5, pp. 51–66. JKMRC monograph series in mining & mineral processing, vol 6. University of Queensland, Australia

Ramdohr P (1980) The ore minerals and their intergrowths, 2nd edn. Pergamon Press, Oxford

Schneiderhöhn H (1952) Erzmikroskopisches Praktikum. E. Schweizerbart'sche Verlagsbuchhandlung, Stuttgart

Schwartz GM (1923) Solving metallurgical problems with the reflecting microscope. Eng Mining J 116:237–238

Schwartz GM (1951) Classification and definitions of textures and mineral structures in ores. Econ Geol 46:578–591

Stanton RL (1972) Ore petrology. McGraw, New York

Taylor R (2009) Ore textures: recognition and interpretation. Springer, Heidelberg

Part II

Automated Ore Analysis

2 Automated Analysis of Ores and Plant Concentrates

Electronics and computers have advanced so rapidly that it is at last becoming feasible to consider image (areal) analysis based on optical properties. This field has great potential and is wide open now that reliable reflectance databases exist (Criddle 1998: 65)

The industrial exploitation of mineral resources nowadays, as shown throughout the text, requires precise and quantified mineralogical information on aspects that condition the behaviour of an ore in plant. This is what is usually understood by a mineralogical characterization of the ore. From its very beginnings reflected light microscopy was an essential technique to obtain this information and soon developed suitable auxiliary instruments, as the point counter or recent automated systems (AMCO).

The application of the point counter, whose use increased steadily during the twentieth century, was an early and important, albeit elementary, step towards automation. For the first time it allowed a substantial quantification of mineralogical information and obtaining objective data on composition (modal analysis), grain size and degree of liberation, among others. However, the irruption of automated digital image analysis techniques in the field of microscopy, towards the last quarter of the twentieth century, was a turning point. It was largely imposed by the industry's evolution, marked by the use of low grade/large tonnage ores and high demand for efficiency in plant processes, both in terms of ore recovery and environmental impact control.

This requirement faces today's plant engineer with the urgent need for massive and continuous mineralogical information, if possible in real time, documenting each stage with quantitative data for control and optimization. Traditional support techniques, based on the point counter coupled to the ore microscope and handled by a manual operator, have demonstrated their historical effectiveness. Nonetheless, the demand for analysis of millions of points, with mathematical data processing and a performance as close as possible to real time, exceeds human capacity for direct observation and measurement, and requires automation.

Automation was first attempted by optical microscopy-based systems, resorting to RGB colour imaging and/or wavelength control of incident (or reflected) beams by means of filters, measuring the grey level as a proxy for reflectance. However, rapid development of automated applications based on scanning electron microscopy combined with energy dispersive X-ray spectroscopy (SEM–EDS)[1] made possible its commercialization as an alternative to optical systems, with a remarkable implementation, although only for large institutions or corporations[2] able to afford the high investment.

For the mineral resources sector as a whole the result of this evolution is far from optimal: only the strongest organisations have access to modern (automated) mineralogical characterisation systems. This is because, to date, the only ones on the market are based on SEM–EDS systems and these, due to their price, are inaccessible to most small- and medium-sized enterprises. Consequently, they must resort to obsolete, or at least slower and more limited, techniques such as the manually operated integrating stage. This involves sending samples for analysis to external laboratories with a consequent delay of typically five weeks in receiving the results, making them of little use, or working without the necessary mineralogical information.

There is one encouraging prospect. In recent years, extensive research on automated optical microscopy techniques and their application to mineralogy indicates that these are a realistic alternative that can be realized in the design of a

[1] There is also a wide range of applicable techniques. The interested reader can find up-to-date documentation by experts and illustrated with numerous case studies in Becker et al. (2016).

[2] In reality, the demand for these SEM–EDS systems is limited by their high price and demanding operational requirements, so total sales figures do not exceed a few hundred worldwide.

R. Castroviejo, *A Practical Guide to Ore Microscopy—Volume 2*,
https://doi.org/10.1007/978-3-031-18954-8_2

commercially viable system. In fact, the developments from the research, together with recent progress in fields such as optics, electronics and computing, have solved many of the problems that limited the usefulness of the first image analysers (e.g. the Quantimet or the successive QTM series of Cambridge Instruments in the 1980s). Not only the characterization of ores from their visible spectra (today well documented in QDF3: Criddle and Stanley 1993) is achieved by applying computer methods based on the digitization of optical parameters for an automated identification (e.g. Bernhardt 1979, 1987; Atkin and Harvey 1979; Kühnel et al. 1980; Gerlitz et al. 1989; Shoji and Kaneda 1994, among others; cf. discussion: Bernhardt 1990): mineralogical recognition by machine vision (Pirard 2004, 2016; Berrezueta and Castroviejo 2007; Castroviejo et al. 2009; Castroviejo et al. 2010; Catalina 2018; Catalina and Castroviejo 2017, among many others) is already a realistic alternative, as well as automated textural classification (Pérez-Barnuevo 2014). An updated and weighted synthesis on these developments is presented by Pirard (2016).

It is striking, however, that this remarkable development has not yet been reflected in the market, as have been the SEM–EDS systems. It is not that there are no objective conditions: already in 1998 Criddle warned of the urgent need to respond to industry demands (mineral resources) by the application of automation based on digital imaging of ore microscopy. He considered this viable more than twenty years ago, with the then-known innovations in electronics and computing, but advances in automated optical microscopy did not translate into the production of finished equipment able to respond to market demand as a commercial offer, ready for use. Overcoming this anomaly is, in our opinion, of interest to the entire mineral resources sector. Since the knowledge and techniques necessary for its production already exist, a high-performance, reliable, competitive and affordable automated system should by now have been launched on the market for small/medium companies, and even universities and geological services.

The AMCO[3] system, developed by a European consortium with the support of EIT RawMaterials, is used for spectral characterization in the VNIR range (visible and near infrared) of the ores described in this text. It is applied to obtain complete information that allows the automated characterization of the ore.

2.1 The AMCO System

The reflectance of the minerals has been measured using the AMCO system prototype of the Applied Microscopy and Image Analysis Laboratory of the Universidad Politécnica de Madrid (LMA-UPM). This is a tool that is highly suitable for the quantitative determination of specular reflectance spectra.

The AMCO system is a multispectral specular reflectance microscopy system that allows the acquisition and processing of images composed of many spectral bands covering at least the visible and near infrared (VNIR) range of the spectrum. It consists of a specialized instrument and two specific software applications.

The instrument comprises a fully-motorized reflected-light microscope, adapted to allow imaging beyond the visible spectrum, on which various auxiliary elements are mounted (a filter wheel with numerous bandpass filters, a monochrome video camera with digital interface and a high-precision motorized XY stage), all connected to the computer running the software. Figure 2.1 shows a schematic view of the instrument, with the auxiliary elements depicted in dark grey.

AMCO system software consists of two independent applications:

- amcoCapture controls all the elements of the instrument and performs the acquisition of series of multispectral specular reflectance images of fields of a polished section. This guarantees the accuracy and reproducibility of the reflectance measurements and perfect registration between all the bands. Images can be captured one by one in manual mode or automatically by the unattended scanning of the area on the polished section specified by the operator.
- amcoAnalysis visualizes the images of a series and performs various types of analysis on them. The operator can choose the image displayed on the screen, view individual bands or linear combinations in true or false colour, zoom in to examine details, adjust levels, apply gamma corrections, and so on. Polygonal zones can be delimited interactively on the images to perform multispectral measurements of the sample's specular reflectance. The mineral corresponding to each zone is identified by

Supported by

eit RawMaterials
Connecting matters

This activity has received funding from the European Institute of Innovation and Technology (EIT), a body of the European Union, under the Horizon 2020, the EU Framework Programme for Research and Innovation

[3] Marketed by Multispectral Microscopic Technologies S.L. (mmt-systems.com).

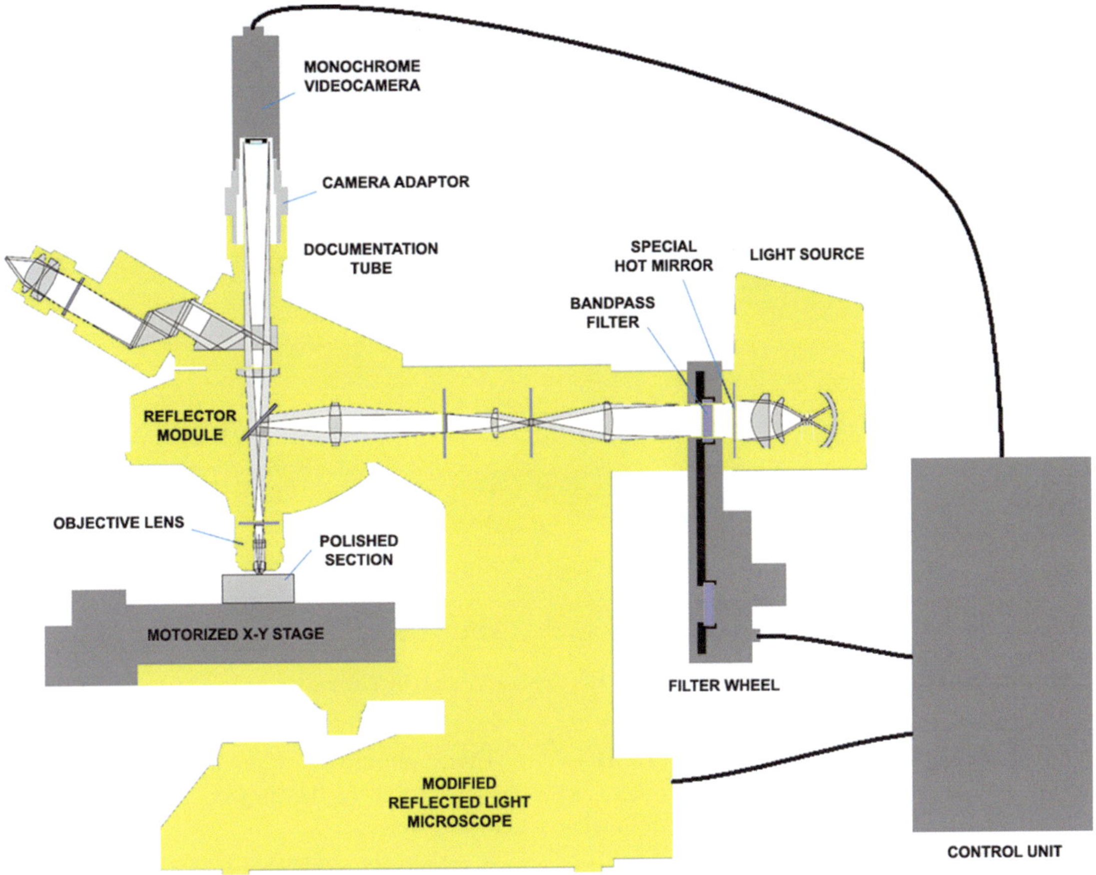

Fig. 2.1 Schematic view of the AMCO system's specialized instrument

comparison to a database of multispectral specular reflectance values of minerals of interest to industry. The program can automatically classify all the pixels of an image, obtaining the modal analysis and a map of the distribution of the mineral phases of the field. It can also automatically process all the images of a series, obtaining the modal analysis of the whole sample. Finally, for polished thin sections of milled ore samples prepared with fluorescent resin, it can perform particle size analysis as well as modal analysis, liberation analysis (both by exposed surface and by composition) and particle composition analysis, all by particle size fractions.

The basic prototype of the AMCO system is based on a Leica DM6000M or DM6M optical microscope, and allows the acquisition of multispectral images composed of up to 20 reflectance bands in the VNIR range (370–1000 nm) plus an optional fluorescence band. The advanced version of the prototype (Fig. 2.2) incorporates an additional camera sensitive to Short-Wave Infrared (SWIR). A second filter wheel with SWIR filters allows the acquisition of up to six additional reflectance bands in the SWIR range between 1100 and 1600 nm, but at a lower resolution.

The AMCO system was specifically conceived to perform an automated microscopic characterization of polished sections of metal ores. It was developed within the AMCO (Automated Microscopic Characterization of Ores) project, a KAVA upscaling research and development project funded by EIT RawMaterials (EIT Project no. 15039, 2016–2018) in the framework of the EU Horizon 2020 programme. This project was carried out by a consortium led by UPM, with the participation of Université de Liège (Belgium), Cobre Las Cruces (Spain), KGHM Polska Miedz (Poland) and ThinSectionLab (France). Its immediate precursor was the CAMEVA system, developed by the UPM in collaboration with AITEMIN in projects CGL2006-13688-C02-01/02

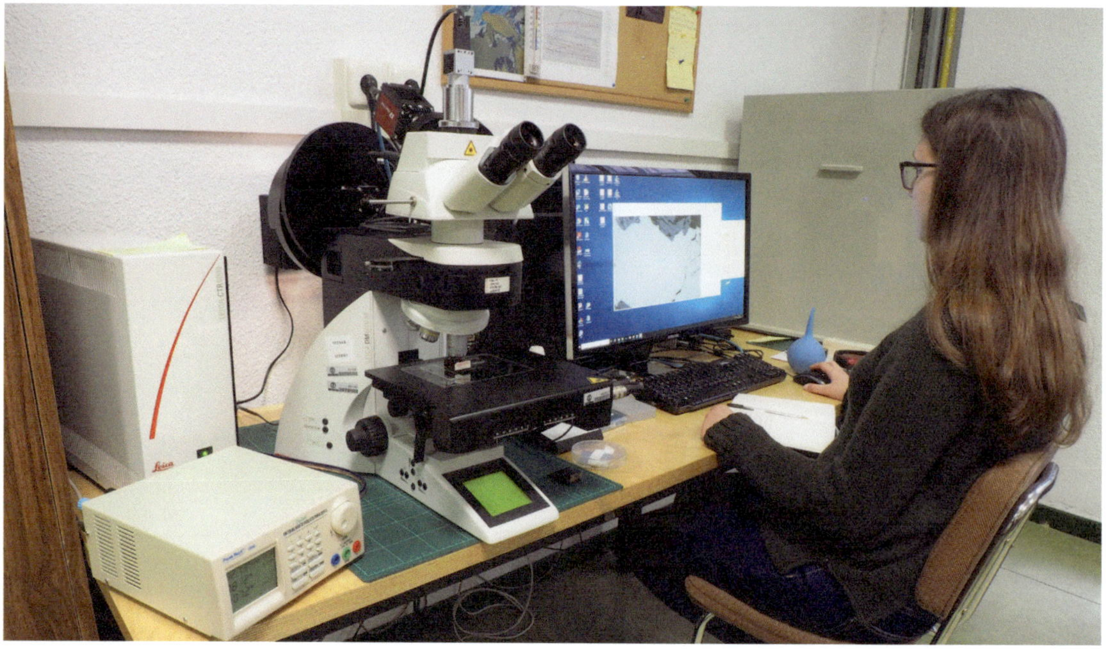

Fig. 2.2 View of the advanced AMCO system prototype at UPM's Applied Microscopy Laboratory

(2007–2009) of the Spanish Plan Nacional de I + D + i (Castroviejo et al. 2009).

The AMCO system was developed with the following objectives:

- To achieve automated identification of the most common ores or those of greatest interest to industry, by means of optical reflection microscopy on polished sections
- To make this process compatible with automated quantification, which involves scanning the sample using a motorized XY stage, thus dispensing with the usual rotating stage
- To build, as an essential resource for identification, a multispectral database covering at least the VNIR range (visible and near infrared, between 400 and 1000 nm) and collecting reliable specular reflectance measurements for each of the ores on randomly oriented grains, thus renouncing polarisation
- Quantify the information of each image pixel by pixel to allow the development of industrial applications (recognition of mineral species, modal, grain size and liberation analysis, textural characterization, etc.)
- Offer an easy-to-use user interface, useful also for university teaching and industrial LLE (life-long education), that facilitates the learning of ore microscopy
- Perform system validation using external data from other analysis systems, statistical analysis and testing of real problems
- Develop a prototype suitable for commercialisation with TRL (Technology Readiness Level) 7–8

These objectives were mostly achieved during the implementation of the project, and were completed in the following years.

A system for automated microscopic characterization of metal ore samples is now available and can be used in an industrial laboratory without the need for an expert in ore microscopy. To carry out particle analysis of milled ore samples it is necessary to use a specific procedure (Grunwald-Romera et al. 2019) developed by the LMA-UPM that uses polished thin sections prepared with fluorescent resin.

2.2 Methodology

The AMCO system is an advanced tool that is particularly suitable for the quantitative determination of specular reflectance spectra of minerals in the VNIR (or VNIR + SWIR) range.

The measurement procedure consists of two phases:

- In the first, the amcoCapture program acquires and stores to disk a series of images of a polished block (PB) or a polished thin section (PTS)
- In the second, the amcoAnalysis program processes the series of images and extracts the desired information.

The reason for separating the phases is that image acquisition requires exclusive use of the microscope and takes much longer than analysis, yet needs to be done only once, whereas analysis of a series of images is usually an iterative process in which the minerals under consideration are judiciously selected and the parameters gradually refined.

Another advantage of this separation is that it allows multiple instances of the analysis program to run simultaneously as it does not make use of the microscope. In this way it is possible to process several samples in parallel, even on different computers.

2.3 Image Acquisition

The amcoCapture program is the application specifically developed to control the microscope and its auxiliary elements and to acquire multispectral specular reflectance images of fields of a polished section. This guarantees the accuracy and reproducibility of the reflectance measurements and obtains perfect registration of the different bands. For this purpose, it implements calibration and correction procedures using two specular reflectance standards and a geometric standard, which were the subject of a patent.

The program can operate in either manual or automatic mode. In the first, the operator scans the preparation at will by manually operating the controls of the motorized stage, searching for areas of interest and capturing, one by one, images on the microscopic field at the desired locations (Fig. 2.3).

In the second, the operator sets the scanning parameters, including the percentage of coverage or scan step in X and Y, and the start and end points of the area to be swept. The operator focuses on several fields spread over the area, then the program automatically performs the indicated scan (Fig. 2.4).

In both cases, the operator must select a disk path and enter a name for the series of images, which will be saved to the specified location in a folder of that name.

The amcoCapture program thus makes it possible to obtain multispectral images in which the grey level of a pixel in one of the bands—multiplied by a defined conversion factor—is equivalent to the value of the specular reflectance of the corresponding point in the polished section, averaged over the spectral range of that band.

Proper calibration of the complete acquisition system (which includes illumination, filters, lenses and camera) with two specular reflectance standards, one of high reflectance (≈90%) and one of low reflectance (≈4.5%) guarantees the accuracy and repeatability of the measurement and, therefore, its reproducibility in different laboratories (Catalina and Castroviejo 2017).

The program incorporates a specific procedure to check that system is correctly calibrated (Fig. 2.5) by acquiring a

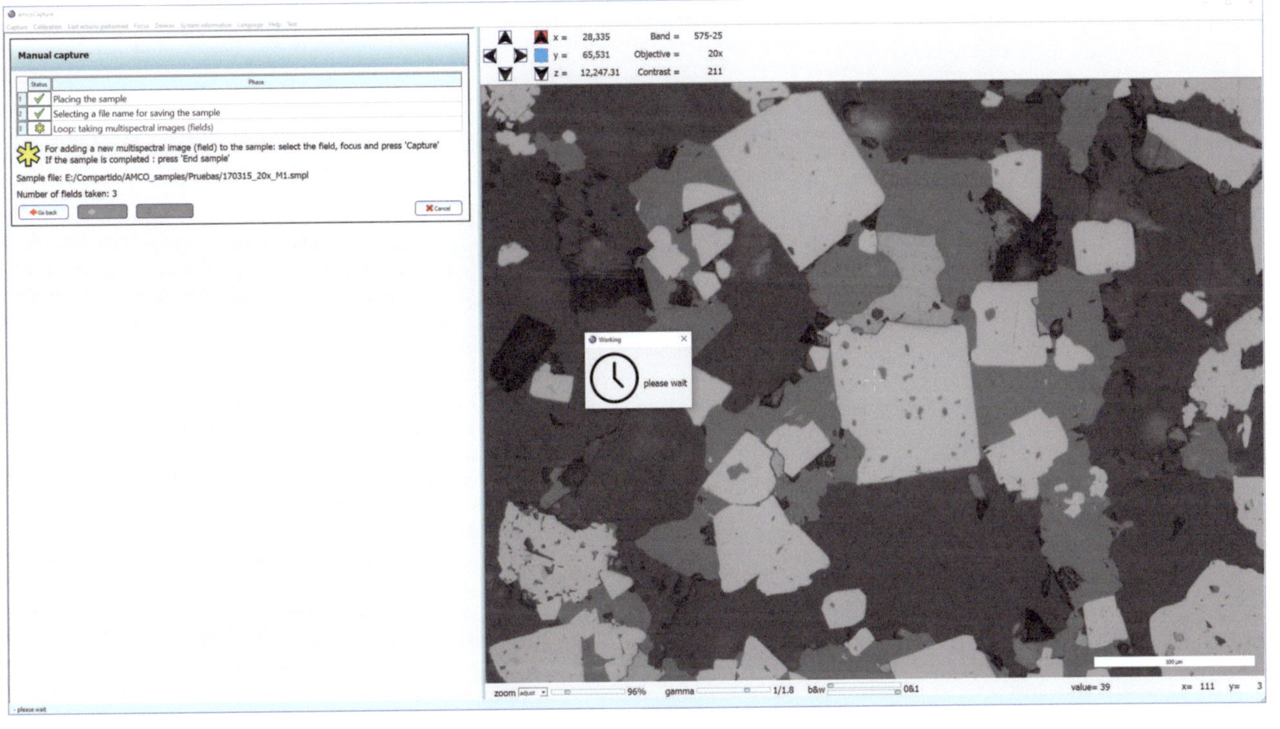

Fig. 2.3 amcoCapture program during manual image acquisition of a polished block

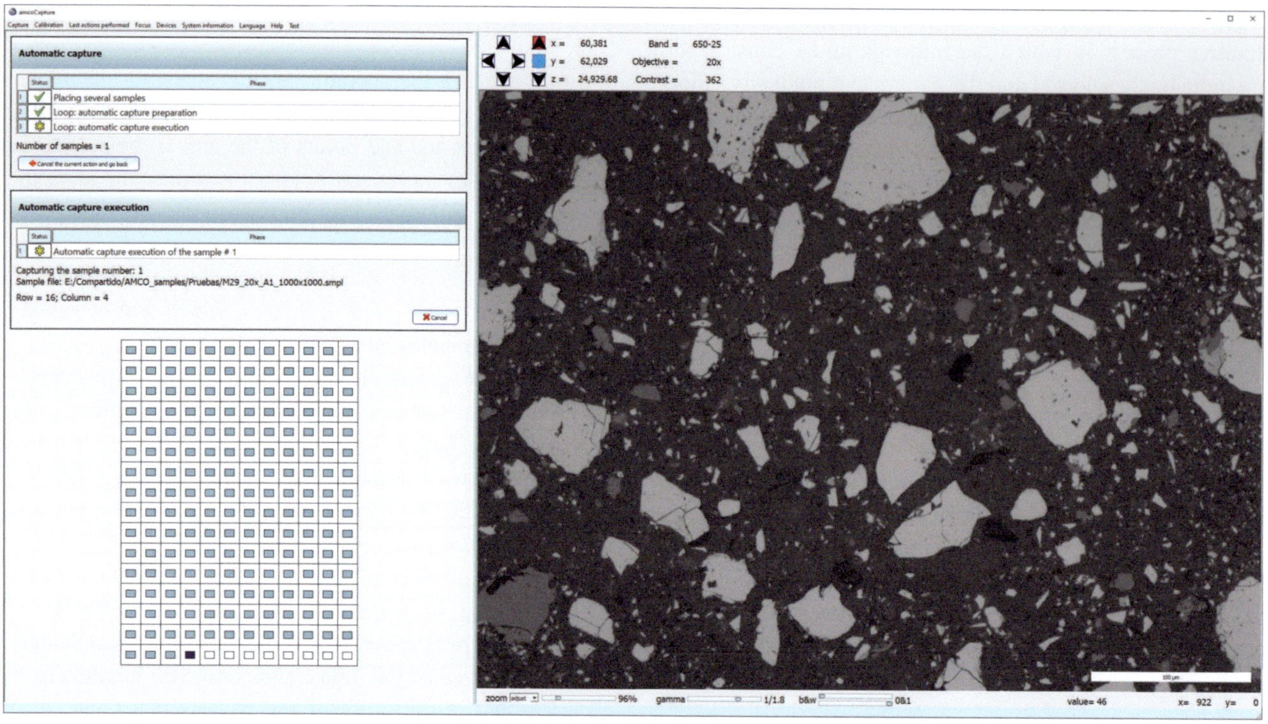

Fig. 2.4 amcoCapture program during an automatic scan (12 × 16 fields) of a polished thin section. The use of a 20× objective and a 1000 μm step in X and Y provide effective coverage of 18.4% of the area

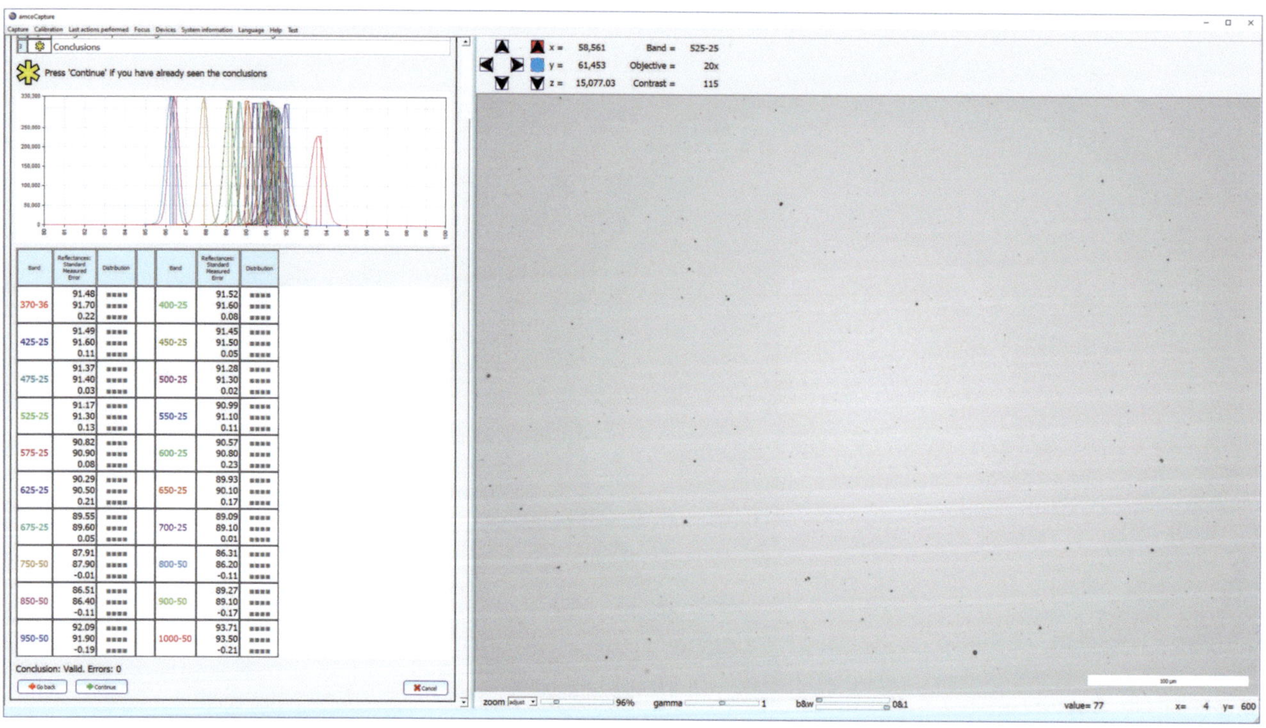

Fig. 2.5 amcoCapture program during calibration check of acquisition system

multispectral image of the high reflectance standard and comparing the average reflectance value of each band to the corresponding nominal reflectance value of the standard. If the absolute value of the difference in any band exceeds a threshold (typically 1%), the program highlights the discrepancy in colour (blue for positive differences and red for negative differences), which indicates to the operator that the system ought to be recalibrated.

Another essential feature of the amcoCapture program is its ability to compensate for the chromatic aberration presented by the microscope lenses when working outside the visible spectrum, applying a specific focus adjustment and geometric correction to each of the bands. This permits perfect registration between them all. For this purpose, geometric calibration of the acquisition system is carried out with a suitable geometric standard.

In this way, it is not necessary to focus—either manually or automatically—on each band separately to achieve a correct focus on all the bands of an image. It is sufficient to focus on any band and, for the others, to add the adjustment corresponding to the focus disparity between the bands, as shown in Fig. 2.6, to the Z focus position obtained.

Figure 2.7 shows the autofocusing process in action on the reference band, which is usually the 525 nm band, as it has the highest definition.

Calibration of the system is lens-specific and must be performed for each objective with which images are to be acquired (it is unnecessary if the lens is only used for visual inspection of the preparation). The normal sequence of the complete calibration procedure for a new objective consists of:

1. Measurement of focus disparities between the bands. The focusing of each band can be done automatically or manually. It is advisable to use a very well-polished section with a high contrast in all bands.
2. Calculation of optimal exposure times. Each band requires a different exposure time, because the intensity of the illumination, the transmissivity of the bandpass filters, the sensitivity of the camera and the reflectance of the patterns depend on the wavelength.
3. Capture of the bright standard. A scan is performed on the high reflectance standard, and the average image is calculated after discarding surface defects in the standard.
4. Capture of the dark standard. The low-reflectance standard is scanned, and the average image is calculated after discarding surface defects.
5. Geometric calibration. A glass slide with a two-dimensional arrangement of reflecting dots, whose diameter and spacing are calibrated, is used as a geometric standard.
6. Measurement of the angle of rotation of the cameras. The geometric standard is also used for this task, taking as angular reference the X axis of the stage.

When the system loses calibration due to ageing of the halogen lamp during normal use, to recalibrate it is sufficient to perform operation (3). Only in the case of significant drift should operations (2), (3) and (4) be performed, and in that order.

Thanks to this calibration procedure, the AMCO system allows the acquisition of multispectral images of 16 bits per

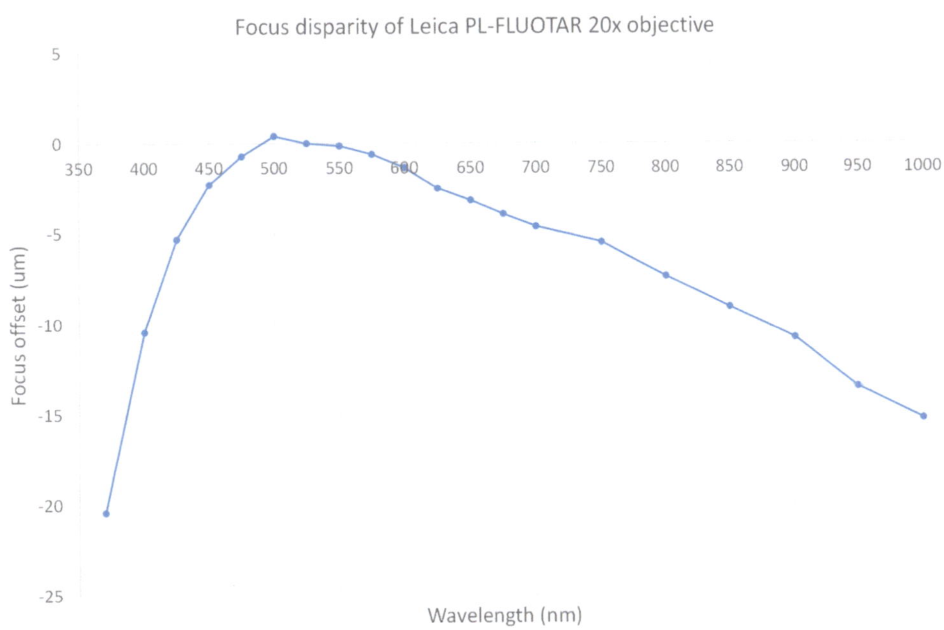

Fig. 2.6 Curve showing the focus disparity between the bands for a given objective

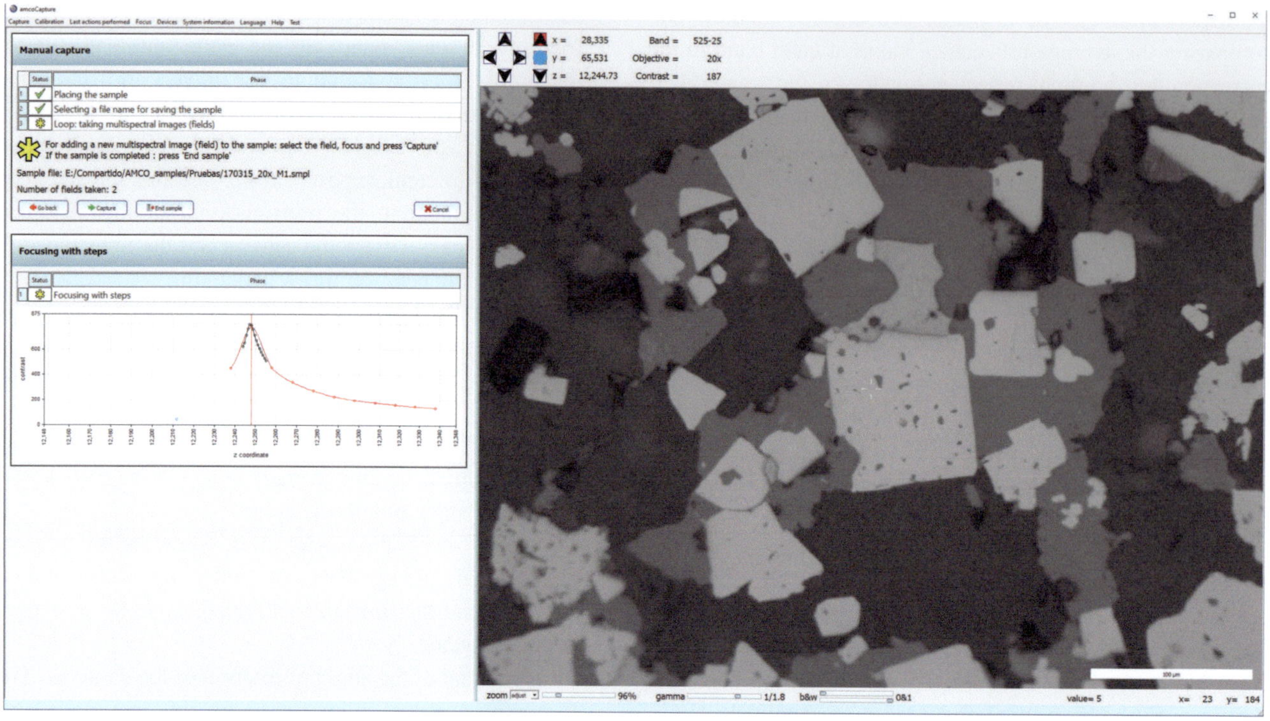

Fig. 2.7 Autofocusing is performed in two stages: first, a fast traverse in the wide vicinity of the initial Z position, then a slow traverse in the close vicinity of the Z of maximum contrast

band to provide a very accurate approximation to the specular reflectance spectrum in the VNIR (or VNIR + SWIR) range of the polished section at the point corresponding to each pixel in an image, enabling the identification of the minerals present in it with a confidence level much higher than that obtained by classical image analysis procedures based on RGB colour images of 8 bits per band.

2.4 Image Analysis

The amcoAnalysis program was developed specifically to visualize and process a series of multispectral images of specular reflectance acquired with the amcoCapture program, then to perform several types of analysis and measurements on them.

Once the selected series of images has been opened, the upper left part of the window gives a graphic representation of the series as a matrix of cells, allowing direct selection with the mouse of the image that the operator wishes to see (Fig. 2.8). In automatic scans with less than 100% coverage, a rectangle is shown within each cell, representing at scale the microscopic field covered by each image (Fig. 2.11). In the case of images acqured manually rather than from an automatic scan, the series of images is represented simply as a horizontal strip of cells.

The operator can change the image display mode at will, choosing between any band of the multispectral image, special views in true or false colour obtained from linear combinations of the bands, or even the classification results. The operator can also zoom the image to observe details, adjust the gamma and modify the black and white levels.

If the series of images corresponds to a ground mineral sample prepared with fluorescent resin according to the specific procedure developed by the LMA-UPM, the first time that the series is opened it is advisable to set a general threshold of binarization for the fluorescence band, so that the resin can be correctly distinguished. For this purpose, it is convenient to compute and display the accumulated histogram of the fluorescence band of the whole series. After this, it is easy to determine a suitable threshold, normally at a value close to the anti-mode that separates the sharp peak corresponding to the mineral particles (non-fluorescent) from the more extended peak corresponding to the resin (fluorescent). Figure 2.9 shows that the binarization threshold has been set to the value corresponding to 12% of the full scale of the fluorescence band.

Binarization of the fluorescence band (Fig. 2.10) to the threshold value set by the operator automatically generates a mask for areas occupied by resin, which effectively delineates the outline of the particles, whether they consist of opaque, gangue or mixed minerals. Figure 2.11 is a true colour view of the same field as Fig. 2.10. It shows in deep

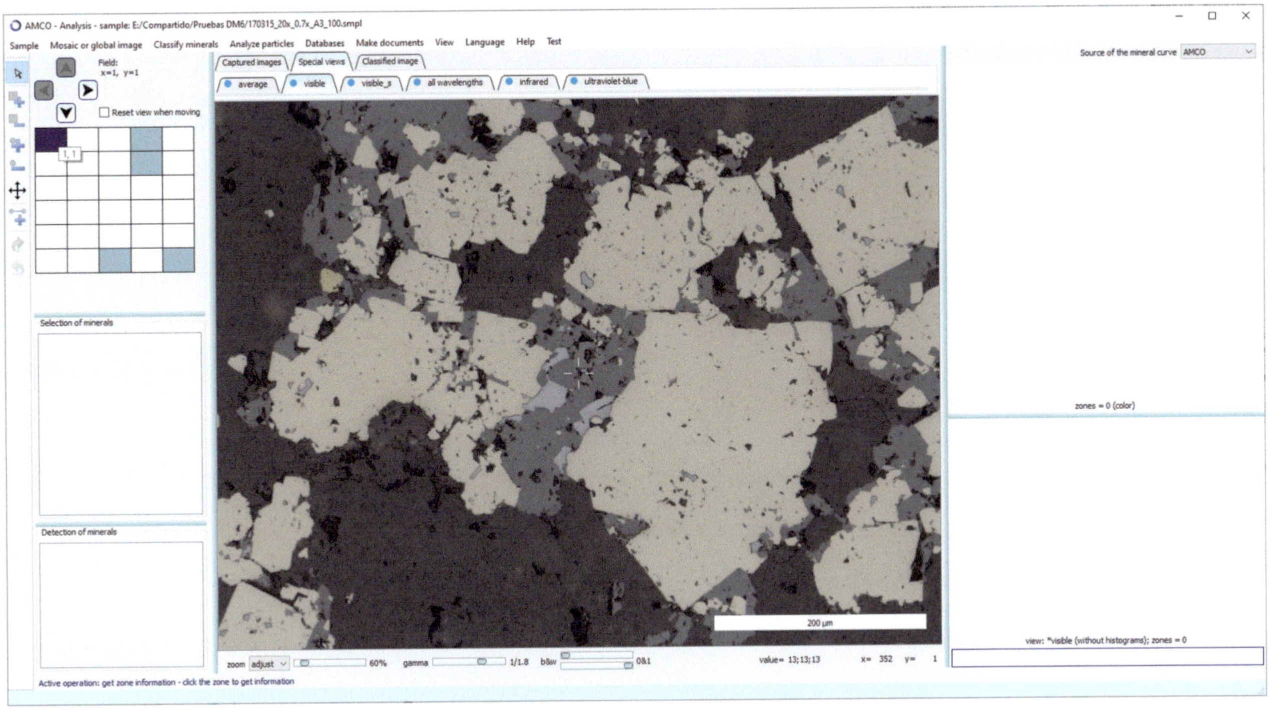

Fig. 2.8 amcoAnalysis program after opening the image series from an automatic scan (5 × 6 fields) of a portion of a polished block with 100% coverage. Special true colour view of field (1,1)

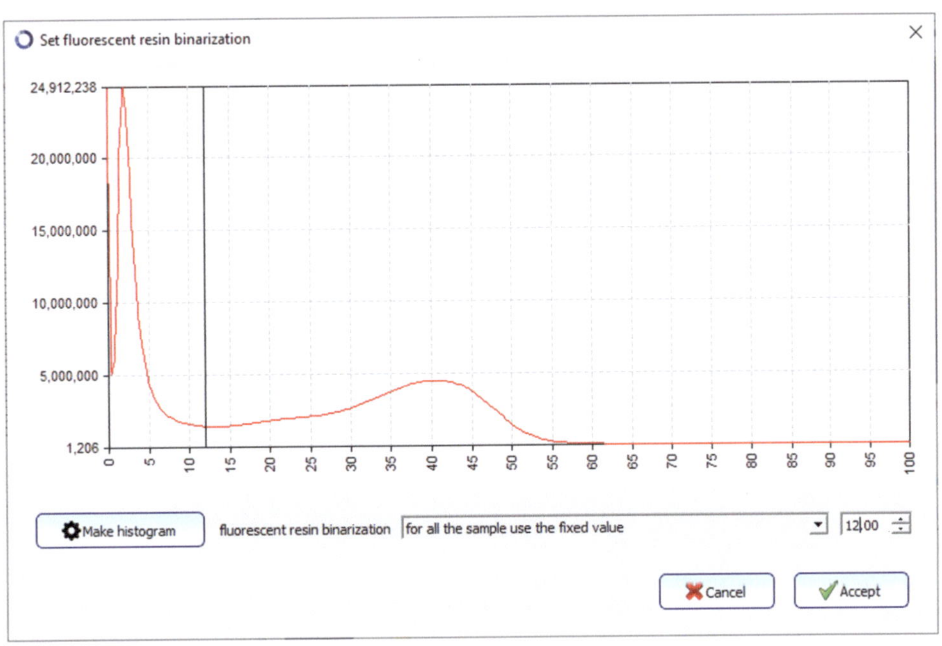

Fig. 2.9 Cumulative histogram of the fluorescence band of a series of images. Determination of binarization threshold to distinguish resin from mineral particles

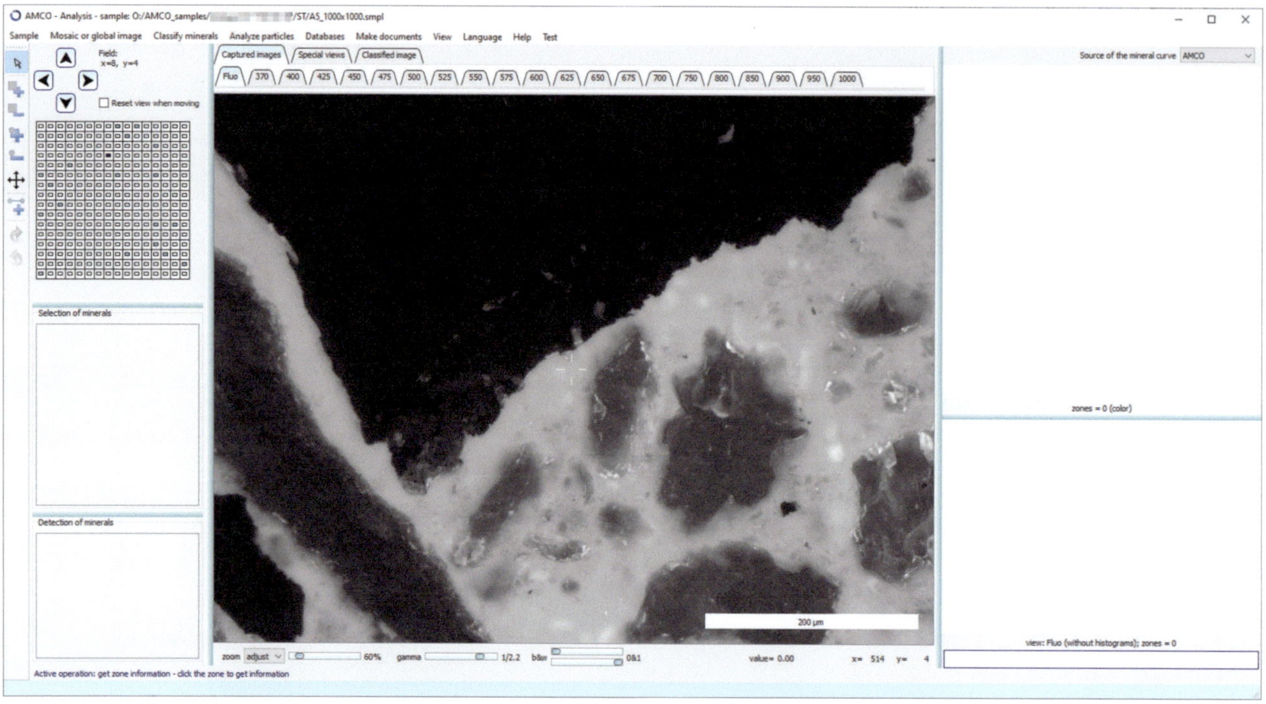

Fig. 2.10 Fluorescence band from a field of a polished thin section made with fluorescent resin

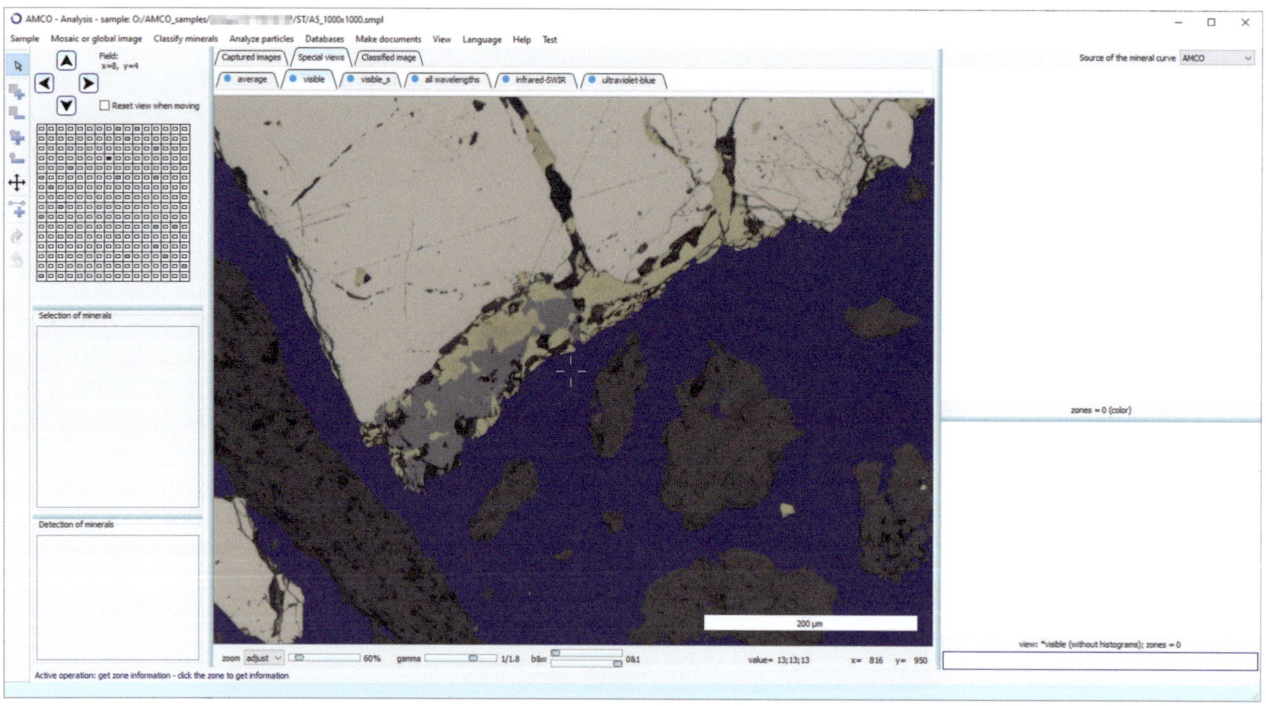

Fig. 2.11 Colour view of the field shown in Fig. 2.10 after setting the threshold value for the resin mask

blue the resin mask obtained after setting the threshold value for the fluorescence band. This mask clearly defines the outline of the various mineral particles that appear in the image, even when made up of gangue minerals that have reflectance spectra very similar to that of the resin, as with the particles at the bottom of the image.

The resin mask allows resin pixels to be discarded automatically when performing modal analysis of a sample, and facilitates segmentation of mineral particles for various characterization analyses based on particle size measurement.

The amcoAnalysis program incorporates a multispectral specular reflectance database in the VNIR range of the main metal ores of interest to industry, as well as their most frequently associated accessory minerals, totalling about a hundred minerals. This database is used by the program to select the mineral species that best match the spectrum of a pixel (or the average spectrum of an image zone). Moreover, the range of minerals can be extended by users by defining additional mineral species in a customised auxiliary database managed by the amcoAnalysis program.

When opening a new series of images, the operator must select the minerals to be considered in the classification, as for reasons of accuracy and efficiency it is best to restrict the options to minerals that may actually be present in the sample. The background of Fig. 2.12 shows an example of the mineral selection window in which the minerals included in the main database are listed alphabetically, along with those in the auxiliary database.

The window incorporates an aid to the initial selection of minerals when the origin of a sample is known. If the operator chooses from a drop-down menu the type of deposit from which it came (Fig. 2.12), the program automatically marks the minerals likely to occur in the sample, which is often a good starting point for analysis. The list of minerals selected for each image series is stored on disk, so it is retained from one session to the next time that the series is opened.

The amcoAnalysis software offers numerous possibilities to approach the study of a sample, from purely manual to fully automated.

In **manual mode**, the program allows the operator to delimit several polygonal zones (Fig. 2.13) in an image interactively. The outlines of the zones are drawn in bright colours to make them easier to locate in the image. The program presents the following types of information about these zones:

1. Average specular reflectance spectrum of each zone (actually, a smooth curve is drawn through the points corresponding to the mean—or the mode—of the specular reflectance values of all pixels within the zone).

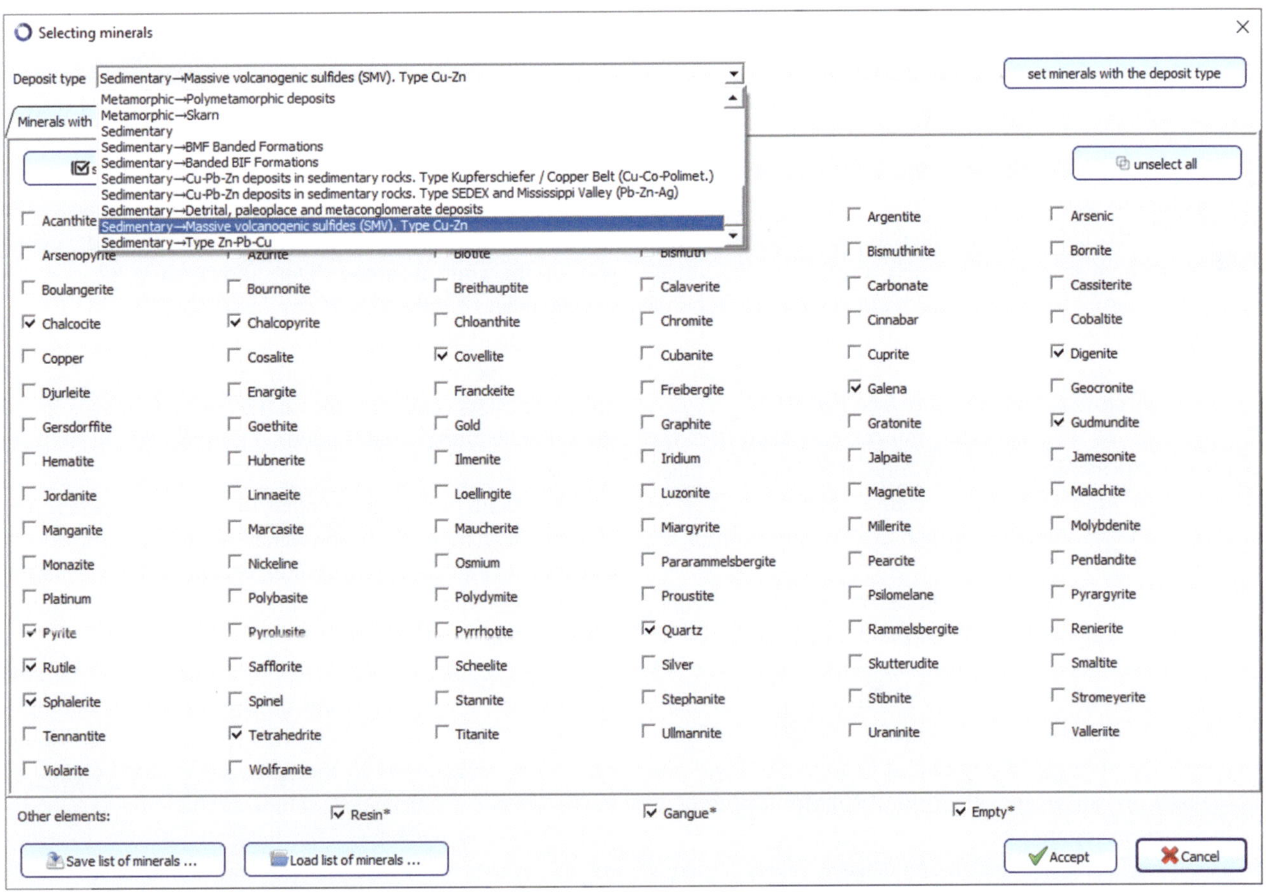

Fig. 2.12 Mineral selection window of the amcoAnalysis program. The Deposit type drop-down menu allows automatic selection of the minerals typical of the deposit type that the sample is from

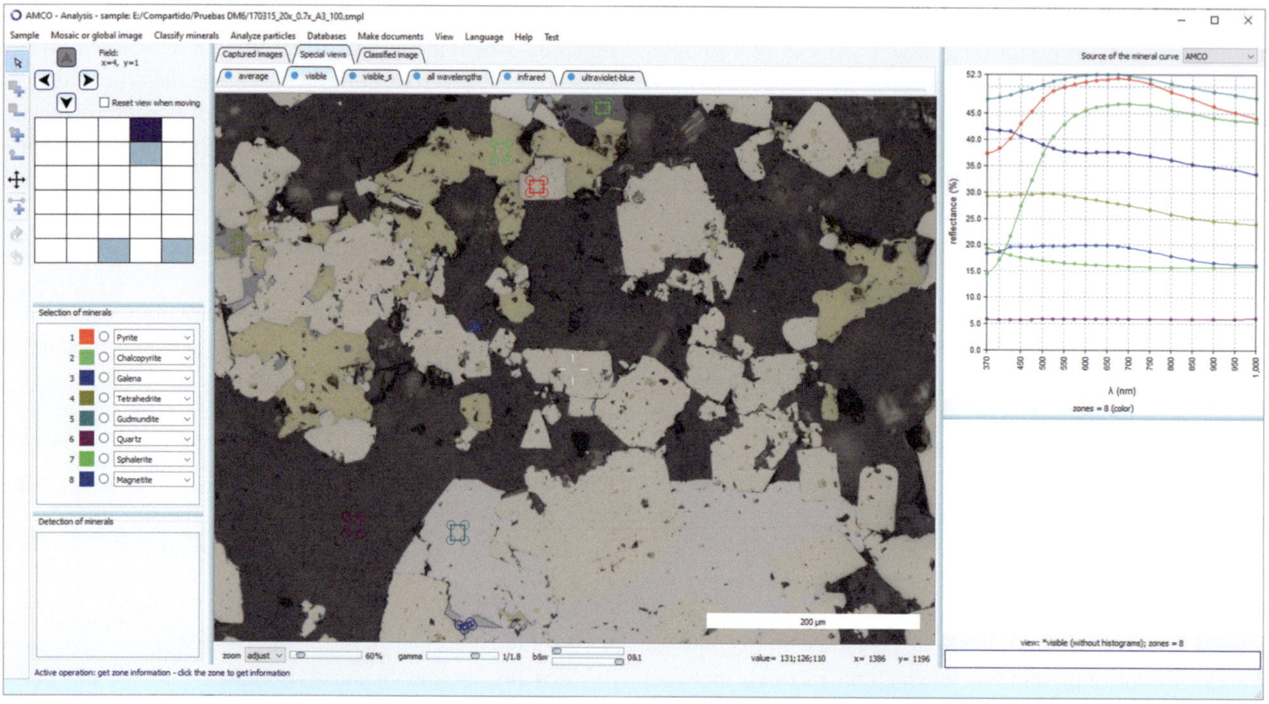

Fig. 2.13 amcoAnalysis program displaying (top right graph) the average specular reflectance spectra of the zones outlined on the selected image (field (4,1) of the polished block scan of Fig. 2.8)

A graph of the spectra of the marked zones is shown in the upper right corner in the same colour as the outline of its zone. The program allows export of a graph of the spectra as an image file (Fig. 2.14), to facilitate its inclusion in reports.

2. Lists of the minerals in the database selected for consideration whose spectra most closely resemble (according to several criteria) the average spectrum of the active marked zone, ordered by increasing distance (equivalent to decreasing likeliness). These lists are displayed in the lower left corner of the window (see Fig. 2.15).
3. Spectrum of the database mineral assigned by the operator to the active marked zone. It is drawn in black in the spectrum graph (upper right corner, Fig. 2.15), so that it can be compared with the average spectrum of the zone.
4. Histogram of specular reflectance of each marked zone. Whenever a band of the multispectral image is displayed, a graph with the histograms of the marked zones for that band is shown in the lower right corner (Fig. 2.15). Each histogram is drawn in the same colour as the outline of its zone. The program allows export of a graph of the histograms as an image file (Fig. 2.16), to facilitate its inclusion in reports.
5. Text files listing the specular reflectance values of pixels belonging to the zones marked on an image (Fig. 2.17) are automatically generated by the program. These are intended for use by the zone database, but can also be used for data processing.

The marked zones can be tagged by the operator according to their nature and characteristics, and automatically become part of the zone database. This allows easy location and management of the zones of any ore (Fig. 2.18).

To expand the list of minerals considered in the classification, the operator has only to define the additional minerals, mark and tag the zones of the images to be used for learning, and include them in the customized database.

The specular reflectance values of pixels from a selection of zones and minerals in the database can be dumped into text files for statistical processing. This provided the starting data for the study of classification techniques reported in López-Benito et al. (2019).

The processing of these files has also made it possible to determine for each mineral the spectra corresponding to the mean, deviation, minimum, maximum and various percentiles represented in the spreadsheet that collects the VNIR specular reflectance values of the minerals in the AMCO system database. This made it possible to generate the graphs of VNIR spectra that are included in this book.

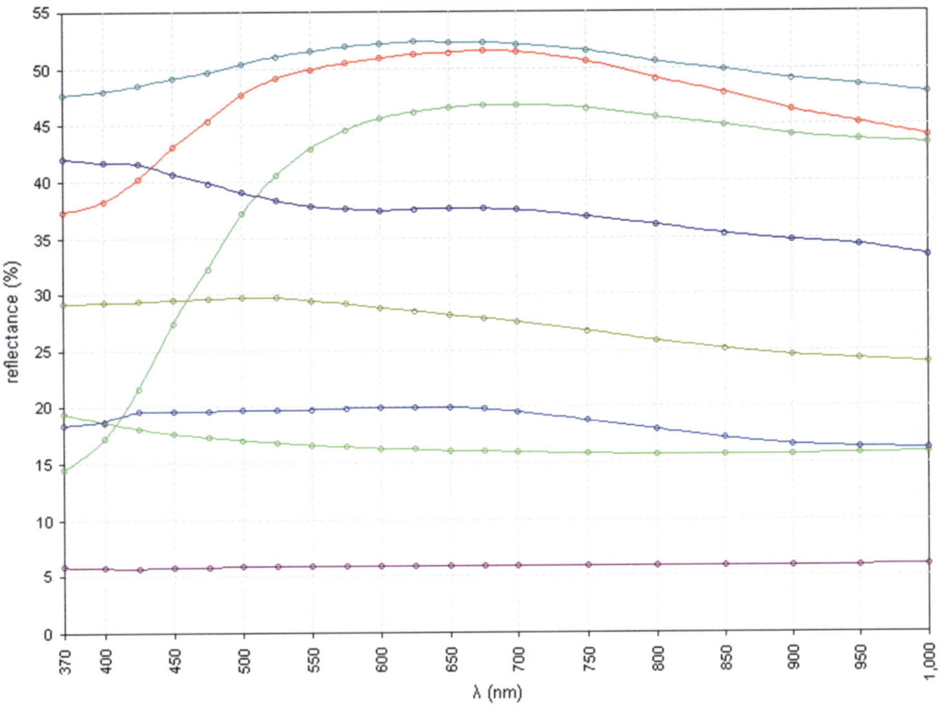

Fig. 2.14 Average specular reflectance spectra of the zones outlined on the image in Fig. 2.13. The colour of each curve matches the outline of the corresponding zone

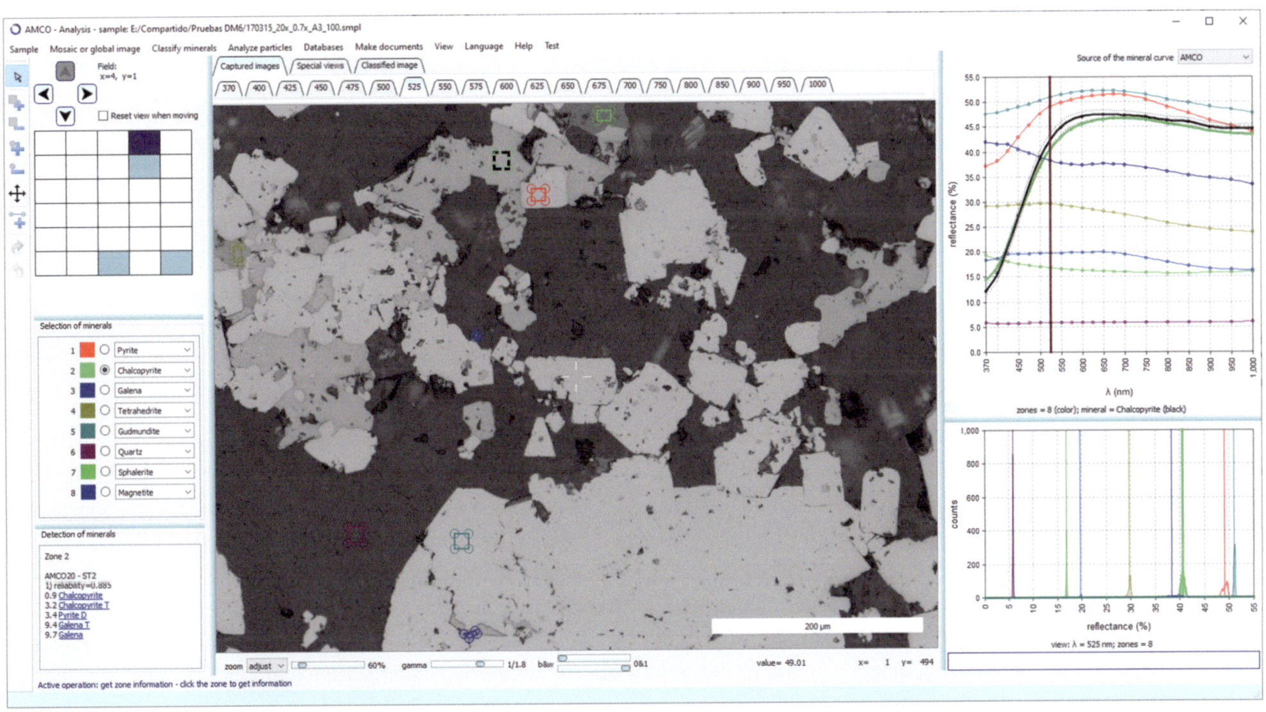

Fig. 2.15 When a marked zone is active, the most likely minerals for the zone are listed (bottom left corner). After the operator assigns it to a mineral, the DB spectrum of that mineral is shown in black (top right corner). When an image band is displayed, the histograms of each zone for that band are shown (lower right corner)

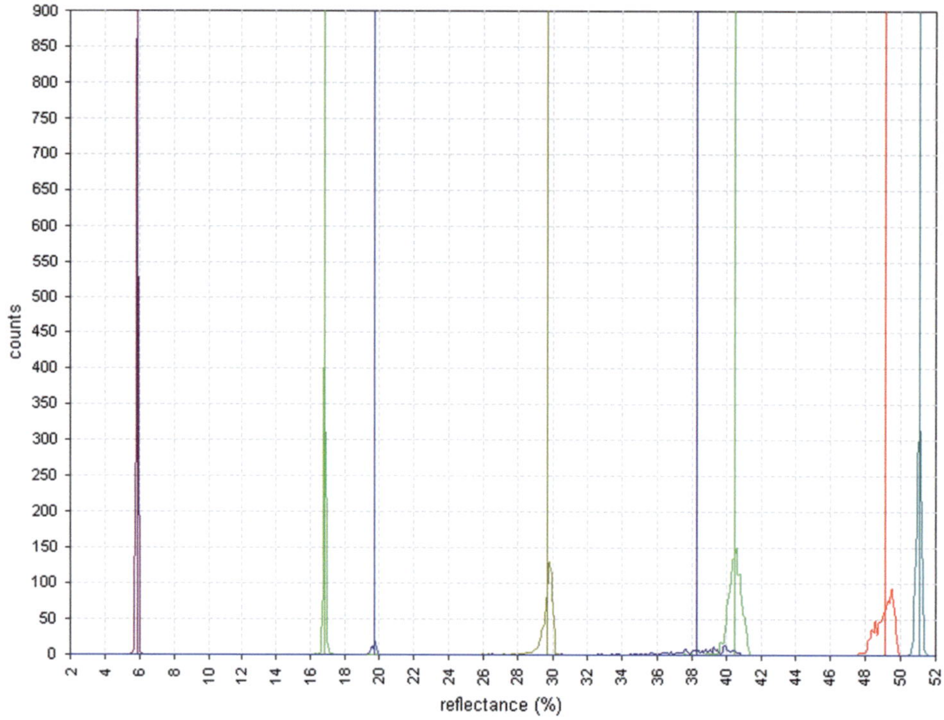

Fig. 2.16 Specular reflectance histograms of the zones marked on Fig. 2.15. The colour of each histogram matches the colour of the outline of the corresponding zone

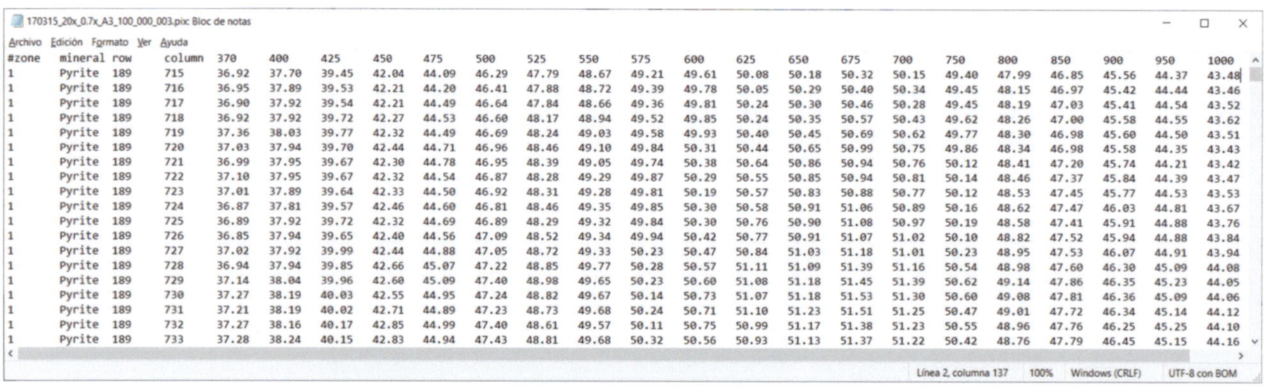

170315_20x_0.7x_A3_100_000_003.pix: Bloc de notas

Archivo Edición Formato Ver Ayuda

#zone	mineral	row	column	370	400	425	450	475	500	525	550	575	600	625	650	675	700	750	800	850	900	950	1000
1	Pyrite	189	715	36.92	37.70	39.45	42.04	44.09	46.29	47.79	48.67	49.21	49.61	50.08	50.18	50.32	50.15	49.40	47.99	46.85	45.56	44.37	43.48
1	Pyrite	189	716	36.95	37.89	39.53	42.21	44.20	46.41	47.88	48.72	49.39	49.78	50.05	50.29	50.40	50.34	49.45	48.15	46.97	45.42	44.44	43.46
1	Pyrite	189	717	36.90	37.92	39.54	42.21	44.49	46.64	47.84	48.66	49.36	49.81	50.24	50.30	50.46	50.28	49.45	48.19	47.03	45.41	44.54	43.52
1	Pyrite	189	718	36.92	37.92	39.72	42.27	44.53	46.60	48.17	48.94	49.52	49.85	50.24	50.35	50.57	50.43	49.62	48.26	47.00	45.58	44.55	43.62
1	Pyrite	189	719	37.36	38.03	39.77	42.32	44.49	46.69	48.24	49.03	49.58	49.93	50.40	50.45	50.69	50.62	49.77	48.30	46.98	45.60	44.50	43.51
1	Pyrite	189	720	37.03	37.94	39.70	42.44	44.71	46.96	48.46	49.10	49.84	50.31	50.44	50.65	50.99	50.75	49.86	48.34	46.98	45.58	44.35	43.43
1	Pyrite	189	721	36.99	37.95	39.67	42.30	44.78	46.95	48.39	49.05	49.74	50.38	50.64	50.86	50.94	50.76	50.12	48.41	47.20	45.74	44.21	43.42
1	Pyrite	189	722	37.10	37.95	39.67	42.32	44.54	46.87	48.28	49.29	49.87	50.29	50.55	50.85	50.94	50.81	50.14	48.46	47.37	45.84	44.39	43.47
1	Pyrite	189	723	37.01	37.89	39.64	42.33	44.50	46.92	48.31	49.28	49.81	50.19	50.57	50.83	50.88	50.77	50.12	48.53	47.45	45.77	44.53	43.53
1	Pyrite	189	724	36.87	37.81	39.57	42.46	44.60	46.81	48.46	49.35	49.85	50.30	50.58	50.91	51.06	50.89	50.16	48.62	47.47	46.03	44.81	43.67
1	Pyrite	189	725	36.89	37.92	39.72	42.32	44.69	46.89	48.29	49.32	49.84	50.30	50.76	50.90	51.08	50.97	50.19	48.58	47.41	45.91	44.88	43.76
1	Pyrite	189	726	36.85	37.94	39.65	42.40	44.56	47.09	48.52	49.34	49.94	50.42	50.77	50.91	51.07	51.02	50.10	48.82	47.52	45.94	44.88	43.84
1	Pyrite	189	727	37.02	37.92	39.99	42.44	44.88	47.05	48.72	49.33	50.23	50.47	50.84	51.03	51.18	51.01	50.23	48.95	47.53	46.07	44.91	43.94
1	Pyrite	189	728	36.94	37.94	39.85	42.66	45.07	47.22	48.85	49.77	50.28	50.57	51.11	51.09	51.39	51.16	50.54	48.98	47.60	46.30	45.09	44.08
1	Pyrite	189	729	37.14	38.04	39.96	42.60	45.09	47.40	48.98	49.65	50.23	50.60	51.08	51.18	51.45	51.39	50.62	49.14	47.86	46.35	45.23	44.05
1	Pyrite	189	730	37.27	38.19	40.03	42.55	44.95	47.24	48.82	49.67	50.14	50.73	51.07	51.18	51.53	51.30	50.60	49.08	47.81	46.36	45.09	44.06
1	Pyrite	189	731	37.21	38.19	40.02	42.71	44.89	47.23	48.73	49.68	50.24	50.71	51.10	51.23	51.51	51.25	50.47	49.01	47.72	46.34	45.14	44.12
1	Pyrite	189	732	37.27	38.16	40.17	42.85	44.99	47.40	48.61	49.57	50.11	50.75	50.99	51.17	51.38	51.23	50.55	48.96	47.76	46.25	45.25	44.10
1	Pyrite	189	733	37.28	38.24	40.15	42.83	44.94	47.43	48.81	49.68	50.32	50.56	50.93	51.13	51.37	51.22	50.42	48.76	47.79	46.45	45.15	44.16

Línea 2, columna 137 100% Windows (CRLF) UTF-8 con BOM

Fig. 2.17 Text file listing pixels' specular reflectance values for the zones marked on an image

In **automatic mode**, the amcoAnalysis program allows to process all the pixels of the image of the currently selected field, performing the following actions:

1. **Classify Minerals in the Current Field** The modal analysis (in % by volume) of the field and an image with the distribution of the various mineral phases in the field are obtained (Fig. 2.19).
2. **Analyse Particles in the Current Field** If the image is of a polished thin section prepared with fluorescent resin according to the specific procedure developed at LMA-UPM for milled ore samples, and the fluorescence band was acquired, the analyses indicated in the particle analysis setup window are performed (Fig. 2.20). However, since a single image is processed, only the particle size analysis and modal analysis by size intervals convey useful information.

Likewise, the amcoAnalysis program allows to process all the images of the series together, performing the following actions:

1. **Classify Minerals in the Entire Sample** All pixels of all the images in a series are classified, obtaining an image of each field with the distribution of the various mineral phases, as well as the overall modal analysis of the

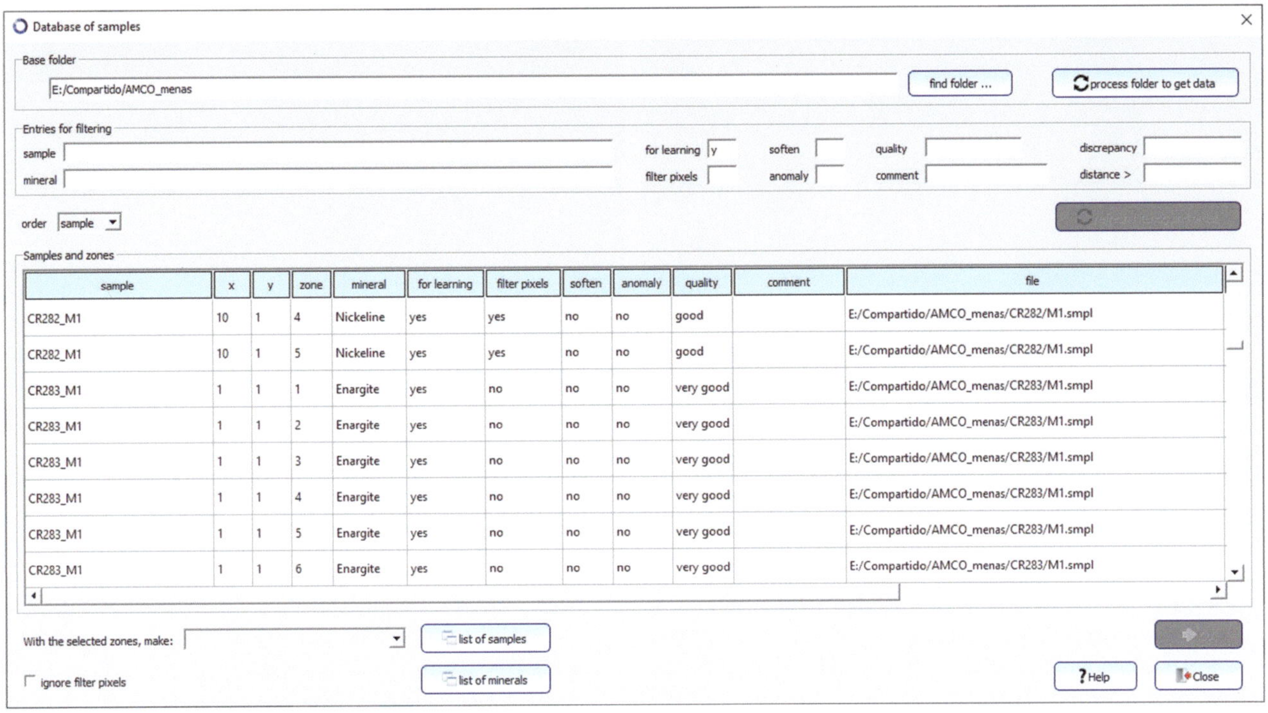

Fig. 2.18 Database window managing the zones delimited in the operator's samples

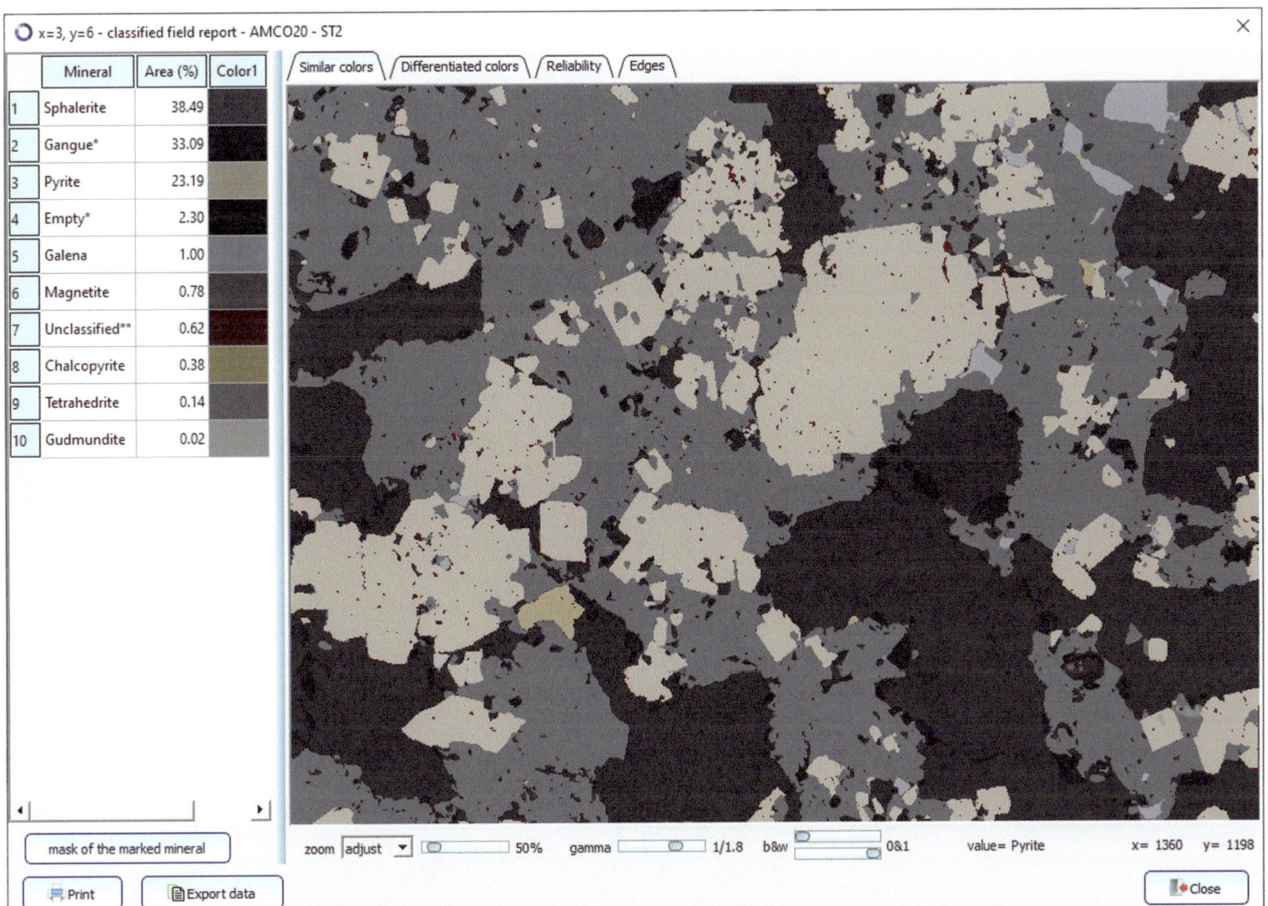

Fig. 2.19 Classification result of field (3,6) of the polished block scan of Fig. 2.8: modal analysis by volume % (left) and classified image (right)

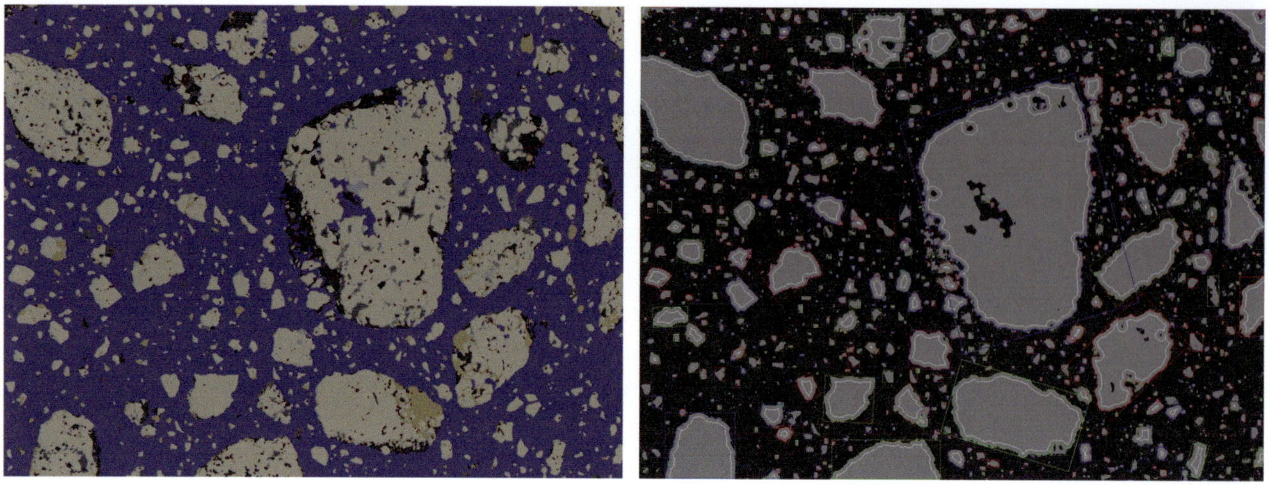

Fig. 2.20 Particle analysis on an image of a polished thin section prepared with fluorescent resin

sample. This quantifies the relative proportion (both in % by volume and % by weight) of each of the mineral species considered in the classification (Fig. 2.21). Pixels assigned to resin are not counted when aggregating the series results. Pixels classified as "Void" or "Not classified" appear in the modal analysis expressed in % by volume, but are discounted when calculating the modal analysis expressed in % by weight.

2. **Analyse Particles in the Entire Sample** If the series is from a polished thin section prepared with fluorescent resin according to the specific procedure developed at LMA-UPM for milled ore samples, and the fluorescence band was acquired, the analyses specified in the particle analysis configuration window are performed. Analyses currently available are:

- Particle size analysis of the sample, expressed as % by weight for each particle size interval. Cumulative values and D80 are also provided.
- Modal analysis by particle size intervals.

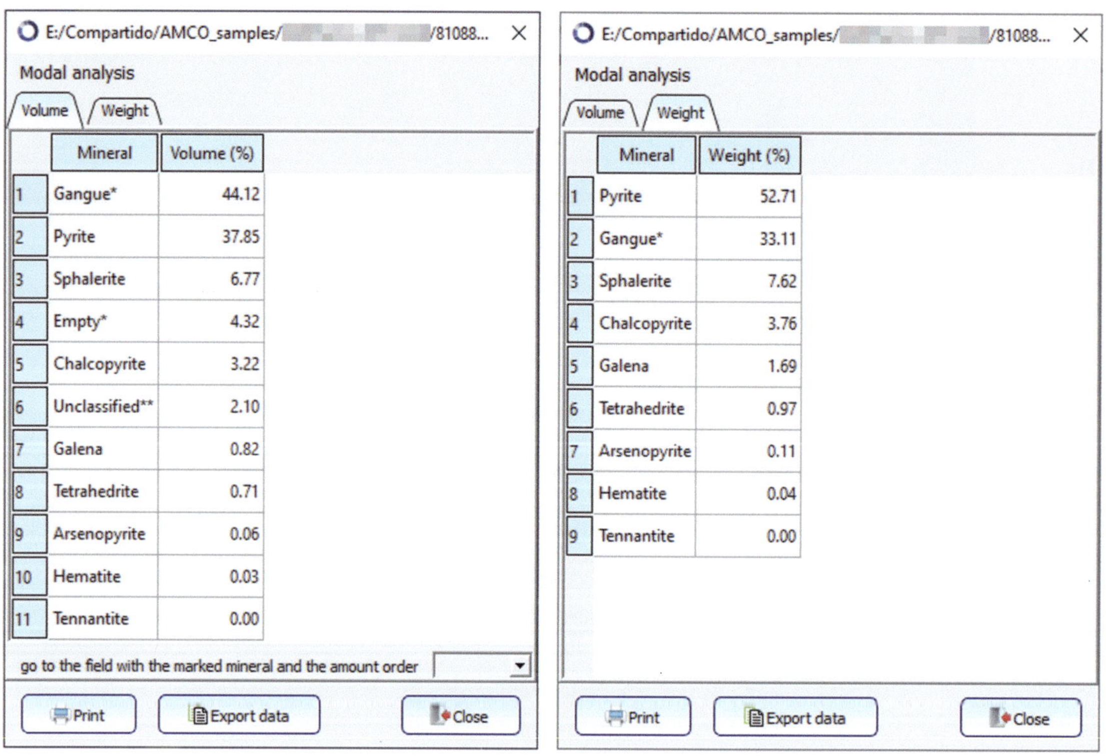

Fig. 2.21 Modal analysis results for the image series from the scan of a polished thin section of a milled ore

- Liberation analysis by % surface area exposed (i.e. degree of surface exposure, calculated as % of the perimeter occupied by the mineral) on particles containing a given mineral—or group of minerals—of interest, distributed by particle size intervals.
- Liberation analysis by weight % composition (i.e. liberation grade, computed from the composition of particles) on particles containing a given mineral—or group of minerals—of interest, distributed by particle size intervals.
- Analysis of the composition (expressed in % by weight) of each of the minerals or mineral groups defined according to the type of particle (single-phase, binary with another mineral group and multi-phase).

The results of these analyses are presented in the form of tables and graphs, organized in a report in HTML format that is automatically created by the program. This can be printed or saved in PDF format. An example of a milled ore report is attached in Case 2 of the Supplement.

2.5 Databases: Construction and Validation

A key element of the AMCO system is the multispectral specular reflectance database, which collects measurements of minerals in each of the reflectance bands captured by the system. These bands cover the visible and near infrared (VNIR) range between 370 and 1000 nm. The polished section containing the sample is placed on a non-rotating stage, so the orientation of each mineral grain is random, which forces it to work with non-polarised light, greatly reducing the effect of bireflectance.

The database was constructed with samples from the research collections of the Applied Microscopy and Image Analysis Laboratory of the Universidad Politécnica de Madrid (LMA-UPM) and other contributions. Selection of the most suitable ores took into account their abundance (e.g. pyrite) and their interest to industry (e.g. Au and PGE), resulting in about 80 species, of which about 40 are of priority, largely coinciding with those described in the microscopy manuals. However, the database is open to incorporation of new species or variants that may be found at certain sites. This is important, because the spectral response can vary with composition, crystallinity, surface alteration or patina, and so on, as well as with the quality of the polish, which must be assured.

Initially, a control microscopic study was carried out to identify the metallic ores, to ensure the suitability of the polished sections (cleanliness, polish quality and texture) and to select suitable fields for measurement. From a total of 589 potentially interesting samples studied, 133 were finally chosen to build the AMCO database. About a hundred minerals were measured, mostly metal ores, among which are species of great abundance, such as pyrite, sphalerite and galena (Fig. 2.22) and others little represented in the classic mineralogy manuals, such as monazite.

Validation of the multispectral specular reflectance database was first undertaken by comparing the spectra to existing databases (Fig. 2.23), such as those of Picot and Johan (1982), Criddle and Stanley (1993) and the precursor project CAMEVA: Castroviejo et al. (2009), or measurements on external systems (H.-J. Bernhardt microspectrometer).

Second, by statistical analysis, various methods of automated classification of the measurements and comparison of the results were applied. The most reliable proved to be the Mahalanobis distance and the Linear Discriminant Analysis, with hit rates usually above 99% (López-Benito et al. 2019). Exceptions have a mineralogical explanation, such as the proximity of cassiterite and chromite spectra or those of sphalerite, manganite and wolframite. These cases can be resolved by additional discrimination criteria, such as paragenetic criteria and extended spectral ranges (VNIR + SWIR).

The final validation was performed by comparing the modal analyses provided by AMCO system to the modal analyses obtained when processing the same polished thin sections on an automated mineralogy system based on SEM–EDS. First, multispectral image series were acquired at LMA-UPM using amcoCapture to perform automatic scans of a set of 18 polished thin sections with a horizontal and vertical scanning step size of 1500 µm, providing an effective coverage of 8.1% of the polished section area (approximately 22 × 32 mm). Each image series was processed with amcoAnalysis to obtain its overall modal analysis. Afterwards, the sections were sent to the Université de Liège, where they were carbon coated and analysed with a ZEISS Mineralogic Mining system (consisting of a ZEISS Sigma 300 field gun emission scanning electron microscope (SEM) fitted with two Bruker xFlash 6|30 X-ray detectors for energy dispersive X-ray spectroscopy (EDS) analysis, controlled by the ZEISS Mineralogic Mining V1.5 software). The mapping mode with 100 µm step size and 50 ms dwell time was typically used to analyse the samples.

The modal analysis results from both systems are reasonably similar, taking into account the completely dissimilar sampling method used by each system and the expected statistical variation in the minority minerals. Some differences in the results of these systems stem from their radically different operating principles and characteristics:

- AMCO cannot reliably distinguish gangue species from each other due to their overlapping spectra, and provides an overall value for the gangue group

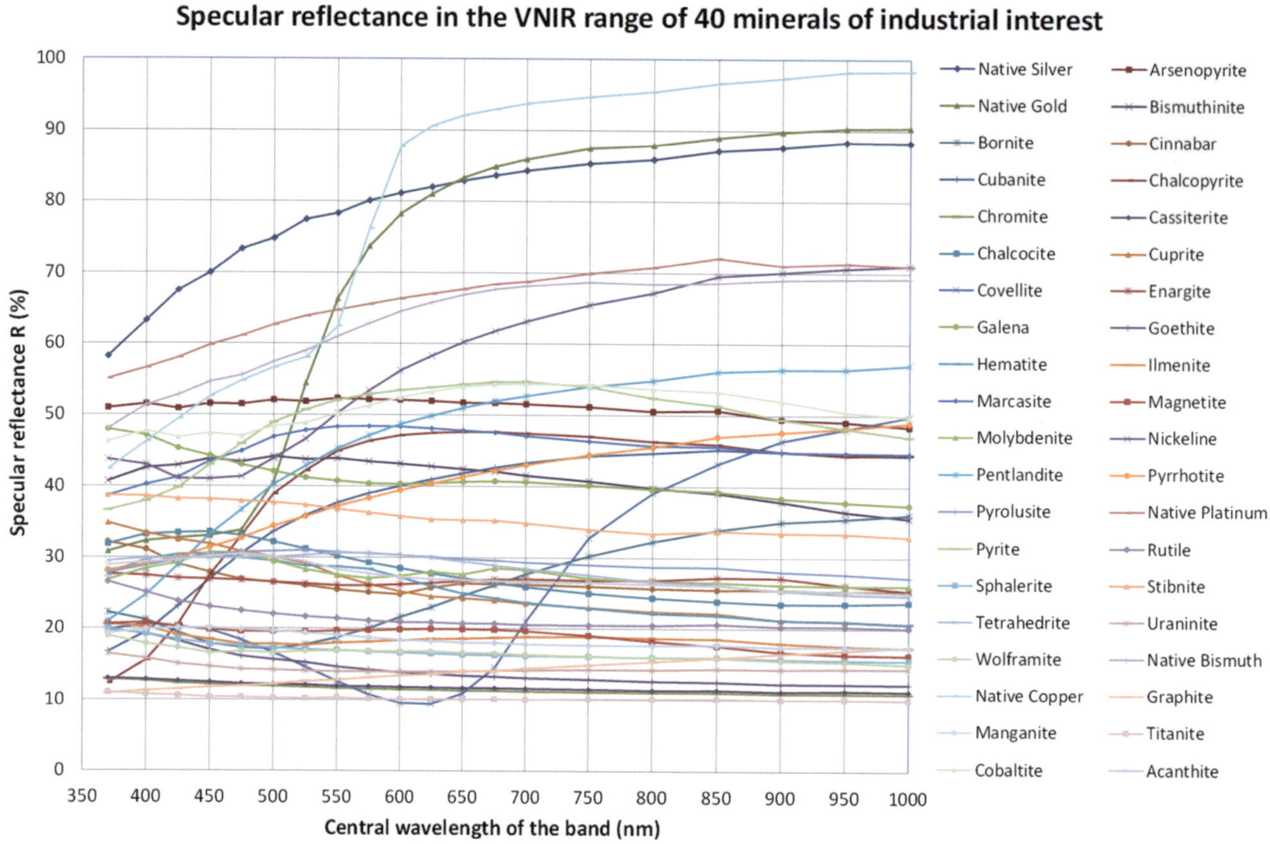

Fig. 2.22 Specular reflectance values in the VNIR range of a selection of 40 ores of interest to industry

- AMCO is able to reliably identify polymorphs and mineral phases having very similar composition that cannot be differentiated by SEM–EDS systems (e.g. pyrite/marcasite, enargite/tennantite/luzonite, digenite/chalcocite/djurleite, magnetite/hematite, etc.).
- AMCO is not expected to distinguish terms of isomorphic series sharing the same optical properties
- AMCO can analyse many more pixels than EDS in a given time
- The percentage of unclassified pixels is usually lower for AMCO than for EDS.

2.6 Results: Automated Identification and Geometallurgical Application

The automated identification of minerals from their multispectral specular reflectance values is particularly effective when the minerals in the sample have distinctly dissimilar specular reflectance spectra. A clear example of this type is sample 170315 from the LMA-UPM. This is a polished block of a massive sulphide ore from the Iberian Pyrite Belt, containing pyrite, sphalerite and gangue as main components, and chalcopyrite, galena, tetrahedrite, magnetite and gudmundite as accessory components.

Figure 2.24 shows the spectra of these minerals in the AMCO system database, which collects specular reflectance values in the VNIR and SWIR ranges (370–1600 nm). It shows that the spectra are sufficiently different in the VNIR range (370–1000 nm) for the minerals to be well distinguished from VNIR images without the need to acquire SWIR bands.

To prove this, the central part of this sample underwent an automatic scan of 5 × 6 contiguous fields (i.e. with 100% coverage of the scanned area) to obtain the corresponding series of multispectral VNIR images. Figures 2.8, 2.13, and 2.19 show several images of this scan.

Case 1 in the Supplements section illustrates the ability of the AMCO system to assist an ore microscopy specialist in the manual analysis of such a sample. It indicates the mineral whose spectrum is closest to that of a manually delimited zone on the image, and can perform automated pixel identification of an image—or the entire image series—providing quantitative results that are virtually impossible to obtain manually.

Moreover, the AMCO system has been designed to carry out automated characterization of milled ore samples, such as plant concentrates. For this purpose, polished thin

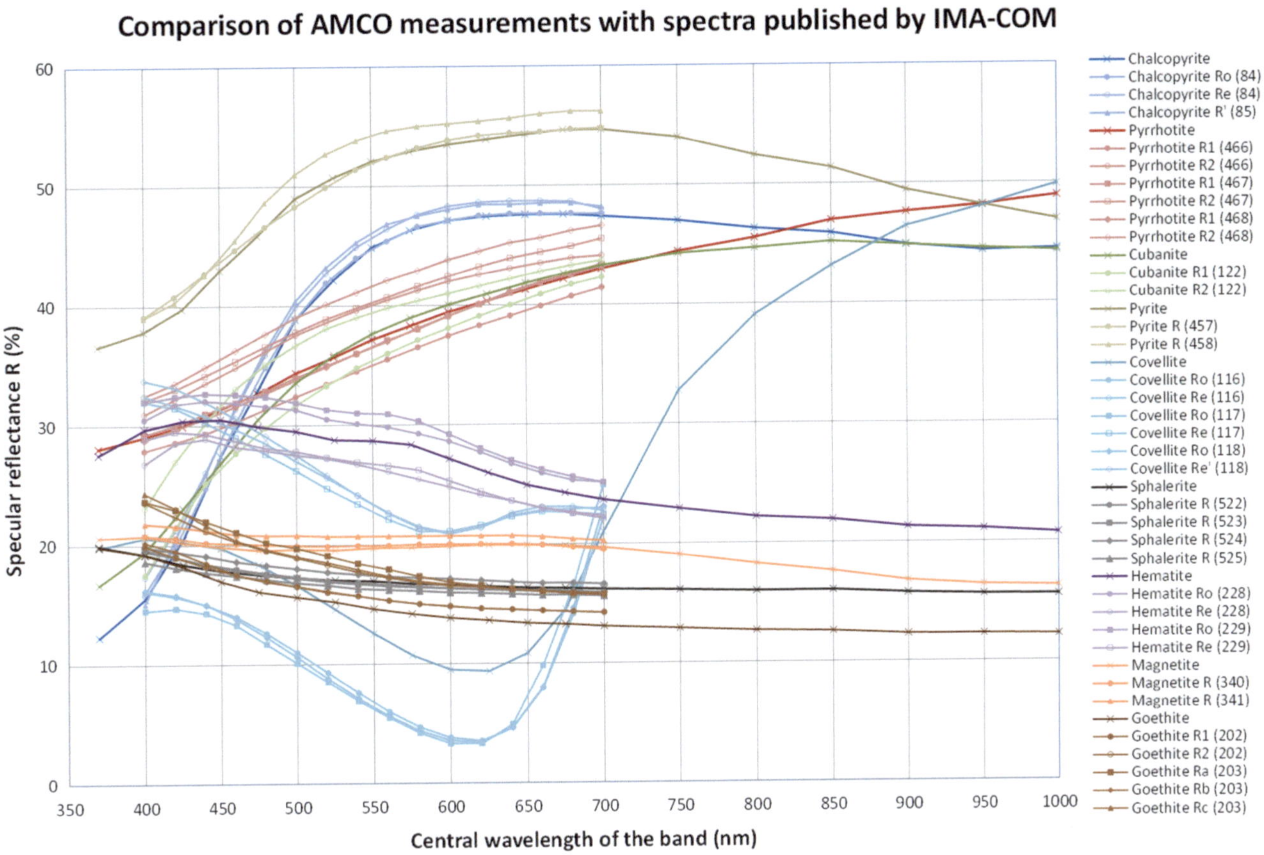

Fig. 2.23 Comparison of AMCO measurements and QDF3 measurements

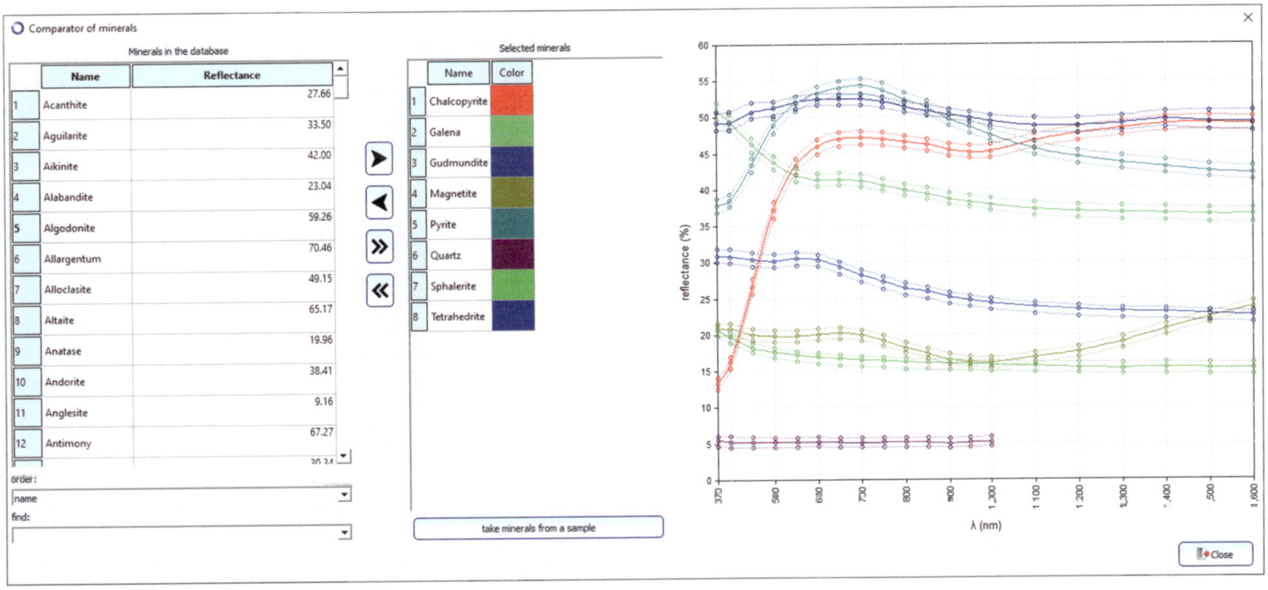

Fig. 2.24 Mineral comparator of the AMCO system database showing the specular reflectance spectra of the eight selected minerals. Each spectrum is composed of three lines: the solid line is the mean value, and the dashed lines are the 16% and 84 percentiles

sections with a nominal thickness of 30 μm are made from the samples, using epoxy resin dyed with a fluorescent dye according to the specific procedure developed by LMA-UPM (Grunwald-Romera et al. 2019). Then, the automatic scanning of a section is performed with the amcoCapture program, using suitable horizontal and vertical step sizes depending on the desired coverage and number of images, making sure to use an adequate exposure time for the acquisition of the fluorescence band. Once the acquisition is finished, the series of images is opened with the amcoAnalysis program. The most appropriate threshold value for the fluorescence band is set and, after selecting the minerals to be considered in the analysis, classification of the mineral phases in the entire sample is carried out, thus obtaining the general modal analysis of the sample. Then, after setting the desired parameters, particle analysis is performed on the entire sample, automatically generating a report in HTML format with numerous tables and graphs that detail the results of all analyses carried out.

Case 2 in the Supplements section contains a typical report of the analysis of a milled ore sample from the Iberian Pyrite Belt. The results show that the AMCO system achieves reliable automated identification and quantification of ores, with a performance comparable to that of the automated mineralogy systems based on scanning electron microscopy with energy-dispersive X-ray spectroscopy (SEM–EDS). It even outperforms them in the distinction of polymorphs (e.g. pyrite/marcasite/melnikovite) and minerals with similar composition, such as some sulphosalts (enargite/tennantite/luzonite), copper sulphides (digenite/chalcocite/djurleite) and iron ores (hematite/magnetite/goethite), and in the identification of graphite.

Automated ore microscopy is not only a useful and high-performance tool for mineralurgical applications but a great aid to learning ore microscopy. The AMCO system represents a real alternative (competing or complementary) to the SEM–EDS systems currently in use. In either case, far from feeling threatened by a black box approach, the expert mineralogist can witness greatly enhanced performance.

2.7 Supplement

Examples of practical application (qualitative and quantitative analysis):

- Case 1 Automated identification (massive sulphide ore)
- Case 2 Analysis and characterization of a plant concentrate.

Case 1 Automated Identification (Massive Sulphide Ore)

This document describes the ability of the AMCO system's amcoAnalysis program to:

- Automatically identify the mineral corresponding to an area manually delimited by the operator on an image
- Automatically identify the mineral corresponding to each of the pixels in an image, or the whole of a series of images.

From the multispectral specular reflectance values.

This capability is of great interest in assisting a microscopy specialist in the manual analysis of a sample, as it allows him/her to obtain quantitative results on its composition that are practically impossible to obtain manually.

To exemplify it, we will use a series of images acquired by performing an automatic scan of 5 × 6 fields over the central part of sample 170315 from the LMA-UPM. This sample is a polished block made from a hand sample of a massive sulphide ore from the Iberian Pyrite Belt, which contains pyrite, sphalerite and gangue as main components, and chalcopyrite, galena, tetrahedrite, magnetite and gudmundite as accessory components.

The series of 30 multispectral VNIR images obtained in the scan can be examined manually by the specialist, by scrolling through the images at will and marking out different areas of interest. The reflectance spectrum of each marked area in an image (calculated as the mean or mode of the spectra of all the pixels in the area) is presented in a graph in the upper right corner of the window, in the same colour assigned to its outline. Comparison of this spectrum to the spectra stored in the database allows the identification of the corresponding mineral, as can be seen in Figure Case 1.1, which presents in the lower left corner a list of the minerals in the database whose spectra most resemble (according to various classification criteria) the spectrum of the active marked zone—sphalerite, in this case—ordered by increasing distance (equivalent to decreasing likeliness).

In this way, the amcoAnalysis program can help the specialist to identify more reliably the minerals present in the sample by quantitatively determining their reflectance spectra.

Once the minerals present in the sample (or those that could possibly appear, knowing its origin) have been identified, the operator can restrict the range of minerals considered in the automatic identification in order to improve the efficiency of the classification process and reduce the possibility of error, marking the minerals that he/she wants the program to take into account in the analysis in the mineral selection window.

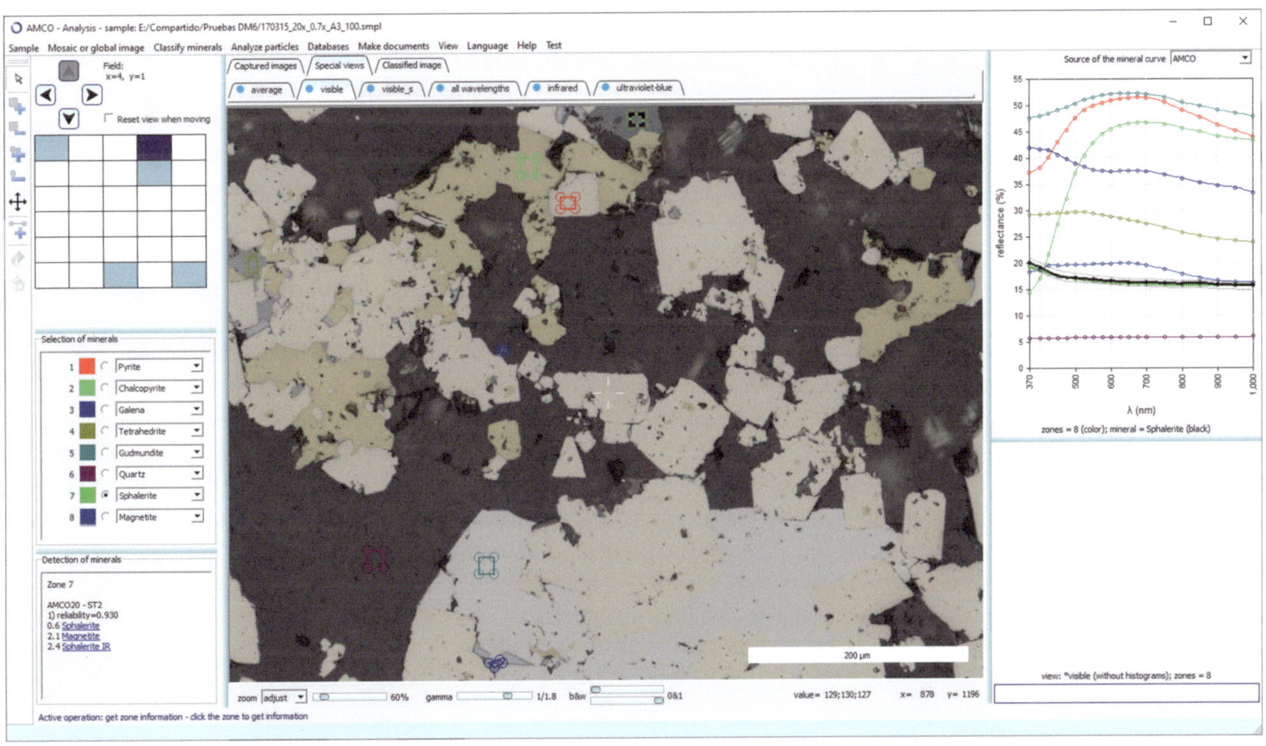

Figure Case 1.1 Manual analysis of field (4,1) of a scan of a polished block. The operator has delimited eight zones of various minerals, which he has tentatively identified. The zone identified as sphalerite is active

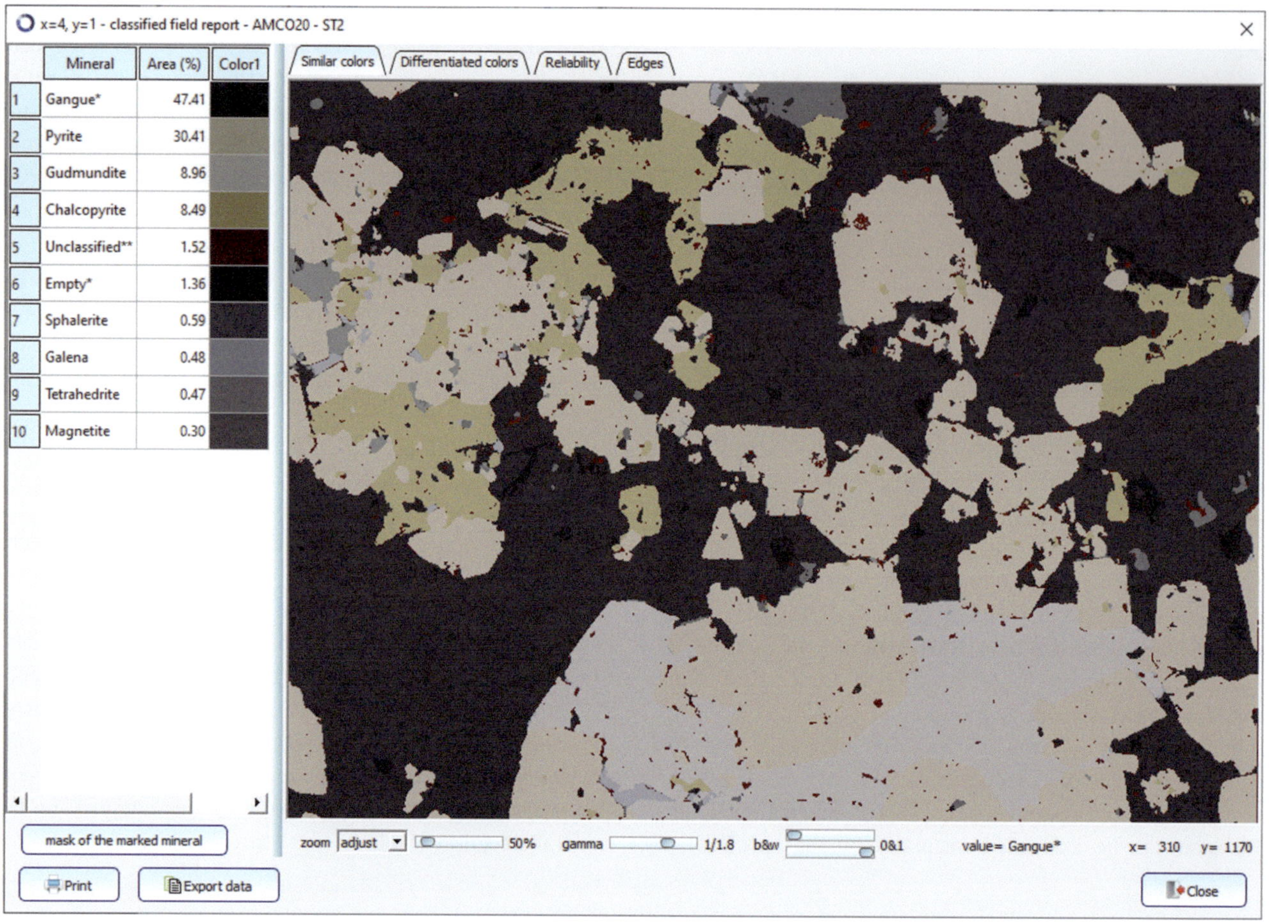

Figure Case 1.2 Window generated by the "Classify minerals in the current field" function. It shows the modal analysis of the field and the result of the classification of all the pixels that compose it

Figure Case 1.2 shows the window generated by applying the function "Classify minerals in the current field", which shows on the left side the modal analysis of all the pixels of the field, expressed in % in volume, and on the right side an image that encodes each pixel with the colour corresponding to the assigned class, so that the different mineral phases appear in uniform colour. It can be seen that the classification of the pixels is very good, with few pixels assigned to the "Not classified" class, which mostly correspond to the edges of particles and grains.

In order that the user can appreciate more easily the location of the different minerals in the image, which can be somewhat confusing in the case of small grains, the program allows to display an image of the distribution of each one of the minerals in its colour, masking the other minerals with a white background.

As an example, the 10 figures from Figures Case 1.3, 1.4, 1.5, 1.6, 1.7, 1.8, 1.9, 1.10, 1.11, 1.12, show the distribution of each of the 10 classes detailed in the modal analysis of the previous field, including pixels assigned to "Empty" (i.e. having no mineral, because grain popping occurred during polishing) and "Not classified" (because the mineral identification is not sufficiently certain to be considered valid).

The function "Classify minerals in the entire sample" allows us to obtain the sample's overall modal analysis, expressed either in % by volume or in % by weight of each mineral species considered in the analysis. Pixels classified as "Empty" or "Not classified" are discounted when calculating the modal analysis expressed in % by weight. Figure case 1.13 shows the overall modal analysis of this sample.

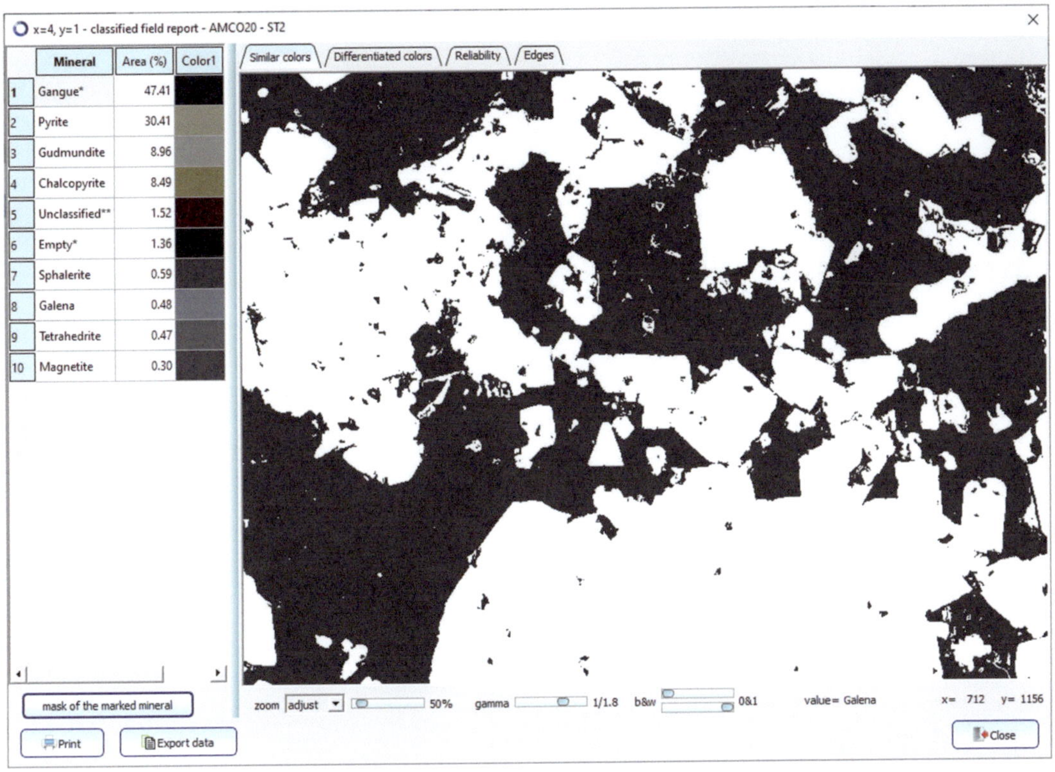

Figure Case 1.3 Distribution of gangue pixels in the current field

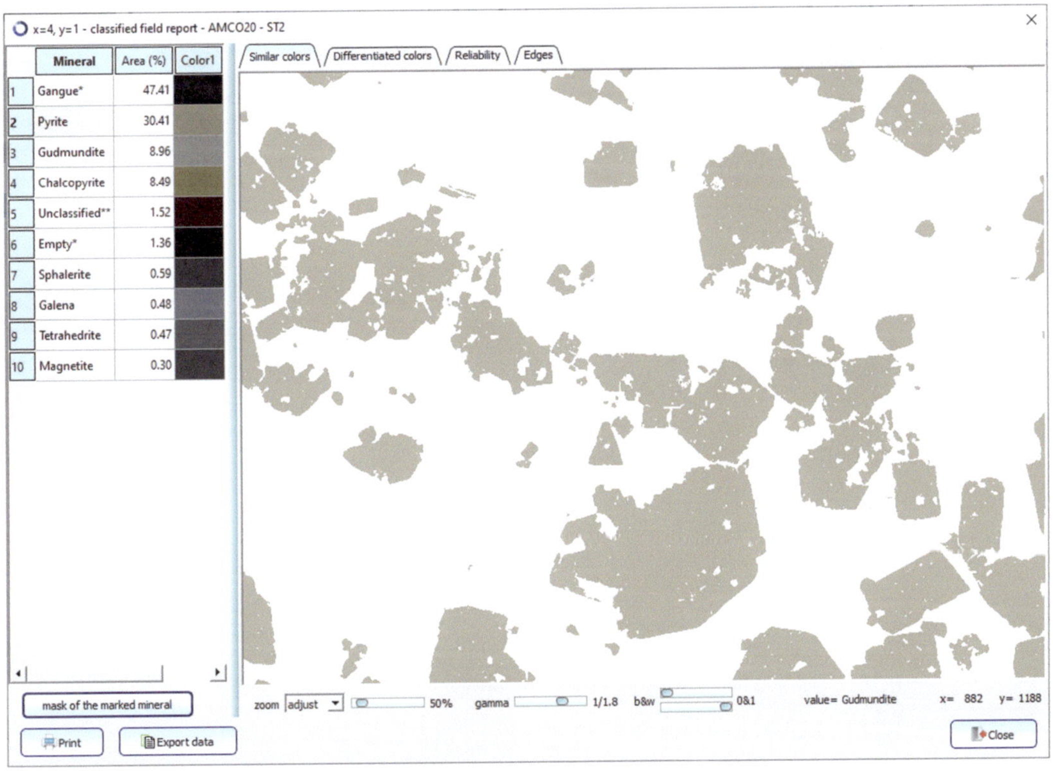

Figure Case 1.4 Distribution of pyrite pixels in the current field

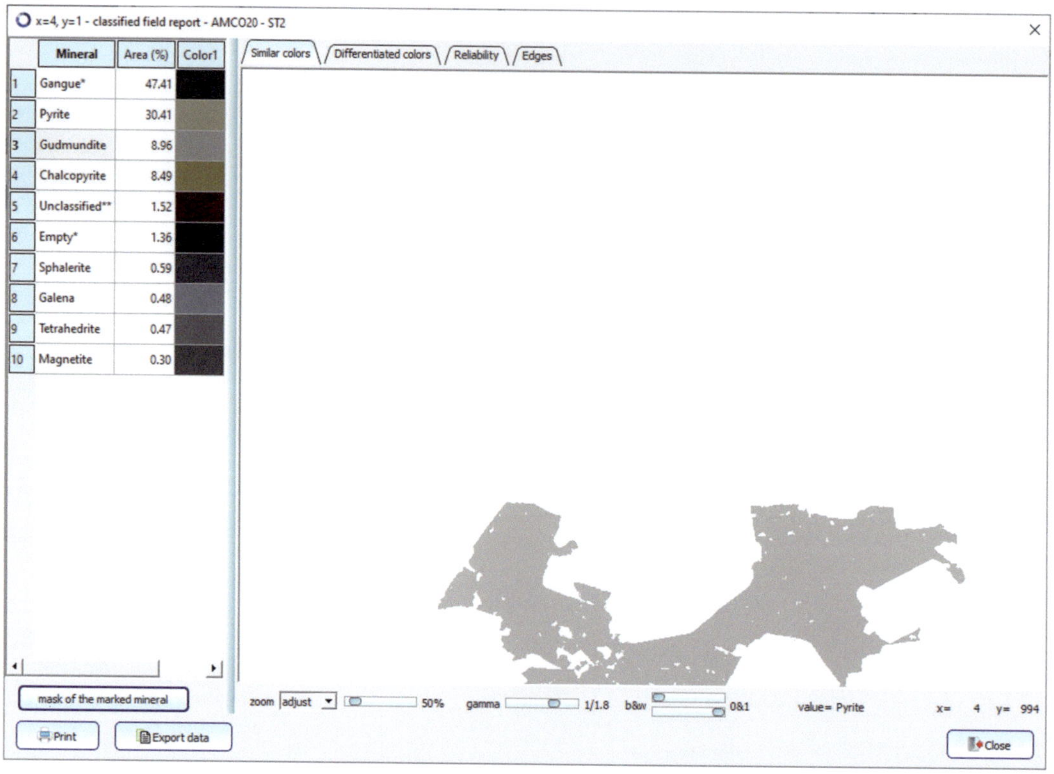

Figure Case 1.5 Distribution of gudmundite pixels in the current field

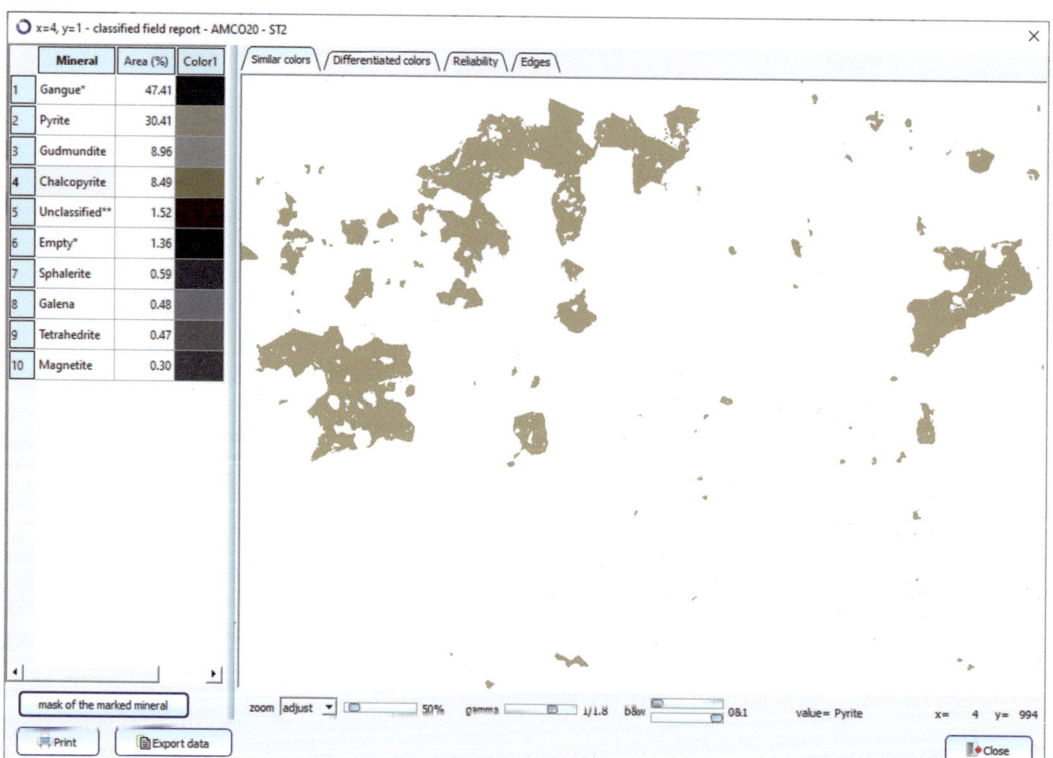

Figure Case 1.6 Distribution of chalcopyrite pixels in the current field

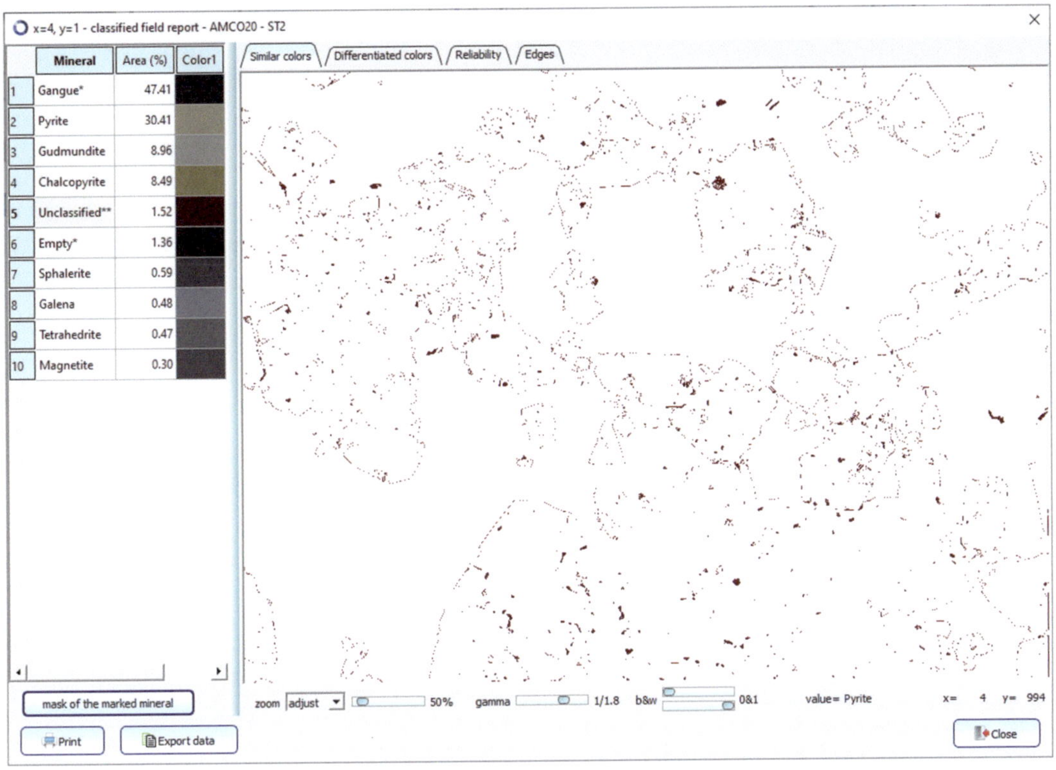

Figure Case 1.7 Distribution of unclassified pixels (due to uncertain identification) in the current field

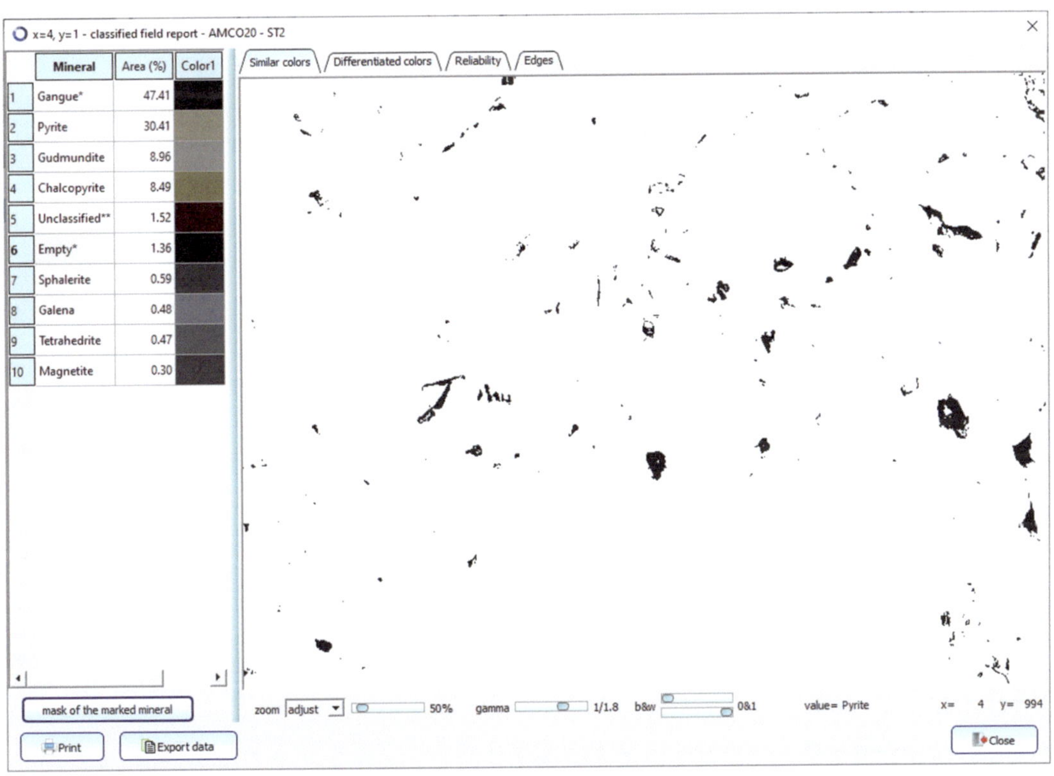

Figure Case 1.8 Distribution of empty pixels (absence of mineral) in the current field

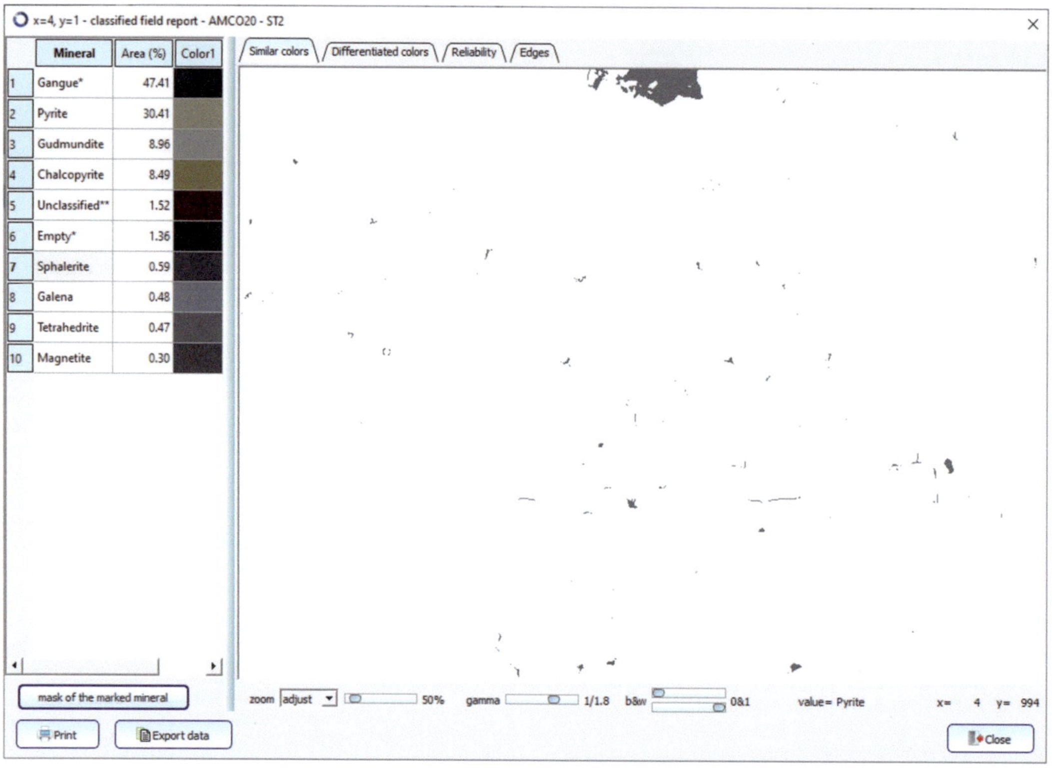

Figure Case 1.9 Distribution of sphalerite pixels in the current field

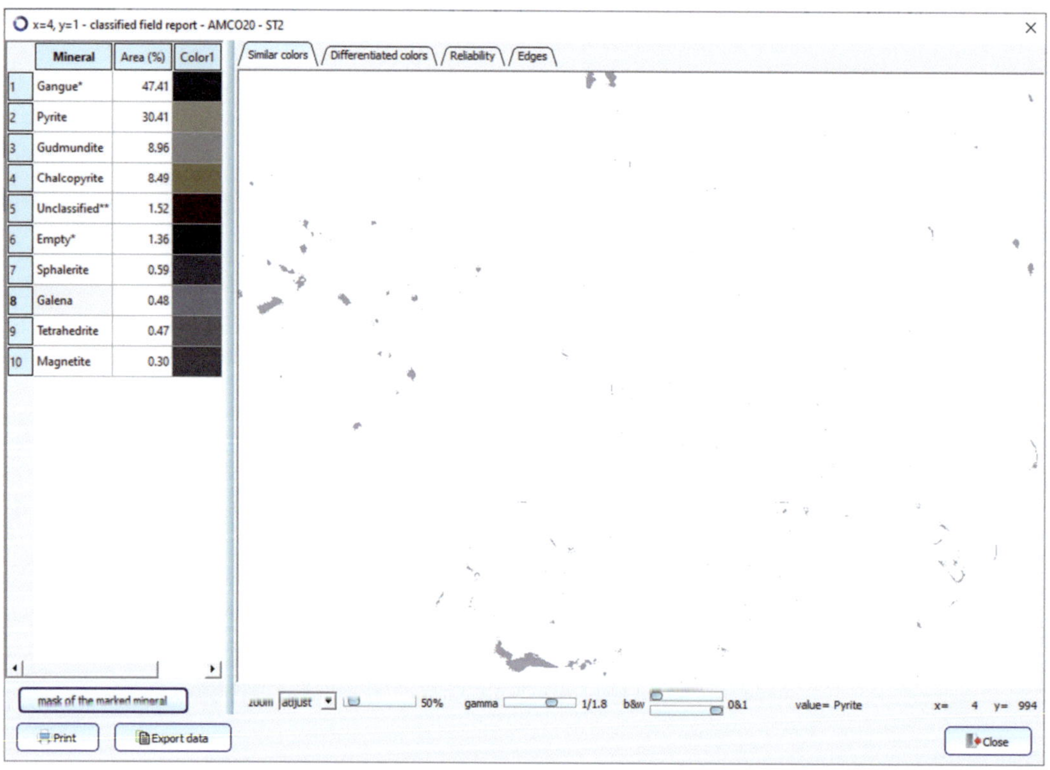

Figure Case 1.10 Distribution of galena pixels in the current field

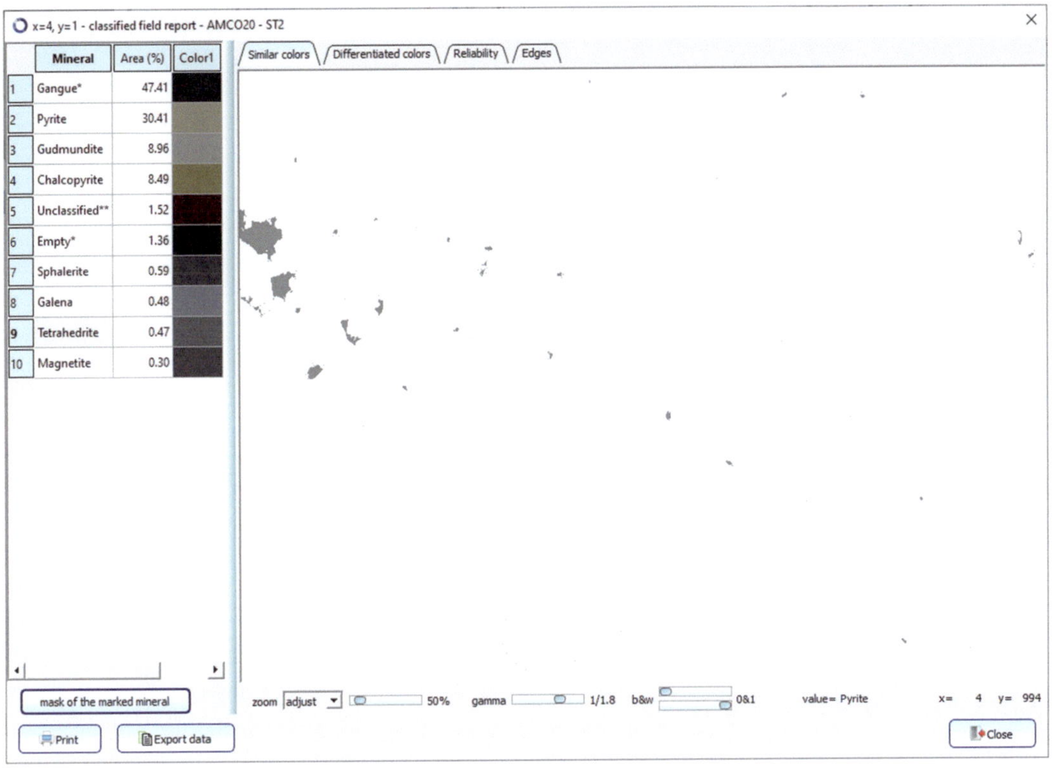

Figure Case 1.11 Distribution of tetrahedrite pixels in the current field

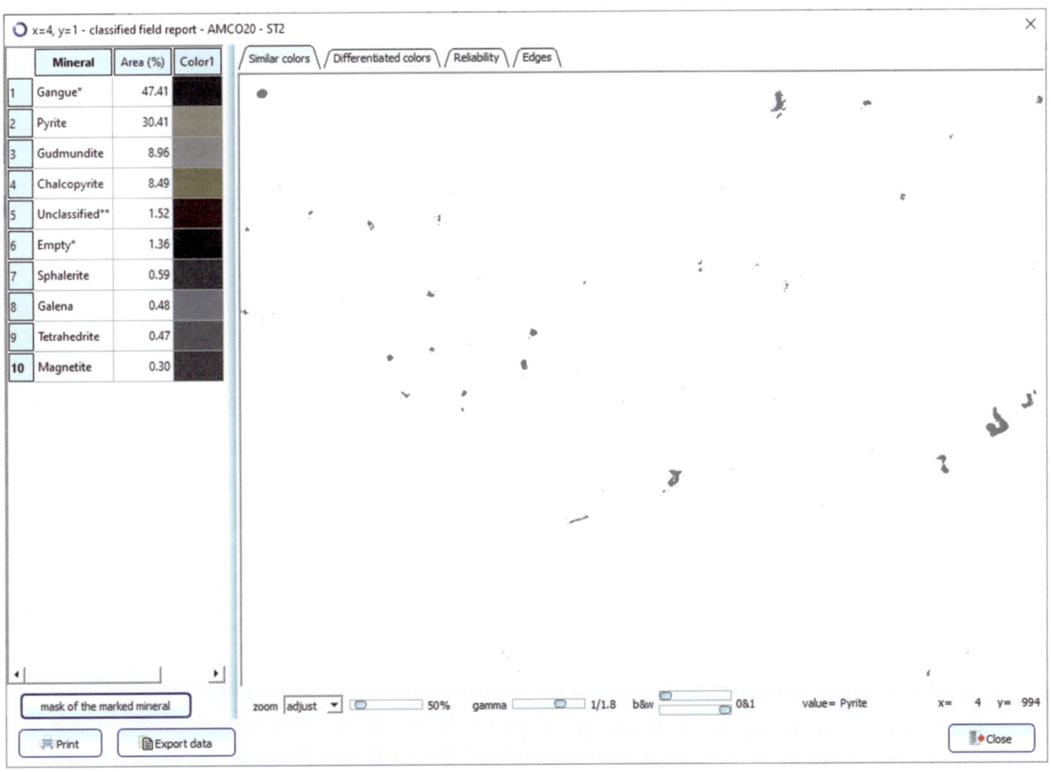

Figure Case 1.12 Distribution of magnetite pixels in the current field

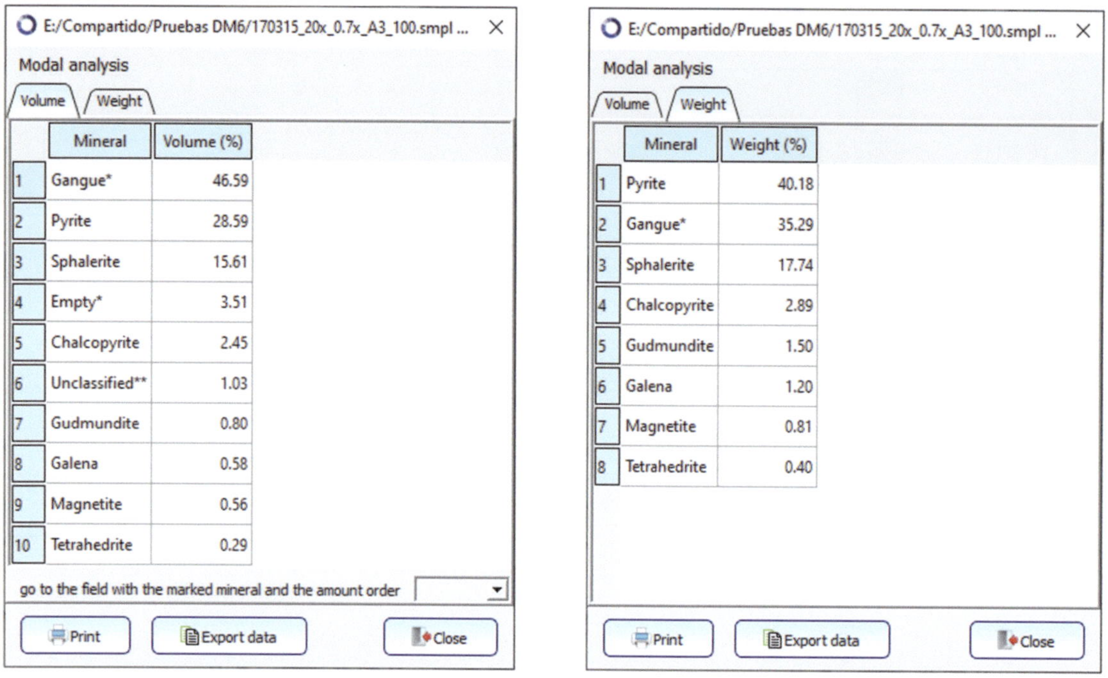

Figure Case 1.13 Overall modal analysis of the image series

Figure Case 1.14 Mosaic of the scan obtained from the true colour view. The region corresponding to field (4,1) is shaded green

Figure Case 1.15 Mosaic of the scan obtained from the classified images: the region corresponding to field (4,1) is shaded green

The "Classify minerals in the entire sample" function also generates the classified image of each of the fields in the image series.

When the image series has been acquired by means of an automatic scan with 100% coverage, as in this series, the amcoAnalysis program can generate a large global image of the scanned area by juxtaposing all the fields as the tesserae of a mosaic, applying the necessary scaling factor so that the resulting image has the indicated size. The image is generated from the currently active view mode, which is usually the true colour view or the classified image.

Figures Case 1.14 and 1.15 show the two "mosaic" images of the scanned region of the section, which has dimensions of 3400 μm (H) × 3007 μm (V). Although the images have been subsampled to a scale of 1 μm/pixel, they are so large that the printing process does not allow the full detail of the image to be appreciated.

In conclusion, the ability of the amcoAnalysis program to identify the minerals present in a multispectral specular reflectance image can be very useful to:

- Provide assistance to the ore microscopy specialist in the identification of rare or questionable minerals when performing manual analysis of a sample.
- Obtain precise quantitative values of the proportion of various minerals in a sample (modal analysis) without the need for a specialist in ore microscopy.
- Help to remove any doubts in the mind of students learning to recognise minerals under the microscope.

Case 2 Analysis and Characterization of a Plant Concentrate

The following pages are a typical example of the amcoAnalysis particle analysis report on the entire series of images from automated scanning of a polished thin section of a milled ore sample. In this case, the sample comes from the crushing of a fragment of a drill core from a mine in the Iberian Pyrite Belt.

The parameters to be applied in the analysis are fully configurable by the user through a configuration window, like the one shown in Figure Case. 2.1.

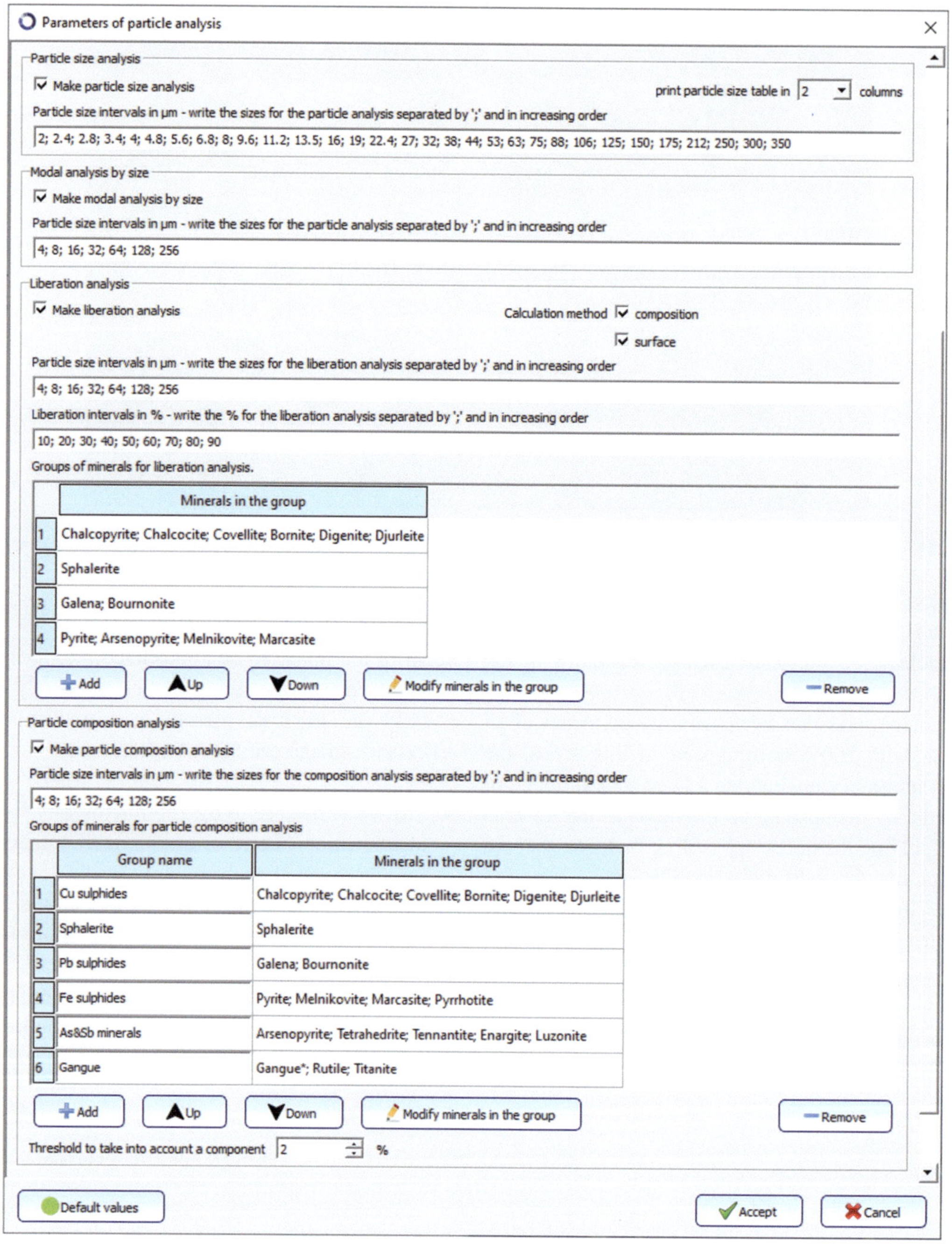

Figure Case 2.1 Capture of the configuration window of particle analysis parameters

Multispectral Microscopic Technologies S.L.

PARTICLE ANALYSIS

Sample:E:/Compartido/AMCO_samples/XXXX/XXXX-M02F/A4_20×_2000 × 2000.smpl
Date time: 2022-02-14 11:10.

Particle Size Distribution

Size interval (µm)	Weight %	Accumulated %
[0; 2)	0.50	0.50
[2; 2.4)	0.50	0.99
[2.4; 2.8)	0.63	1.62
[2.8; 3.4)	1.10	2.72
[3.4; 4)	1.21	3.93
[4; 4.8)	1.72	5.65
[4.8; 5.6)	1.76	7.41
[5.6; 6.8)	3.12	10.53
[6.8; 8)	3.86	14.39
[8; 9.6)	3.61	18.00
[9.6; 11.2)	2.47	20.47
[11.2; 13.5)	2.94	23.40
[13.5; 16)	2.89	26.30
[16; 19)	3.04	29.34
[19; 22.4)	3.30	32.64
[22.4; 27)	3.73	36.37

Size interval (µm)	Weight %	Accumulated %
[27; 32)	4.01	40.38
[32; 38)	4.80	45.18
[38; 44)	3.81	48.99
[44; 53)	5.24	54.23
[53; 63)	5.41	59.64
[63; 75)	5.11	64.76
[75; 88)	5.59	70.34
[88; 106)	7.21	77.56
[106; 125)	4.49	82.04
[125; 150)	6.32	88.37
[150; 175)	4.11	92.48
[175; 212)	3.99	96.47
[212; 250)	2.18	98.64
[250; 300)	0.84	99.48
[300; 350)	0.52	100.00
[350; ∞)	0.00	100.00

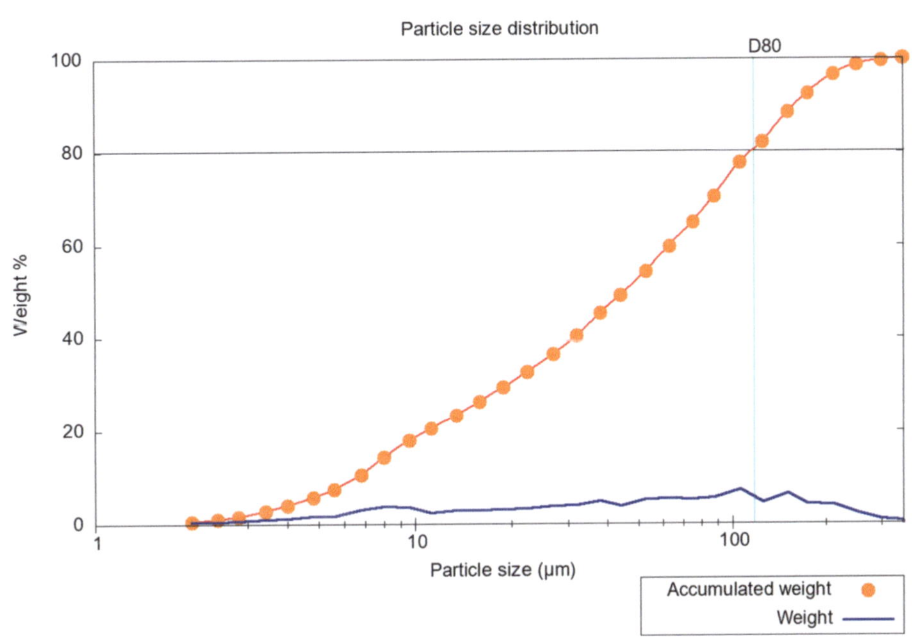

Modal Analysis by Particle Size Intervals

Modal analysis by particle size intervals Sum of all cells (size interval, mineral) is 100%									
Weight %		Size interval (µm)							
		[0; 4)	[4; 8)	[8; 16)	[16; 32)	[32; 64)	[64; 128)	[128; 256)	[256; ∞)
Minerals	**Total**	3.93	10.46	11.91	14.08	19.92	22.91	15.56	1.22
Pyrite	57.91	0.38	4.07	6.57	8.61	12.59	14.74	10.19	0.77
Marcasite	20.43	2.54	4.33	3.05	2.64	3.02	2.95	1.76	0.14
Chalcopyrite	10.02	0.66	1.17	1.18	1.46	2.01	2.02	1.41	0.10
Gangue*	5.10	0.11	0.29	0.40	0.57	0.97	1.57	1.06	0.13
Sphalerite	4.12	0.15	0.40	0.48	0.52	0.80	1.04	0.69	0.05
Tetrahedrite	1.16	0.06	0.09	0.08	0.13	0.23	0.30	0.26	0.02
Galena	0.94	0.02	0.09	0.11	0.11	0.23	0.22	0.14	0.01
Bornite	0.22	0.00	0.01	0.02	0.03	0.05	0.06	0.04	0.00
Covellite	0.08	0.00	0.01	0.01	0.01	0.01	0.02	0.01	0.00
Arsenopyrite	0.03	0.00	0.01	0.01	0.01	0.00	0.00	0.00	0.00
Tennantite	0.00	0.00	0.00	0.00	0.00	0.00	0.00	0.00	0.00

Modal analysis by particle size intervals Sum of every column (size interval) is 100%									
Weight %		Size interval (µm)							
Minerals	**Total**	[0; 4)	[4; 8)	[8; 16)	[16; 32)	[32; 64)	[64; 128)	[128; 256)	[256; ∞)
Pyrite	57.91	9.78	38.89	55.14	61.16	63.20	64.31	65.45	63.03
Marcasite	20.43	64.51	41.37	25.59	18.75	15.18	12.86	11.33	11.54
Chalcopyrite	10.02	16.85	11.17	9.94	10.34	10.10	8.83	9.05	8.43
Gangue*	5.10	2.70	2.78	3.34	4.04	4.86	6.86	6.82	10.94
Sphalerite	4.12	3.76	3.82	4.02	3.69	4.03	4.53	4.41	4.09
Tetrahedrite	1.16	1.63	0.82	0.70	0.91	1.15	1.30	1.64	1.24
Galena	0.94	0.55	0.90	0.93	0.78	1.16	0.97	0.92	0.51
Bornite	0.22	0.09	0.10	0.17	0.20	0.24	0.26	0.28	0.21
Covellite	0.08	0.11	0.08	0.08	0.06	0.07	0.07	0.09	0.01
Arsenopyrite	0.03	0.03	0.08	0.09	0.07	0.00	0.00	0.00	0.00
Tennantite	0.00	0.00	0.00	0.00	0.00	0.00	0.00	0.00	0.00

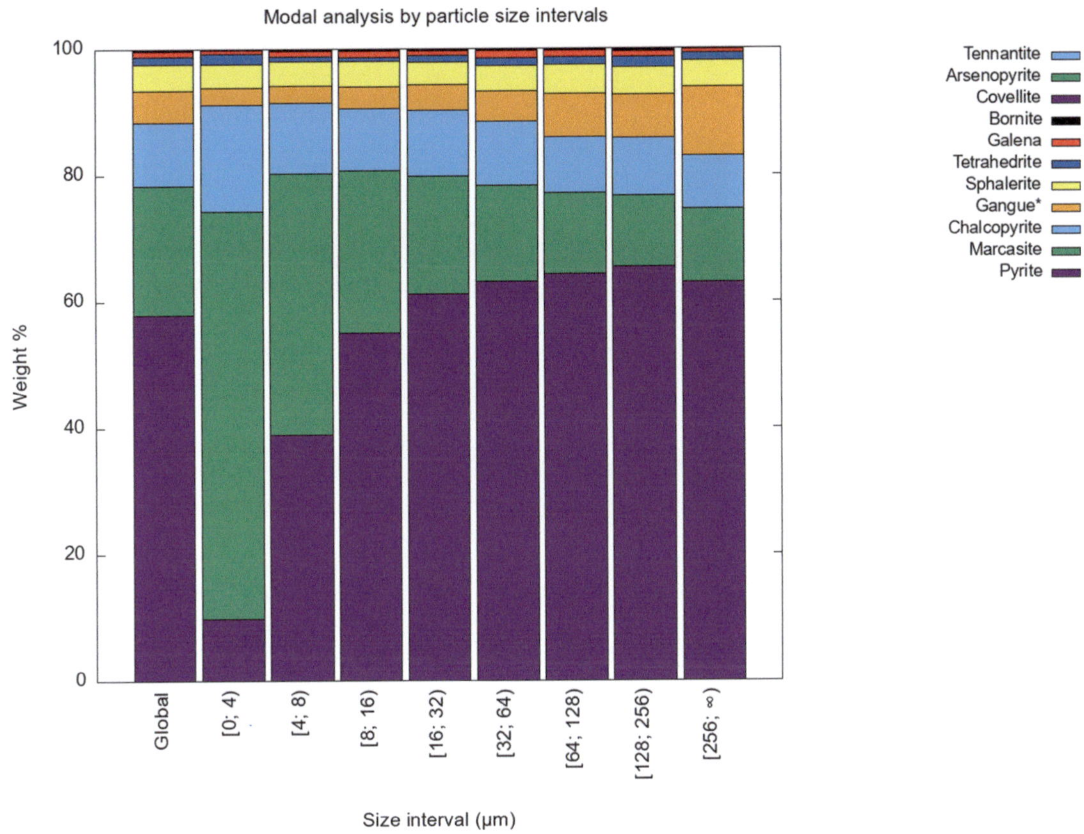

Liberation of Mineral Groups by Particle Size Intervals

Liberation by composition of chalcopyrite + chalcocite + covellite + bornite + digenite + djurleite									
Weight % distribution by size and liberation intervals									
Liberation (%)	**Size interval (μm)**								**Total**
	[0; 4)	[4; 8)	[8; 16)	[16; 32)	[32; 64)	[64; 128)	[128; 256)	[256; ∞)	
[0; 10)	0.10	1.13	2.30	3.57	5.78	7.65	3.96	0.66	25.15
[10; 20)	0.31	2.09	3.13	4.69	7.31	8.75	7.37	0.37	34.02
[20; 30)	0.49	1.69	1.88	2.37	3.78	2.90	2.50	0.00	15.61
[30; 40)	0.51	1.19	1.19	1.40	1.92	0.79	0.40	0.00	7.40
[40; 50)	0.47	0.92	0.73	0.84	0.40	0.29	0.00	0.00	3.65
[50; 60)	0.41	0.63	0.42	0.48	0.62	0.00	0.00	0.00	2.56
[60; 70)	0.32	0.52	0.28	0.45	0.07	0.00	0.00	0.00	1.64
[70; 80)	0.24	0.42	0.30	0.20	0.15	0.00	0.00	0.00	1.31
[80; 90)	0.24	0.41	0.35	0.29	0.00	0.00	0.00	0.00	1.29
[90; 100]	3.42	2.51	1.18	0.18	0.08	0.00	0.00	0.00	7.37

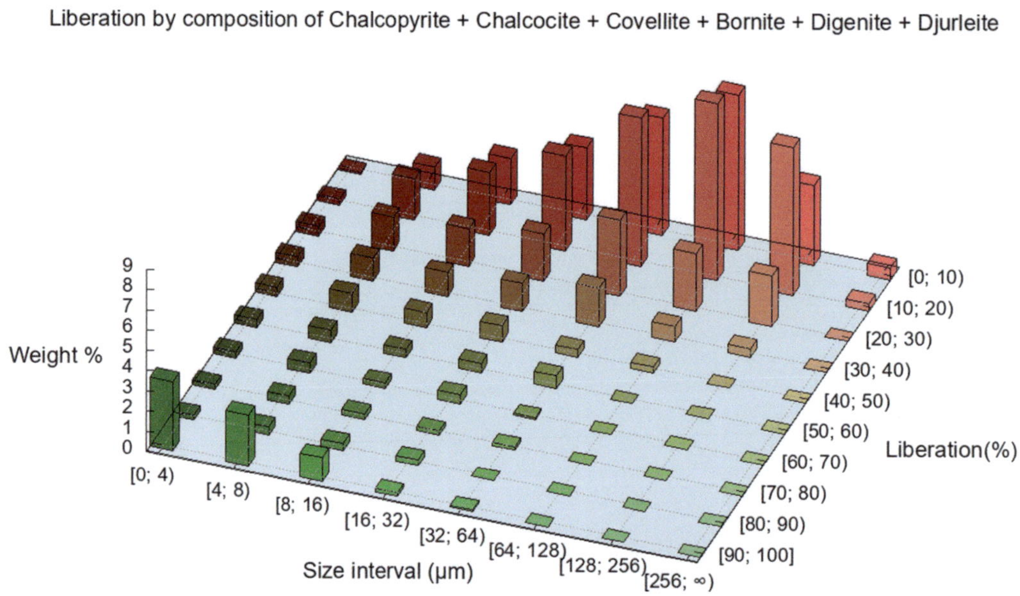

Liberation by exposed surface of chalcopyrite + chalcocite + covellite + bornite + digenite + djurleite									
Weight % distribution by size and liberation intervals									
Liberation (%)	Size interval (µm)								Total
	[0; 4)	[4; 8)	[8; 16)	[16; 32)	[32; 64)	[64; 128)	[128; 256)	[256;∞)	
[0; 10)	0.05	0.24	0.54	0.54	0.66	1.15	1.03	0.08	4.29
[10; 20)	0.10	0.82	1.42	1.87	3.16	7.39	5.49	0.84	21.08
[20; 30)	0.20	1.23	1.95	3.20	5.80	5.94	5.40	0.00	23.73
[30; 40)	0.30	1.39	1.98	2.92	5.29	3.22	1.85	0.10	17.04
[40; 50)	0.39	1.34	1.75	2.52	3.05	1.76	0.47	0.00	11.29
[50; 60)	0.53	1.33	1.14	1.60	1.39	0.81	0.00	0.00	6.80
[60; 70)	0.54	1.08	0.81	0.97	0.41	0.10	0.00	0.00	3.91
[70; 80)	0.50	0.84	0.63	0.41	0.23	0.00	0.00	0.00	2.60
[80; 90)	0.43	0.66	0.49	0.28	0.12	0.00	0.00	0.00	1.98
[90; 100]	3.46	2.59	1.06	0.18	0.00	0.00	0.00	0.00	7.29

Liberation by surface of Chalcopyrite + Chalcocite + Covellite + Bornite + Digenite + Djurleite

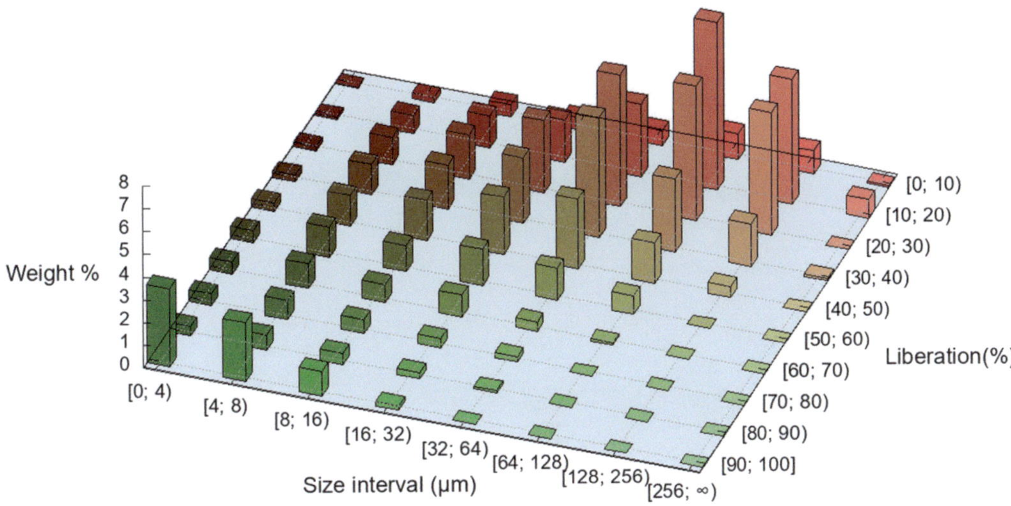

Liberation by composition of sphalerite									
Weight % distribution by size and liberation intervals									
Liberation (%)	Size interval (μm)								Total
	[0; 4)	[4; 8)	[8;16)	[16; 32)	[32; 64)	[64; 128)	[128; 256)	[256; ∞)	
[0; 10)	0.06	0.53	1.67	4.20	9.18	14.36	10.63	1.21	41.84
[10; 20)	0.08	0.81	1.48	2.71	4.81	7.20	5.23	0.00	22.32
[20; 30)	0.10	0.63	1.05	1.18	1.36	1.74	0.79	0.00	6.86
[30; 40)	0.09	0.55	0.86	0.69	1.30	0.96	0.00	0.00	4.45
[40; 50)	0.09	0.39	0.58	0.88	0.45	0.34	0.00	0.00	2.73
[50; 60)	0.11	0.37	0.56	0.55	0.36	0.57	0.00	0.00	2.52
[60; 70)	0.10	0.40	0.50	0.20	0.35	0.00	0.00	0.00	1.55
[70; 80)	0.10	0.46	0.37	0.10	0.07	0.00	0.00	0.00	1.09
[80; 90)	0.20	0.63	0.82	0.75	0.36	0.00	0.00	0.00	2.76
[90; 100]	2.66	4.91	3.73	1.33	1.25	0.00	0.00	0.00	13.88

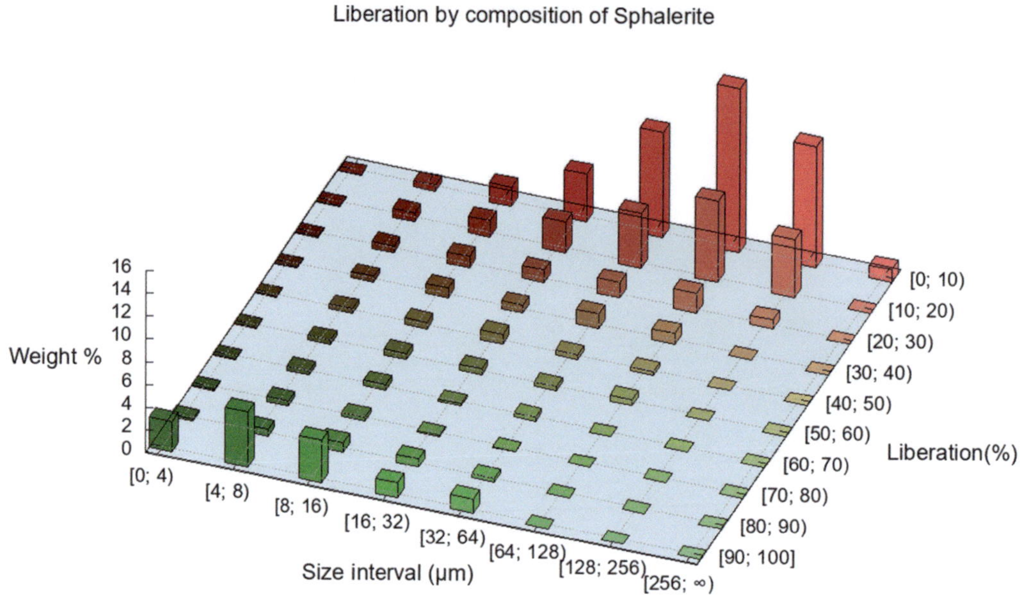

Liberation by exposed surface of sphalerite									
Weight % distribution by size and liberation intervals									
Liberation (%)	**Size interval (μm)**								**Total**
	[0; 4)	[4; 8)	[8; 16)	[16; 32)	[32; 64)	[64; 128)	[128; 256)	[256; ∞)	
[0; 10)	0.04	0.26	0.97	3.10	7.06	12.53	9.97	1.21	35.12
[10; 20)	0.06	0.58	1.68	2.97	5.52	9.12	4.11	0.00	24.05
[20; 30)	0.10	0.75	1.26	1.75	2.24	2.52	2.57	0.00	11.19
[30; 40)	0.11	0.68	0.91	1.05	1.12	0.74	0.00	0.00	4.62
[40; 50)	0.11	0.62	0.80	1.00	1.52	0.27	0.00	0.00	4.33
[50; 60)	0.14	0.51	0.78	0.48	0.98	0.00	0.00	0.00	2.88
[60; 70)	0.13	0.53	0.73	0.56	0.36	0.00	0.00	0.00	2.31
[70; 80)	0.15	0.74	0.77	0.44	0.19	0.00	0.00	0.00	2.29
[80; 90)	0.22	0.89	1.01	0.67	0.38	0.00	0.00	0.00	3.17
[90; 100]	2.52	4.13	2.69	0.58	0.12	0.00	0.00	0.00	10.05

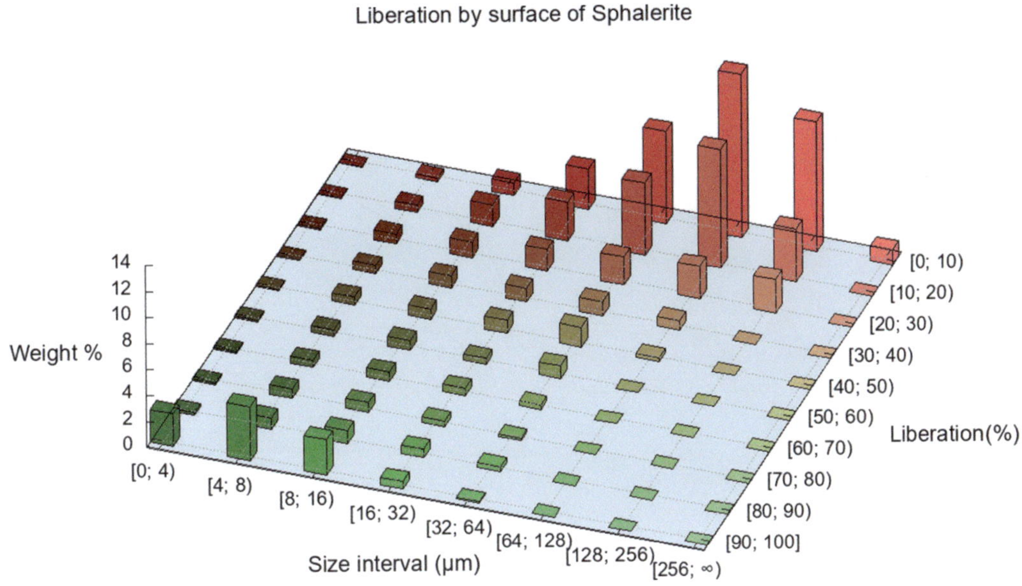

Liberation by composition of galena + bournonite									
Weight % distribution by size and liberation intervals									
Liberation (%)	**Size interval (µm)**								Total
	[0; 4)	[4; 8)	[8; 16)	[16; 32)	[32; 64)	[64; 128)	[128; 256)	[256; ∞)	
[0; 10)	0.04	0.59	2.27	5.95	13.98	15.40	13.31	0.66	52.21
[10; 20)	0.16	0.84	1.52	2.10	4.05	1.91	1.84	0.00	12.41
[20; 30)	0.22	0.77	1.35	1.99	2.22	6.36	0.00	0.00	12.91
[30; 40)	0.28	1.01	0.41	0.00	0.85	0.00	0.00	0.00	2.55
[40; 50)	0.34	1.03	0.84	0.67	0.53	0.00	0.00	0.00	3.42
[50; 60)	0.34	1.02	0.37	0.58	3.02	0.00	0.00	0.00	5.34
[60; 70)	0.29	1.60	0.60	0.00	0.00	0.00	0.00	0.00	2.49
[70; 80)	0.25	1.28	1.01	0.06	0.00	0.00	0.00	0.00	2.60
[80; 90)	0.15	1.11	2.40	0.37	0.00	0.00	0.00	0.00	4.02
[90; 100]	0.23	0.80	1.03	0.00	0.00	0.00	0.00	0.00	2.07

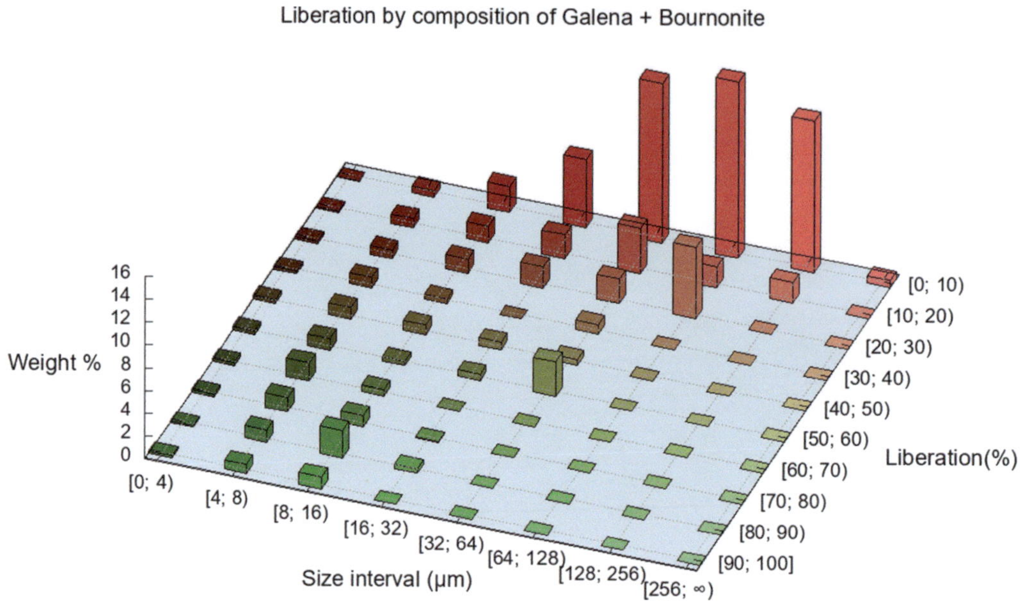

Liberation by exposed surface of Galena + Bournonite									
Weight % distribution by size and liberation intervals									
Liberation (%)	Size interval (µm)								Total
	[0; 4)	[4; 8)	[8; 16)	[16; 32)	[32; 64)	[64; 128)	[128; 256)	[256; ∞)	
[0; 10)	0.56	2.81	4.61	10.08	21.93	23.22	15.15	0.66	79.02
[10; 20)	0.35	1.75	1.59	0.63	1.24	0.45	0.00	0.00	6.01
[20; 30)	0.35	1.46	1.48	0.65	1.48	0.00	0.00	0.00	5.41
[30; 40)	0.22	1.20	1.70	0.37	0.00	0.00	0.00	0.00	3.50
[40; 50)	0.20	1.05	0.71	0.00	0.00	0.00	0.00	0.00	1.96
[50; 60)	0.15	0.57	0.73	0.00	0.00	0.00	0.00	0.00	1.44
[60; 70)	0.12	0.37	0.51	0.00	0.00	0.00	0.00	0.00	1.00
[70; 80)	0.08	0.51	0.18	0.00	0.00	0.00	0.00	0.00	0.77
[80; 90)	0.08	0.22	0.30	0.00	0.00	0.00	0.00	0.00	0.59
[90; 100]	0.19	0.11	0.00	0.00	0.00	0.00	0.00	0.00	0.31

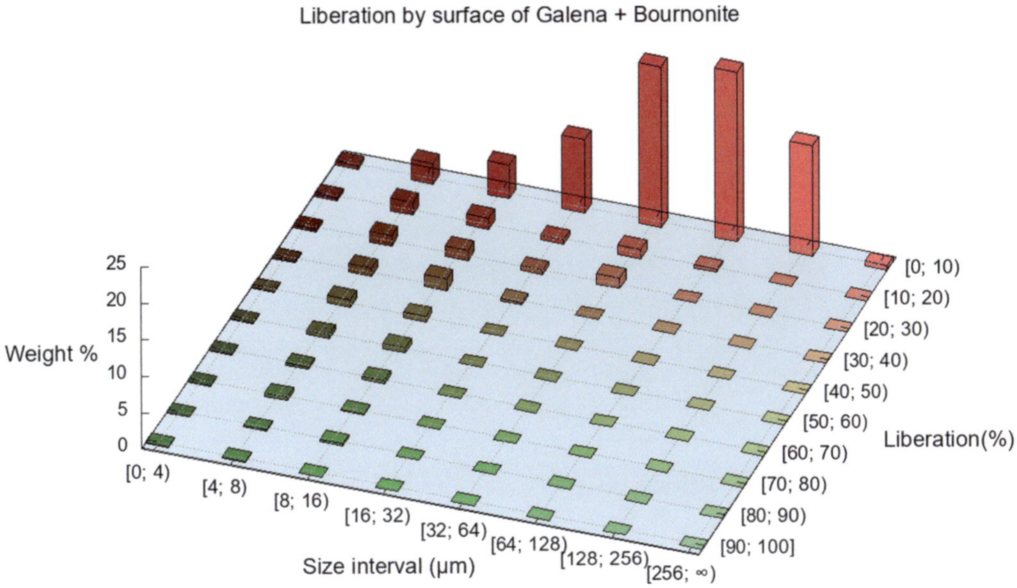

Liberation by composition of pyrite + arsenopyrite + melnikovite + marcasite									
Weight % distribution by size and liberation intervals									
Liberation (%)	Size interval (μm)								Total
	[0; 4)	[4; 8)	[8; 16)	[16; 32)	[32; 64)	[64; 128)	[128; 256)	[256; ∞)	
[0; 10)	0.00	0.01	0.01	0.01	0.01	0.00	0.00	0.00	0.03
[10; 20)	0.01	0.03	0.04	0.04	0.03	0.04	0.00	0.00	0.18
[20; 30)	0.02	0.06	0.09	0.08	0.11	0.06	0.14	0.00	0.56
[30; 40)	0.03	0.10	0.13	0.19	0.19	0.45	0.37	0.11	1.57
[40; 50)	0.05	0.15	0.21	0.32	0.47	0.79	0.21	0.00	2.20
[50; 60)	0.10	0.26	0.39	0.70	1.17	2.04	1.31	0.00	5.96
[60; 70)	0.16	0.45	0.75	1.42	2.90	3.21	2.29	0.49	11.68
[70; 80)	0.22	0.86	1.28	2.44	5.21	5.52	5.58	0.17	21.29
[80; 90)	0.28	1.82	2.65	4.38	6.73	8.92	5.34	0.39	30.51
[90; 100]	2.86	6.98	6.73	4.79	3.12	1.53	0.00	0.00	26.02

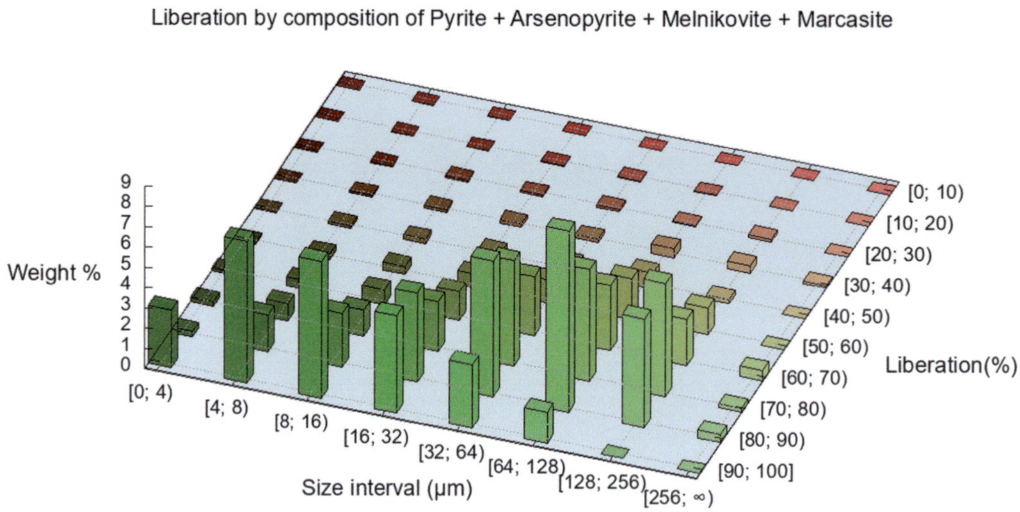

Liberation by exposed surface of pyrite + arsenopyrite + melnikovite + marcasite									
Weight % distribution by size and liberation intervals									
Liberation (%)	Size interval (μm)								Total
	[0; 4)	[4; 8)	[8; 16)	[16; 32)	[32; 64)	[64; 128)	[128; 256)	[256; ∞)	
[0; 10)	0.02	0.07	0.12	0.21	0.55	0.84	0.35	0.00	2.16
[10; 20)	0.04	0.15	0.27	0.78	2.40	4.39	4.24	0.95	13.21
[20; 30)	0.07	0.27	0.48	1.44	3.97	7.80	7.56	0.00	21.59
[30; 40)	0.11	0.39	0.74	2.05	4.99	5.64	2.25	0.21	16.38
[40; 50)	0.13	0.56	1.04	2.45	3.43	2.60	0.68	0.00	10.89
[50; 60)	0.18	0.84	1.46	2.29	2.45	0.97	0.18	0.00	8.37
[60; 70)	0.18	1.07	1.71	2.03	1.29	0.30	0.00	0.00	6.56
[70; 80)	0.15	1.28	1.88	1.61	0.56	0.02	0.00	0.00	5.49
[80; 90)	0.14	1.38	2.10	1.02	0.25	0.00	0.00	0.00	4.90
[90; 100]	2.71	4.71	2.47	0.50	0.05	0.00	0.00	0.00	10.45

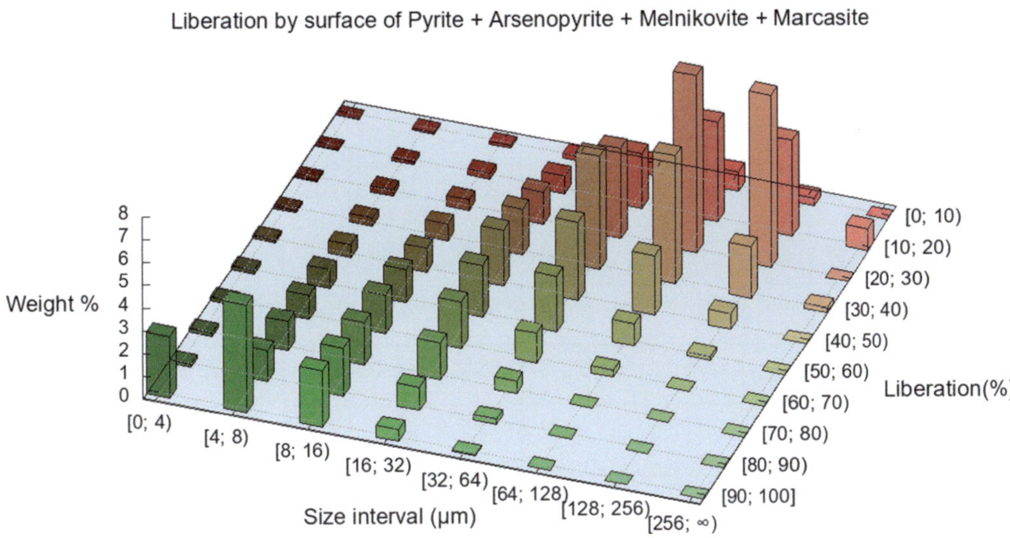

Composition of Particles by Mineral Groups and Particle Size Intervals

Definition of groups	
Group	**Minerals**
[1] Cu sulfides	chalcopyrite + chalcocite + covellite + bornite + digenite + djurleite
[2] Sphalerite	sphalerite
[3] Pb sulfides	galena + bournonite
[4] Fe sulfides	pyrite + melnikovite + marcasite + pyrrhotite
[5] As and Sb minerals	arsenopyrite + tetrahedrite + tennantite + enargite + luzonite
[6] Gangue	gangue* + rutile + titanite

Composition of particles by mineral groups – weight %								
Group	Monophase	Binary						Multiphase
		[1]	[2]	[3]	[4]	[5]	[6]	
[1] Cu sulfides	6.44	☒	0.23	0.01	22.48	0.03	1.56	69.26
[2] Sphalerite	9.66	0.65	☒	0.10	2.08	0.35	6.43	80.72
[3] Pb sulfides	0.69	0.04	0.09	☒	0.84	10.16	0.00	88.18
[4] Fe sulfides	12.05	23.24	0.56	0.09	☒	0.18	1.05	62.83
[5] As and Sb minerals	4.09	0.22	0.77	3.91	2.55	☒	0.71	87.74
[6] Gangue	2.11	1.09	0.86	0.00	1.86	0.05	☒	94.04

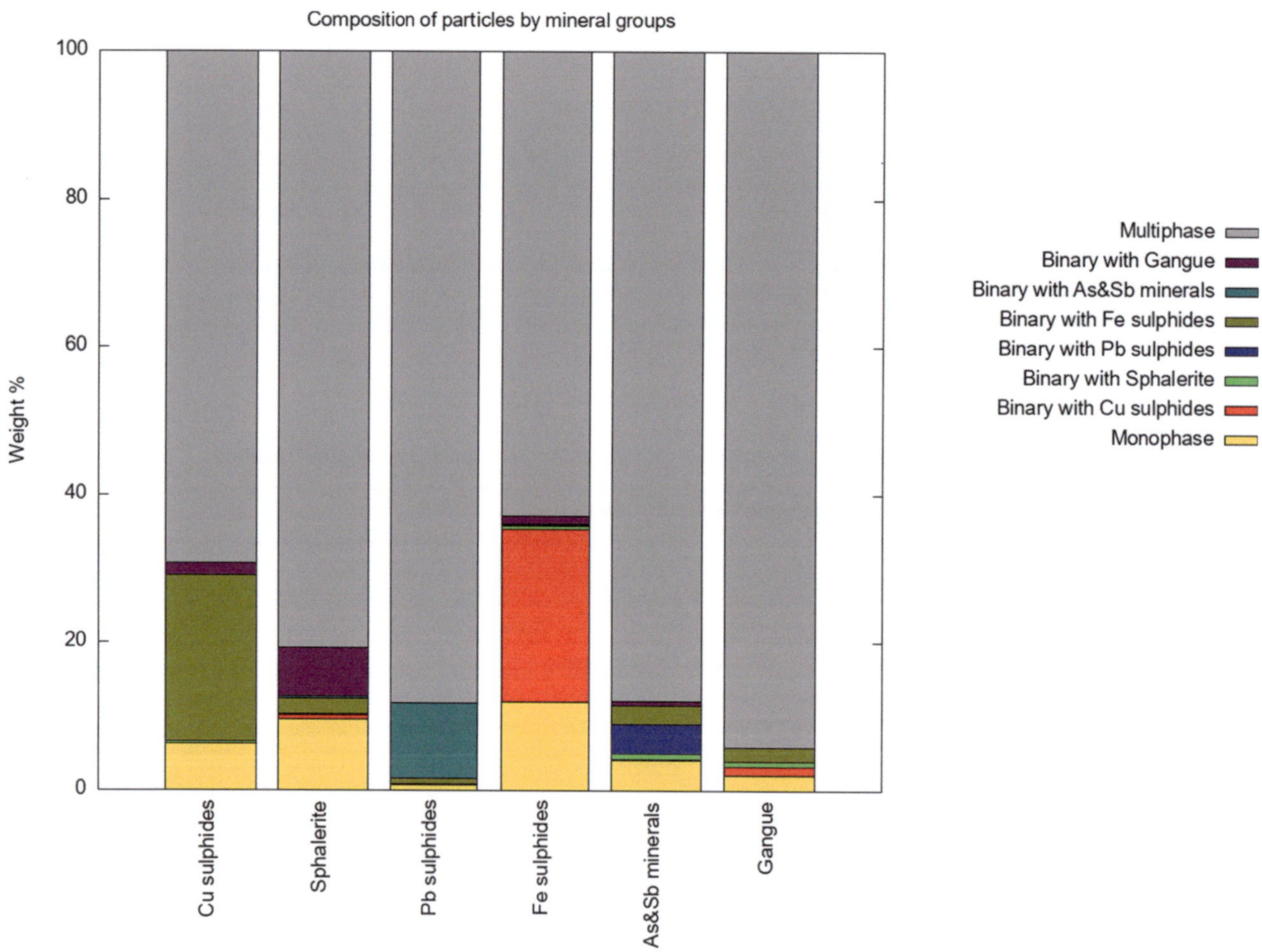

[1] Cu sulfides								
Size interval (μm)	Monophase	Binary						Multiphase
		[1]	[2]	[3]	[4]	[5]	[6]	
[0; 4)	3.17	☒	0.06	0.00	2.57	0.02	0.35	0.37
[4; 8)	2.19	☒	0.08	0.00	5.88	0.01	0.61	2.77
[8; 16)	0.90	☒	0.08	0.00	4.83	0.01	0.40	5.45
[16; 32)	0.10	☒	0.01	0.00	3.73	0.00	0.20	10.38
[32; 64)	0.08	☒	0.00	0.00	3.30	0.00	0.00	16.76
[64; 128)	0.00	☒	0.00	0.00	2.06	0.00	0.00	18.37
[128; 256)	0.00	☒	0.00	0.00	0.10	0.00	0.00	14.13
[256; ∞)	0.00	☒	0.00	0.00	0.00	0.00	0.00	1.03

[2] Sphalerite								
Size interval (μm)	Monophase	Binary						Multiphase
		[1]	[2]	[3]	[4]	[5]	[6]	
[0; 4)	2.54	0.07	☒	0.00	0.20	0.06	0.57	0.34
[4; 8)	4.07	0.21	☒	0.00	0.66	0.19	1.90	3.15
[8; 16)	2.49	0.22	☒	0.04	0.60	0.10	1.65	6.94
[16; 32)	0.47	0.16	☒	0.06	0.25	0.00	1.08	10.42
[32; 64)	0.09	0.00	☒	0.00	0.08	0.00	1.23	17.70
[64; 128)	0.00	0.00	☒	0.00	0.29	0.00	0.00	24.44
[128; 256)	0.00	0.00	☒	0.00	0.00	0.00	0.00	16.48
[256; ∞)	0.00	0.00	☒	0.00	0.00	0.00	0.00	1.26

[3] Pb sulfides								
Size interval (µm)	Monophase	Binary						Multiphase
		[1]	[2]	[3]	[4]	[5]	[6]	
[0; 4)	0.20	0.01	0.04	☒	0.08	2.02	0.00	0.38
[4; 8)	0.22	0.03	0.00	☒	0.22	4.95	0.00	6.47
[8; 16)	0.27	0.00	0.01	☒	0.29	3.12	0.00	10.10
[16; 32)	0.00	0.00	0.04	☒	0.20	0.07	0.00	12.16
[32; 64)	0.00	0.00	0.00	☒	0.05	0.00	0.00	25.32
[64; 128)	0.00	0.00	0.00	☒	0.00	0.00	0.00	21.43
[128; 256)	0.00	0.00	0.00	☒	0.00	0.00	0.00	12.32
[256; ∞)	0.00	0.00	0.00	☒	0.00	0.00	0.00	0.00

[4] Fe sulfides								
Size interval (µm)	Monophase	Binary						Multiphase
		[1]	[2]	[3]	[4]	[5]	[6]	
[0; 4)	2.62	0.90	0.03	0.00	☒	0.02	0.06	0.09
[4; 8)	4.60	4.39	0.17	0.01	☒	0.06	0.27	1.22
[8; 16)	2.95	5.33	0.17	0.02	☒	0.08	0.23	3.49
[16; 32)	1.18	5.04	0.08	0.03	☒	0.03	0.18	7.83
[32; 64)	0.51	4.54	0.08	0.02	☒	0.00	0.08	14.70
[64; 128)	0.19	2.97	0.04	0.00	☒	0.00	0.23	19.14
[128; 256)	0.00	0.06	0.00	0.00	☒	0.00	0.00	15.20
[256; ∞)	0.00	0.00	0.00	0.00	☒	0.00	0.00	1.16

■ [5] As and Sb minerals								
Size interval (µm)	■ Monophase	Binary						■ Multiphase
		■ [1]	■ [2]	■ [3]	■ [4]	■ [5]	■ [6]	
[0; 4)	3.52	0.13	0.41	1.81	0.52	☒	0.50	0.96
[4; 8)	0.34	0.08	0.31	1.76	1.06	☒	0.19	7.56
[8; 16)	0.23	0.01	0.05	0.33	0.85	☒	0.02	9.24
[16; 32)	0.00	0.00	0.00	0.02	0.12	☒	0.00	12.63
[32; 64)	0.00	0.00	0.00	0.00	0.00	☒	0.00	18.28
[64; 128)	0.00	0.00	0.00	0.00	0.00	☒	0.00	20.36
[128; 256)	0.00	0.00	0.00	0.00	0.00	☒	0.00	18.15
[256; ∞)	0.00	0.00	0.00	0.00	0.00	☒	0.00	0.56

■ [6] Gangue								
Size interval (µm)	■ Monophase	Binary						■ Multiphase
		■ [1]	■ [2]	■ [3]	■ [4]	■ [5]	■ [6]	
[0; 4)	1.02	0.29	0.17	0.00	0.28	0.03	☒	0.38
[4; 8)	0.79	0.54	0.43	0.00	0.79	0.02	☒	3.24
[8; 16)	0.27	0.19	0.18	0.00	0.41	0.01	☒	6.61
[16; 32)	0.02	0.06	0.05	0.00	0.14	0.00	☒	10.42
[32; 64)	0.00	0.00	0.04	0.00	0.05	0.00	☒	18.48
[64; 128)	0.00	0.00	0.00	0.00	0.18	0.00	☒	30.82
[128; 256)	0.00	0.00	0.00	0.00	0.00	0.00	☒	21.33
[256; ∞)	0.00	0.00	0.00	0.00	0.00	0.00	☒	2.75

References

Atkin BP, Harvey PK (1979) Nottingham interactive system for opaque mineral identification: NISOMI. Trans Inst Mining Metall 88:1324–1327

Becker M, Wightman EM, Evans CL (eds) (2016) Process mineralogy—JKMRC monograph series in mining and mineral processing, vol. 6. University of Queensland, Australia

Bernhardt HJ (1979) Computer-gestuetzte Erzmineral-Diagnose mittels Reflexionsspektren in sichtbarem Licht. N Jb Miner Mh 9:403–407

Bernhardt HJ (1987) A simple, fully-automated system for ore mineral identification. Mineral Petrol 36:241–245

Bernhardt HJ (1990) Microscopic identification, and identification schemes, of ore minerals. In: Jambor JL, Vaughan DJ (eds) Advanced microscopic studies of ore minerals: short course handbook, 17. Mineralogical Association of Canada, Ottawa, pp 189–212

Berrezueta E. & Castroviejo R. (2007) Reconocimiento automatizado de menas metálicas mediante análisis digital de imagen: Un apoyo al proceso mineralúrgico. I: Ensayo metodológico. Rev Metalurgia 43(4):294–309

Castroviejo R, Brea C, Pérez-Barnuevo L, Catalina JC, Segundo F, Bernhardt HJ, Pirard E (2009) Using computer vision for microscopic identification of ores with reflected light: preliminary results. Williams et al. (eds) In: Proceedings of the 10th Biennial SGA meeting. Townsville, Australia, pp 682–684. ISSN 9780980558685

Castroviejo R, Catalina JC, Bernhardt HJ, Pirard E, Segundo F, Brea C, Pérez-Barnuevo L (2010) A fully automated system for multispectral ore microscopy. In: IMA2010 (20th general meeting international mineralogical association). Acta mineralogica petrographica

—Abstract series (HU ISSN 0324-6523/HU ISSN 1589–4835), Szeged, Budapest, p. 281 (Proceedings of CD_ROM, session EG53_geometallurgy and process mineralogy, code 4NF15)

Catalina JC, Castroviejo R (2017) Multispectral reflectance microscopy: application to the automated recognition of metallic ores. J Metall 53–54:1–20. https://doi.org/10.3989/revmetalm.107

Catalina JC (2018) Desarrollo de un sistema multiespectral para microscopía de menas. Doctoral thesis, Escuela Técnica Superior de Ingenieros de Minas y Energía. Universidad Politécnica de Madrid. https://doi.org/10.20868/UPM.thesis.52452, http://oa.upm.es/52452/

Criddle AJ, Stanley CJ (eds) (1993) Quantitative data file for ore minerals, 3rd edn. Chapman & Hall, London

Criddle AJ (1998) Ore microscopy and photometry (1890–1998) In: Cabri LJ, Vaughan DJ (eds) Modern approaches to ore and environmental mineralogy. COM/IMA short course series, vol 27, Ottawa, Ontario, 421 p

Gerlitz CN, Leonard BF Criddle AJ (1989) QDF database system, version 1.0: Reflectance of ore minerals—A search-and-match identification system for IBM and compatible microcomputers using the IMA/COM quantitative data file for ore minerals, 2nd issue. U. S. Geological Survey. Open-File Report 89-0306A-E

Grunwald-Romera U, Catalina JC, Alarcón D, López-Benito A, Castroviejo R (2019) A reliable method for the automated distinction of quartz gangue and epoxy resin with reflected light microscopy and its application to digital image analysis. In: Proceedings of 15th SGA Biennial meeting. Glasgow, Scotland (August)

Kühnel RA, Prins JJ, Roorda HJ (1980) The 'Delft' system for mineral identification. Delft University Press, Delft, The Netherlands, p 204

López-Benito A, Catalina JC, Alarcón D, Grunwald-Romera U, Romero P, Castroviejo R (2019) Automated ore microscopy based on multispectral measurements of specular reflectance. I—A comparative study of some supervised classification techniques. Miner Eng 146:106–136, ISSN 0892-6875. https://doi.org/10.1016/j.mineng.2019.106136, https://www.sciencedirect.com/science/article/pii/S0892687519305473

Pérez-Barnuevo L (2014) Caracterización automatizada de texturas de menas mediante análisis digital de imagen para su aplicación geometalúrgica. Ph.D. thesis, Escuela Técnica Superior de Ingenieros de Minas y Energía, Universidad Politécnica de Madrid, 298 p

Picot P, Johan Z (1982) Atlas of ore minerals. Bureau de Recherches Géologiques et Minières (BGRM). Elsevier

Pirard E (2004) Multispectral imaging of ore minerals in optical microscopy. Mineral Mag 68(2):323–333

Pirard E (2016) Optical microscopy. In: Becker M, Wightman EM, Evans CL (eds) Process mineralogy. JKMRC Monograph series in mining and mineral processing, vol 6, chap. 5. University of Queensland, Australia, pp 51–66

Shoji T, Kaneda H (1994) An interactive system to assist mineral identification in ore microscopy. Math Geol 26(8):961–972

Mineral Indices, Annexes

Analytical Index of Minerals Described*

Name	Abbreviation	Explanation
Acanthite	ac	+
Adularia	adu	gg
Aguilarite	agl	+
Aikinite	aik	cf. bm
Alabandite	alb	+
Albite	ab	gg
Allargentum	aag	+
Alloclasite	alc	cf. cob
Amalgam	am (amal)	+
Amblygonite	amb	cf. Phos
Amphibole (Group)	Anf, Amp	gg
Anatase	ana	+
Anglesite	ang	gg
Anhydrite	anh	gg
Anilite	anil	cf. cc, dg
Ankerite	ank	gg
Annabergite	anbg	gg
Anorthite	an	gg
Antimony	Sb	+
Apatite	ap	gg
Argentite	arg	+
Argentopyrite	agp	+
Arsenic	As	+
Arsenopyrite	asp	+
Aurian silver Ag–Au	*ausi*	cf. *el*
Azurite	az	+
Baddeleyite	bad	+
Barite	bar	gg
Bastnäsite	bst, (bas)	+
Berthierite	brt	+
Berzelianite	bzl	cf. dg
Biotite	bio	gg
Bismite	bim	gg

(continued)

R. Castroviejo, *A Practical Guide to Ore Microscopy—Volume 2*,
https://doi.org/10.1007/978-3-031-18954-8

(continued)

Name	Abbreviation	Explanation
Bismuth	Bi	+
Bismuthinite	bm	+
Bismutite	bis	gg
Bornite	bn	+
Boulangerite	bl	+
Bournonite	bnn	+
Braggite	brg	cf. erl
Braunite	bra (br)	+
Bravoite	brv	+
Breithauptite	bth	+
Briartite	bri	cf. ren
Brookite	bk	gg
Calaverite	clv	+
Calciotantite	cta	cf. mlt, pyc
Calcite	calc	gg
Carrollite	car	cf. ln
Cassiterite	cst (cs)	+
Cattierite	ctt	cf. brv
Cerargyrite	crg	cf. cag
Cerussite	crs	cf. gn
Cervantite	cerv	cf. ochres Sb
Chalcedony	cld	cf. q
Chalcocite	cc	+
Chalcopyrite	ccp	+
Chert	ch	gg
Chloanthite (disc.)	cloa	cf. sk
Chlorargyrite	cag	+
Chlorite (Series)	Chl	gg
Chromite	chr	+
Chrysocolla	crc	+
Cinnabar	cin	+
Clausthalite	clt (cla)	+
Clay	Arc	gg
Clay minerals (Gr.)	CM, MArc	gg
Coal—organic matter	COM	cf. gg & gra
Cobaltite	cob	+
Cohenite	coh	cf. Fe
Coloradoite	colo (cld)	+
Columbite	cb	cf. cbt
Columbotantalite	cbt	+
Cooperite	cpr	cf. erl
Copper	Cu	+
Coronadite	crn	cf. rom, psi
Corundum	cor	gg
Cosalite	cos	+
Covellite	cv	+
Cryptomelane	cry	+
Cubanite	cbn	+
Cuprite	cup	+
Cylindrite	cyl	cf. F 41.04

(continued)

(continued)

Name	Abbreviation	Explanation
Digenite	dg	+
Djurleite	dju (dj)	cf. cc, dg
Dolomite	dol	gg
Dyscrasite	dy	+
Electrum	el	+
Enargite	en	+
Epidote (Group)	Ep	gg
Erlichmanite	erl	+
Erythrite	ery	gg
Eugenite	eug	cf. am, aag
Euxenite	eux	cf. REE
Falkmanite	flk	cf. bl
Famatinite	fm	+
Feldspar (Group)	Fs, Felds	gg
Ferberite	fb	cf. wf
Ferroselite	fse	cf. clt, um
Fletcherite	fle	cf. ln
Fluorite	fl	gg
Franklinite	fkl (fk)	+
Freibergite	fbg	cf. td
Freieslebenite	flb	+
Galena	gn	+
Gangue (undiff.)	gg	gg
Garnet (Group)	grt	gg
Geocronite	gc	+
Germanite	ger	cf. ren
Gersdorffite	gf	+
Glaucodot	gld	+
Glauconite	gln	gg
Goethite	gth	+
Gold	Au	+
Gossan	gos	cf. lim
Graphite	gra	+
Gratonite	gtt	+
Greenockite	gk	gg
Greigite	grg	cf. ln
Gudmundite	gud (gd)	+
Hematite	ht	+
Hessite	hss	
Hollandite	hol	cf. psi
Hornblende	hb	gg
Horobetsuite	hrb	cf. bm
Hübnerite	hub	cf. wf
Illite	ili	gg
Ilmenite	il	+
Ilmenorutile	ilr	+
Iridium	Ir	+
Iridosmine (discr.)	iros	cf. Os, Ir
Iron	Fe	+
Isocubanite	isc	+

(continued)

(continued)

Name	Abbreviation	Explanation
Ixiolite	ix	cf. rt (F13-14)
Jacobsite	jac	+
Jalpaite	jl	cf. Sulf
Jamesonite	jm	+
Jarosite	jar	cf. lim
Jasper	jas	gg
Jordanite	jd	+
K Feldspar	Kfs	gg
Kermesite	ker, (km)	+
Kësterite	ks	cf. stn
Krennerite	krn	+
Laurite	lrt	cf. erl
Lepidocrocite	lpc	+
Leucoxeno	lcx	cf. ana (gg)
Limonite	lim, lm	+
Linnaeite	ln	+
Löllingite	lo	+
Ludwigite	ldw	cf. von
Luzonite	lz	+
Mackinawite	mck	+
Maghemite	mgh	+
Magnesite	mag	cf. Carb, calc
Magnetite	mt, (mg)	+
Malachite	mch	+
Manganite	mng	+
Marcasite	mc	+
Maucherite	mau	+
Melnicovite	mln	cf. py
Mercury/Quicksilver	Hg	cf. cin
Metacinnabar	mcin	cf. cin
Metastibnite	msbt	cf. sbt
Miargyrite	my	+
Mica (Group)	Mica	gg
Microlite	mlt	cf. pyc
Millerite	ml	+
Molybdenite	mol	+
Monazite	mz	+
Moschellandsbergite	mobg	cf. am
Muscovite	mv, mus	gg
Naumannite	nau	+
Nickeline, Niccolite	nc, (nk)	+
Ochre, Ocher	Ochre	cf. Oxides
Olivin (Series)	ov	gg
Organic matter	OM	gg
Orpiment	orp, op	+
Orthoclase	or	gg
Orthopyroxene (Gr.)	Opx	gg
Osmium	Os	+
Oxycalciomicrolite	ocm	cf. mlt, bim
Palladium	Pd	cf. PGE

(continued)

(continued)

Name	Abbreviation	Explanation
Pararammelsbergite	prm	+
Pearceite	pc	+
Pentlandite	pn	+
Perovskite	pvk	gg
Phosphate	Phos	gg
Phyllosil. (Subclass)	FSlc, PSlc	gg
Piemontite	pie	cf. bra
Pitchblende	pch	cf. urn
Plagioclase (Series)	Plag	gg
Platinum	Pt	+
Platinum Group Elements	PGE	cf. Pt, Os, Ir
Platinum Group Minerals	PGM	cf. spy
Polianite	poli	cf. prl
Polybasite	plb	+
Polydymite	pdm	cf. ln
Proustite	pru	+
Psilomelane	psi	+
Pyrargyrite	prg	+
Pyrite	py	+
Pyrochlore	pyc (pyl)	+
Pyrolusite	prl	+
Pyromorphite	pymp	gg
Pyrostilpnite	pyst	cf. prg
Pyroxene (Group)	Px, Pyx	gg
Pyrrhotite	po	+
Quartz	q, qz…	gg
Rammelsbergite	rm	+
Ramsdellite	rdl	+
Realgar	rl	+
Renierite	ren	+
Rhodochrosite	rhc	gg
Rhodonite	rhn	gg
Romanechite	rom	+
Rutheniridosmine	ruiro	cf. Os, Ir
Rutile	rt	+
Rynersonite	ryn	cf. mlt, pyc
Safflorite	sf	+
Schachnerite	snr	cf. am, F stm8
Scheelite	sch	+
Scorodite	sco	gg (cf. asp)
Selenides	Sel	cf. nau, um
Senarmontite	sen	cf. ochres Sb
Sericite	ser	gg
Serpentine	srp	gg
Siderite	sid (sd)	cf. Carb
Siegenite	sg	cf. ln
Silica	Sic	cf. q

(continued)

(continued)

Name	Abbreviation	Explanation
Silicate (Class)	Slc	gg
Silver	Ag	+
Skutterudite	sk	+
Smaltite (≈sk)	smal	cf. sk
Smithsonite	sms	gg
Souxite	sx	cf. cst
Sperrylite	spy	+
Sphalerite	sp	+
Spinel	spn	+
Spionkopite—Bbl.cv	spk	cf. cv, F dg
Stannite	stn	+
Stephanite	stp	+
Sternbergite	stb	+
Stibarsen	sba	+
Stibioluzonite	sbl	cf. fm
Stibnite	sbt	+
Stromeyerite	stm	+
Sulfur	S	gg
Sylvanite	syl (sv)	cf. Tel
Talnakhita	tlk	cf. ccp
Tantalite	tan (tt)	cf. cbt
Tapiolite	tp, (tap)	+
Tedradymite	tdy	cf. Tel
Tellurium	Te	+
Tennantite	tn	cf. td
Tetrahedrite	td	+
Titanite	tit (sph)	+
Tourmaline	tur	gg
Troilite	tr (tro)	cf. po, Fe
Ullmannite	ul	+
Ulvöspinel (Ulvite)	ulv	+
Umangite	um	+
Uraninite	urn	+
Vaesite	vae	cf. ctt
Valentinite	val	cf. ochres Sb
Valleriite	vll	+
Vanadinite	van	+
Vanadinite	van	gg
Violarite	vio	cf. ln
Vonsenite	von	gg
Willemite	wil	cf. Slc
Wolframite	wf	+
Wurtzite	wz	+
Xanthoconite	xan	cf. pru
Xenotime	xtm	cf. Phos
Zeolite (Group)	Zeo	gg

(continued)

(continued)

Name	Abbreviation	Explanation
Zincite	znc	gg (cf. fkl)
Zinkenite	zk	cf. Ss
Zircon	zir	gg
Other name—Synonym		Observations
Allemontite		cf. sba
Antimonite		cf. sbt
Blende		cf. sp
Chalcopyrrhotite		cf. isc
Fahlore		cf. td
Hollandite		cf. rom, psi
Martite		cf. ht, mt
Mispickel		cf. asp
Polianite		cf. prl
Schalenblende		cf. sp
Sphene		cf. tit
Steel		cf. Fe
Strüverite		cf. rt
Ulvite		cf. ulv

*__Key__ to locating the descriptions of ores and minerals in the text

+	Ore is fully described in the chapter under its own name, symbol or abbreviation
***cf m1, m2*, etc.**	Ore or mineral described/commented on in the chapter under mineral *m1, m2*, etc.
gg	Described in the corresponding chapter in Part III (Gangue)

Selected Ore Minerals—Learning Priority 1

Name (EN)	Abbreviation	Nombre (ES)	Observations	Priority
Acanthite	**ac**	**Acantita**	+	**1**
Antimony	**Sb**	**Antimonio**	+	**1**
Arsenopyrite	**asp**	**Arsenopirita**	+	**1**
Bastnäsite	**bst (bas)**	**Bastnäsita**	+	**1**
Bornite	**bn**	**Bornita**	+	**1**
Cassiterite	**cst (cs)**	**Casiterita**	+	**1**
Chalcocite	**cc**	**Calcocita = Calcosina**	+	**1**
Chalcopyrite	**ccp**	**Calcopirita**	+	**1**
Chromite	**chr**	**Cromita**	+	**1**
Cinnabar	**cin**	**Cinabrio**	+	**1**
Cobaltite	**cob**	**Cobaltita**	+	**1**
Columbotantalite	**cbt**	**Columbotantalita**	+	**1**
Copper	**Cu**	**Cobre**	+	**1**
Covellite	**cv**	**Covellita**	+	**1**
Cuprite	**cup**	**Cuprita**	+	**1**
Digenite	**dg**	**Digenita**	+	**1**
Galena	**gn**	**Galena**	+	**1**
Goethite	**gth**	**Goethita**	+	**1**
Gold	**Au**	**Oro**	+	**1**
Graphite	**gra**	**Grafito**	+	**1**
Hematite	**ht**	**Hematites**	+	**1**
Ilmenite	**il**	**Ilmenita**	+	**1**
Iron	**Fe**	**Hierro**	+	**1**

(continued)

(continued)

Name (EN)	Abbreviation	Nombre (ES)	Observations	Priority
Limonite	**lim, lm**	**Limonita**	+	**1**
Linnaeite	**ln**	**Linneíta**	+	**1**
Magnetite	**mt, (mg)**	**Magnetita**	+	**1**
Manganite	**mng**	**Manganita**	+	**1**
Marcasite	**mc**	**Marcasita**	+	**1**
Miargyrite	**my**	**Miargirita**	+	**1**
Millerite	**ml**	**Millerita**	+	**1**
Molybdenite	**mol**	**Molibdenita**	+	**1**
Nickeline, Niccolite	**nc, (nk)**	**Niquelita, Niquelina**	+	**1**
Pentlandite	**pn**	**Pentlandita**	+	**1**
Platinum	**Pt**	**Platino**	+	**1**
Proustite	**pru**	**Proustita**	+	**1**
Psilomelane	**psi**	**Psilomelana**	+	**1**
Pyrargyrite	**prg**	**Pirargirita**	+	**1**
Pyrite	**py**	**Pirita**	+	**1**
Pyrolusite	**prl**	**Pirolusita**	+	**1**
Pyrrhotite	**po**	**Pirrotita (Pirrotina)**	+	**1**
Rammelsbergite	**rm**	**Rammelsbergita**	+	**1**
Romanechite	**rom**	**Romanechita**	+	**1**
Rutile	**rt**	**Rutilo**	+	**1**
Safflorite	**sf**	**Safflorita**	+	**1**
Silver	**Ag**	**Plata**	+	**1**
Skutterudite	**sk**	**Skutterudita**	+	**1**
Sphalerite	**sp**	**Esfalerita**	+	**1**
Stibnite	**sbt**	**Estibina, Antimonita**	+	**1**
Tetrahedrite	**td**	**Tetraedrita-Tennant**	+	**1**
Uraninite	**urn**	**Uraninita**	+	**1**
Wolframite	**wf**	**Wolframita**	+	**1**

Selected Ore Minerals—Learning Priority 2

Name (EN)	Abbreviation	Nombre (ES)	Observations	Priority
Alabandite	*alb*	*Alabandita*	+	*2*
Allargentum	*aag*	*Allargentum*	+	*2*
Amalgam	*am, (amal)*	*Amalgama*	+	*2*
Argentopyrite	*agp*	*Argentopirita*	+	*2*
Arsenic	*As*	*Arsénico*	+	*2*
Aurian silver Ag–Au	*ausi*	*Plata áurea Ag–Au*	cf. *el*	*2*
Azurite	*az*	*Azurita*	+	*2*
Berthierite	*brt*	*Berthierita*	+	*2*
Bismuth	*Bi*	*Bismuto*	+	*2*
Bismuthinite	*bm*	*Bismutinita*	+	*2*
Boulangerite	*bl*	*Boulangerita*	+	*2*
Braunite	*bra (br)*	*Braunita*	+	*2*
Bravoite	*brv*	*Bravoíta*	+	*2*
Breithauptite	*bth*	*Breithauptita*	+	*2*
Calaverite	*clv*	*Calaverita*	+	*2*

(continued)

(continued)

Name (EN)	Abbreviation	Nombre (ES)	Observations	Priority
Chlorargyrite	*cag*	*Clorargirita*	+	2
Cohenite	*coh (chn)*	*Cohenita*	cf. *Fe*	2
Cosalite	cos	Cosalita	+	2
Cryptomelane	*cry*	*Criptomelana*	+	2
Cubanite	*cbn*	*Cubanita*	+	2
Dyscrasite	*dy*	*Discrasita*	+	2
Electrum	*el*	*Electrum*	+	2
Enargite	*en*	*Enargita*	+	2
Famatinite	*fm*	*Famatinita*	+	2
Franklinite	*fk (fkl)*	*Franklinita*	+	2
Gersdorffite	*gf*	*Gersdorffita*	+	2
Glaucodot	*gld*	*Glaucodoto*	+	2
Gudmundite	*gud (gd)*	*Gudmundita*	+	2
Ilmenorutile	*ilr*	*Ilmenorutilo*	+	2
Iridium	*Ir*	*Iridio*	+	2
Isocubanite	*isc*	*Isocubanita*	+	2
Jamesonite	jm	Jamesonita	+	2
Lepidocrocite	*lpc*	*Lepidocrocita*	+	2
Löllingite	*lo*	*Löllingita*	+	2
Luzonite	*lz*	*Luzonita*	+	2
Mackinawite	*mck*	*Mackinawita*	+	2
Maghemite	*mgh*	*Maghemita*	+	2
Malachite	*mch*	*Malaquita*	+	2
Maucherite	*mau*	*Maucherita*	+	2
Monazite	*mz*	*Monazita*	+	2
Orpiment	*orp, op*	*Oropimente*	+	2
Osmium	*Os*	*Osmio*	+	2
Pararammelsbergite	*prm*	*Pararamelsbergita*	+	2
Pearcite	*pc*	*Pearceíta*	+	2
Platinum Group Minerals	*PGM*	*Minls del Grupo del Pt*	cf. *spy*	2
Polybasite	*plb*	*Polibasita*	+	2
Pyrochlore	*pyc (pyl)*	*Pirocloro*	+	2
Renierite	*ren*	*Renierita*	+	2
Scheelite	*sch*	*Scheelita*	+	2
Sperrylite	*spy*	*Sperrylita*	+	2
Spinel	*spn*	*Espinela*	+	2
Stannite	*stn*	*Estannita*	+	2
Stephanite	*stp*	*Stephanita*	+	2
Stibarsen	*sba*	*Stibarsen*	+	2
Stromeyerite	*stm*	*Stromeyerita*	+	2
Tapiolite	*tp, (tap)*	*Tapiolita*	+	2
Tellurium	*Te*	*Teluro*	+	2
Titanite	*tit (sph)*	*Titanita*	+	2
Umangite	*um*	*Umangita*	+	2
Valleriite	*vll*	*Vallerita*	+	2
Wurtzite	*wz*	*Wurtzita*	+	2

Most Common Gangue Minerals

Name	Abbreviation	Observations
Adularia	adu	cf. Slc, Fs
Albite	ab	cf. Slc, Fs
Amblygonite	amb	cf. Phos
Amphibole (Group)	Anf, Amp	cf. Slc
Anglesite	ang	cf. Sulfates
Anhydrite	anh	cf. Sulfates
Ankerite	ank	cf. Carb
Annabergite	anbg	Arsenates
Anorthite	an	cf. Slc, Fs
Apatite	ap	cf. Phos
Barite	bar	cf. Sulfates
Biotite	bio	cf. Slc, FSlc
Bismite	bim	cf. Bi ochre
Bismutite	bis	cf. Bi ochre
Brookite	bk	Oxides, cf. lcx
Calcite	calc	cf. Carb
Carbonate (Class)	Carb	cf. Carb, gg
Cerussite	crs	cf. Carbonates
Cervantite	cerv	cf. Sb ochre
Chalcedony	cld	cf. q
Chert	ch	cf. Slc, Sic
Chlorite (Series)	Chl	cf. Slc, FSlc
Chrysocolla	crc	cf. Slc
Clay	Arc	cf. FSlc
Clay Minerals	CM, MArc	cf. Slc, FSlc
Coal—organic matter	COM	cf. Org.Matter
Corundum	cor	
Dolomite	dol	cf. Carb
Epidote (Group)	Ep	cf. Slc
Erythrite	ery	Arsenates
Feldspar (Group)	Fs, Felds	cf. Slc
Fluorite	fl	cf. Halides
Gangue (undiff.)	gg	
Garnet (Group)	grt	cf. Slc
Glauconite	gln	cf. Slc, FSlc
Hornblende	hb	cf. Slc, Inoslc
Illite	ili	cf. Slc, FSlc
Jarosite	jar	cf. lim
Jasper	jas	cf. q
K Feldspar	Kfs	cf. Slc, Fs
Leucoxene	lcx	cf. tit, ana… +
Ludwigite	ldw	cf. von
Magnesite	mag	cf. Carb
Mica (Group)	Mica	cf. Slc, FSlc
Muscovite	mv, mus	cf. Mica
Ochre, Ocher	Ocre	cf. Oxides
Olivine (Series)	ov	cf. Slc
Organic matter	OM	
Orthoclase	or	cf. Slc, Fs

(continued)

(continued)

Name	Abbreviation	Observations
Perovskite	pvk	Oxides
Phosphate	Phos	gg
Phyllosilicates (Subclass)	FSlc, PSlc	cf. Slc
Piemontite	pie	cf. bra, Ep
Plagioclase (Series)	Plag	cf. Slc, Fs
Pyromorphite	pymp	cf. Phos
Pyroxene (Group)	Px, Pyx	cf. Slc, Inoslc
Quartz	q (qz, qtz)	cf. Slc, Sic
Rhodochrosite	rhc	cf. Carb
Rhodonite	rhn	cf. rdl F 1-4
Scorodite	sco	cf. asp-Arsenates
Senarmontite	sen	cf. Sb ochre
Sericite	ser	cf. Slc, FSlc
Serpentine	srp	cf. Slc, FSlc
Siderite	sid (sd)	cf. Carb
Silica	Sic	cf. Slc, q
Silicate (Class)	Slc	gg
Smithsonite	sms	cf. Carb
Souxite	sx	cf. cst
Sulfur	S	gg
Tourmaline	tur	cf. Slc
Valentinite	val	cf. Sb ochre
Vanadinite	van	cf. Vanadates
Vonsenite	von	cf. Carb-Borates
Willemite	wil	cf. Slc
Xenotime	xtm	cf. Phos
Zeolite (Group)	Zeo	cf. Slc
Zincite	znc	Oxides
Zircon	zir	cf. Slc
Other name (synonym)		**Observations**
Sphene		cf. tit

Annex 1: Abbreviations and Criteria

The following list of mineral abbreviations is based on that used at the Laboratory of Applied Microscopy and Image Analysis (LMA) of the Madrid School of Mines, Universidad Politécnica de Madrid (UPM). It responds to the practical need to annotate quickly, accurately and easily the minerals identified in images, graphs, technical reports and texts, such as those used in the daily mineralogy work of the LMA. It refers mainly to the most common ores, without excluding certain rock-forming minerals (transparent) that stand out for their importance and frequency either as companions of metallic ores or for their interest to industry. The criteria for mineral selection in the text (§ 1.2, vol. 1) have been taken into account. Given the practical and necessarily restrictive character of these criteria, we cannot aim to include all possible species: the most common are selected, not the scarcer or highly specific ones from peculiar deposit types, for instance many rare and as yet not well-known mineral species of the groups PGE (Platinum Group Elements) or REE (Rare Earth Elements).

As there is no unanimously accepted and complete international proposal containing at least all those considered as common ores, the list has been drawn up according to our own criteria.[1] In the first place, the classical or most widespread proposals (especially for **metallic ores**) in international circles have been consulted and, as far as reasonable, taken into account. *Vbgr.* Chace 1956; Kretz 1983; Fontboté 2006; Whitney and Evans 2010; Neumann 2019; or those applied unilaterally by journals, texts or institutions specialized in the field of mineral resources (such as that published just for names "sanctioned by the International Mineralogical Association" by *The Canadian Mineralogist*, cf. https://canminportal.files.wordpress.com/2018/03/symbols2.pdf , visit. 2020/12/31). These proposals, generally disparate from each other, do not entirely satisfy the criteria appropriate to our needs, which are summarized below. Nevertheless, Chace's approaches (1956) have proved a particularly useful reference, and many of the abbreviations that he proposes for minerals have been incorporated into the list.

These are the criteria followed to elaborate the list of abbreviations used in this text (cf. Tables of Selected Minerals and Abbreviations, **Annexes 1.1, 1.2** and **1.3 a**, **m**):

1. As far as possible, advantage should be taken of pre-existing contributions and criteria whose rationality is supported by their use in the international literature.
3. Any confusion should be avoided, so that the abbreviation-mineral correspondence is, as far as possible, bi-univocal.
4. In order to speed up its use and to maintain reference to the international literature, the direct correspondence between the mineral symbol and the names of the mineral in English and Spanish is established by means of a simple database. This becomes a table with three columns: Abbreviation, Name (EN) and Nombre (ES), thus offering instant information from any of the three.
5. Any coincidence of an abbreviation or symbol of a mineral with a symbol for a chemical element of the Periodic System should be avoided (unless the mineral species is a native element, in which case it is obligatory to use the same symbol).
6. This confusion is ideally avoided by introducing a differential criterion, such as the requirement to write the abbreviation of the mineral species in lower case (unless it is an element); moreover, it is convenient that there is no literal coincidence. For example, Ag, Au or S are valid for native silver, gold or sulfur, respectively; yet *fe* for ferberite, *mo* for molybdenite, *ba* for barite or *cr* for chromite are **not** valid.

[1] After finishing the book, a new proposal (Warr 2021) was published, which unfortunately could not be considered here for obvious scheduling reasons.

R. Castroviejo, *A Practical Guide to Ore Microscopy—Volume 2*,
https://doi.org/10.1007/978-3-031-18954-8

7. Exceptionally, the coincidence of mineral abbreviations sanctioned by a tradition of extended use in the mineralogical community with symbols of elements is admitted when nature excludes any real risk of confusion. Among others, this happens with minerals such as *po* (pyrrhotite), *ac* (acanthite), *mc* (marcasite) or *mt* (magnetite), which no minimally informed reader could confuse with the elements *Po* (Polonium), *Ac* (Actinium), *Mc* (Moscovium) or *Mt* (Meitnerium), respectively. These are radioactive and extremely rare (ppb) in nature, have no known mineralogical expression and are only found as traces in radioactive ores (uraninite), are the product of the decay of a radioactive series or are known only as a synthetic product in the laboratory—some of them for only a few decades.
8. To distinguish mineral species from groups, the former use only lower-case letters and the latter have a capital initial letter. For example: ab = albite, but Plag = plagioclase (group); hb = hornblende, but Anf = amphibole (group); likewise, Pt = Platinum, PGM = *Platinum Group Minerals*, and so on.
9. When a mineral has several names, the preference is for defining the abbreviation to the most commonly used term (usually in English) and accepted by IMA/COM (International Mineralogical Association, Commission on Ore Mineralogy) to facilitate as far as possible equivalence with international usage.
10. As long as they are compatible with the other criteria, the usual pre-existing usages must be respected, for instance py = pyrite, po = pyrrhotite. Exceptionally, the possibility of more than one symbol for the same species is allowed when their use is very widespread (for example: q or qz for quartz); then the recommended name is given as the first option and, as the second, and in brackets, the other.

In compiling this Table of Abbreviations we have sought the greatest possible consensus and maximum simplicity, avoiding any proliferation of superfluous symbols that would generate confusion. Discussions and exchanges with colleagues and students over the years have been a welcome contribution to the development of this table, and the input of Dr. César Cánepa (Lima, Peru) deserves special recognition. If the result meets our aims, these colleagues share any merit; if not, the errors are mine alone.

For ease of use, the table is presented in two versions, depending on the column chosen as an entry (in alphabetical order): **Index of Minerals** (labelled **m**); or **Index of Abbreviations** (labelled **a**). Each can be consulted through three search options: **all minerals** (**Annex 1.1**); only **ores** (**Annex 1.2**); or only **gangue** (**Annex 1.3**).

Finally, the Meanings of the Acronyms, Abbreviations and Symbols used are summarized in Annex 1.4 MAAS Glossary, to make the explanations in the text and tables easier.

Selected Minerals and Use of the Table of Abbreviations The species chosen (*common minerals*, collected in the Abbreviations table) have been selected by applying the criteria explained in the text (§ 1.2, vol. 1), taking into account the interest in the ore, according to its abundance or its importance to industry (§ 1.2.1, vol. 1). As for the gangue (silicates, carbonates, oxides, etc.), the species chosen are those most frequently found in common ore bodies (§ 1.2.2 and 133.2, vol. 1). They result in about 300 entries, of which ≈ 230 correspond to ores and ≈80 to gangue.

Since their significance can vary widely, the entries have been further sorted and grouped by importance. The main ores are marked + in the Remarks column of the table, meaning that they are described fully. They are further classified in the *Priority* column into a key group (Priority 1: fundamental ores) and a collateral group (Priority 2: important, but subordinate). This is intended to guide the beginner in following the text, to ensure orderly progress and effective learning. This will always rely on the practitioner's personal observations under the microscope.

- The ≈ 50 ores whose learning is considered essential for the beginner or junior practitioner to get started are identified as Priority 1.
- For a more advanced level, another ≈ 60 ores follow (Priority 2), with which it is recommended to become familiar.
- Finally, the remaining species described are marked+, and are also worthy of attention. They might not be in the priority groups because they are less frequent or seldom cited in the literature (actually, some are sometimes confused or uncited simply because they are unknown to the observer: their supposed scarcity in nature seems an information gap).

Apart from the microscopic descriptions, most of the ores in the table, due to their importance or industrial interest, have been investigated for their automated identification by artificial vision (CAMEVA and AMCO systems: cf. Part II, vol. 2). Therefore, their VNIR specular reflectance spectra (Visible and Near Infrared ranges) have been measured and incorporated into the AMCO database created at the LMA_UPM. These data are also available in this text as Tables and R-λ plots to support the description of each ore.

Annex 1.1.a All Described Minerals' Symbols or Abbreviations and Comments

Abbreviation	Name (EN)	Nombre (ES)	Observations
aag	Allargentum	Allargentum	+
ab	Albite	Albita	gg
ac	Acanthite	Acantita	+
adu	Adularia	Adularia	gg
Ag	Silver	Plata	+
agl	Aguilarite	Aguilarita	+
agp	Argentopyrite	Argentopirita	+
aik	Aikinite	Aikinita	cf. bm
alb	Alabandite	Alabandita	+
alc	Alloclasite	Aloclasita	cf. cob
alt	Altaite	Altaíta	
am (amal)	Amalgam	Amalgama	+
amb	Amblygonite	Ambligonita	cf. Phos
an	Anorthite	Anortita	gg
ana	Anatase	Anatasa	+
anbg	Annabergite	Annabergita	gg
Anf, Amp	Amphibole (Group)	Anfíbol (Grupo)	gg
ang	Anglesite	Anglesita	gg
anh	Anhydrite	Anhidrita	gg
anil	Anilite	Anilita	cf. cc, dg
ank	Ankerite	Ankerita	gg
ap	Apatite	Apatito	gg
Arc	Clay	Arcilla	gg
arg	Argentite	Argentita	+
As	Arsenic	Arsénico	+
asb	Asbolane	Asbolana	
asp	Arsenopyrite	Arsenopirita	+
Au	Gold	Oro	+
ausi	*Aurian silver Ag–Au*	*Plata áurea Ag–Au*	cf. *el*
az	Azurite	Azurita	+
bad	Baddeleyite	Baddeleyita	+
bar	Barite	Barita	gg
baux	Bauxite	Bauxita	
Bi	Bismuth	Bismuto	+
bim	Bismite	Bismita	gg
bio	Biotite	Biotita	gg
bis	Bismutite	Bismutita	gg
bk	Brookite	Brookita	gg
bl	Boulangerite	Boulangerita	+
bm	Bismuthinite	Bismutinita	+
bn	Bornite	Bornita	+
bnn	Bournonite	Bournonita	+
bra (br)	Braunite	Braunita	+
brg	Braggite	Braggita	cf. erl
bri	Briartite	Briartita	cf. ren
brt	Berthierite	Berthierita	+
brv	Bravoite	Bravoíta	+
bst (bas)	Bastnäsite	Bastnäsita	+

(continued)

(continued)

Abbreviation	Name (EN)	Nombre (ES)	Observations
bth	Breithauptite	Breithauptita	+
buk	Bukovite	Bukovita	
bzl	Berzelianite	Berzelianita	cf. dg
cag	Chlorargyrite	Clorargirita	+
calc	Calcite	Calcita	gg
car	Carrollite	Carrolita	cf. ln
Carb	Carbonate (Class)	Carbonato (Clase)	
cb	Columbite	Columbita	cf. cbt
cbn	Cubanite	Cubanita	+
cbt	Columbotantalite	Columbotantalita	+
cc	Chalcocite	Calcocita = Calcosina	+
ccp	Chalcopyrite	Calcopirita	+
cerv	Cervantite	Cervantita	cf. ocres Sb
ch	Chert	Chert	gg
Chl	Chlorite (Series)	Clorita (Serie)	gg
chr	Chromite	Cromita	+
cin	Cinnabar	Cinabrio	+
cld	Chalcedony	Calcedonia	cf. q
cloa	Chloanthite (disc. IMA)	Cloantita	cf. sk
clt (cla)	Clausthalite	Clausthalita	+
clv	Calaverite	Calaverita	+
CM, MArc	Clay minerals (Group)	Minerales Arcillosos (Grupo)	gg
cob	Cobaltite	Cobaltita	+
coh	Cohenite	Cohenita	cf. Fe
colo (cld)	Coloradoite	Coloradoíta	+
COM	Coal—organic matter	Carbón-Materia Orgánica	cf. gg & gra
cor	Corundum	Corindón	gg
cos	Cosalite	Cosalita	+
cpr	Cooperite	Cooperita	cf. erl
crc	Chrysocolla	Crisocola	+
crg	Cerargyrite	Cerargirita	cf. cag
crn	Coronadite	Coronadita	cf. rom, psi
crs	Cerussite	Cerusita	cf. gn
cry	Cryptomelane	Criptomelana	+
cst (cs)	Cassiterite	Casiterita	+
cta	Calciotantite	Calciotantita	cf. mlt, pyc
ctt	Cattierite	Cattierita	cf. brv
Cu	Copper	Cobre	+
cup	Cuprite	Cuprita	+
cv	Covellite	Covellita	+
cyl	Cylindrite	Cilindrita	cf. F 41.04
dg	Digenite	Digenita	+
dju (dj)	Djurleite	Djurleíta	cf. cc, dg
dol	Dolomite	Dolomita	gg
dom	Domeykite	Domeiquita	
dy	Dyscrasite	Discrasita	+
el	Electrum	Electrum	+
en	Enargite	Enargita	+
Ep	Epidote (Group)	Epidota (Grupo)	gg
erl	Erlichmanite	Erlichmanita	+

(continued)

(continued)

Abbreviation	Name (EN)	Nombre (ES)	Observations
ery	Erythrite	Eritrina	gg
eug	Eugenite	Eugenita	cf. am, aag
eux	Euxenite	Euxenita	cf. REE
fb	Ferberite	Ferberita	cf. wf
fbg	Freibergite	Freibergita	cf. td
Fe	Iron	Hierro	+
fkl (fk)	Franklinite	Franklinita	+
fl	Fluorite	Fluorita	gg
flb	Freieslebenite	Freieslebenita	+
fle	Fletcherite	Fletcherita	cf. ln
flk	Falkmanite	Falkmanita	cf. bl
fm	Famatinite	Famatinita	+
Fs, Felds	Feldspar (Group)	Feldespato (Grupo)	gg
fse	Ferroselite	Ferroselita	cf. clt, um
FSlc, PSlc	Phyllosilicates (Subclass)	Filosilicatos (Subclase)	gg
gal	Gallite	Galita	
gc	Geocronite	Geocronita	+
ger	Germanite	Germanita	cf. ren
gf	Gersdorffite	Gersdorffita	+
gg	Gangue (undifferentiated)	Ganga (indiferenciada)	gg
gk	Greenockite	Greenockita	gg
gld	Glaucodot	Glaucodoto	+
gln	Glauconite	Glauconita, Glauconia	gg
gn	Galena	Galena	+
gnb	Galenobismuthite	Galenobismutina	
gos	Gossan	Gossan	cf. lim
gra	Graphite	Grafito	+
grg	Greigite	Greigita	cf. ln
grt	Garnet (Group)	Granate (Grupo)	gg
gth	Goethite	Goethita	+
gtt	Gratonite	Gratonita	+
gud (gd)	Gudmundite	Gudmundita	+
hb	Hornblende	Hornblenda	gg
Hg	Mercury (Quicksilver)	Mercurio	cf. cin
hol	Hollandite	Hollandita	cf. psi
hrb	Horobetsuite	Horobetsuita	cf. bm
hs	Hausmannite	Hausmanita	
hss	Hessite	Hessita	
ht	Hematite	Hematites	+
htg	Heterogenite	Heterogenita	
hub	Hübnerite	Hübnerita	cf. wf
il	Ilmenite	Ilmenita	+
ili	Illite	Illita	gg
ilr	Ilmenorutile	Ilmenorutilo	+
Ir	Iridium	Iridio	+
iros	Iridosmine (discredited)	Osmiridio (desacreditado)	cf. Os, Ir
isc	Isocubanite	Isocubanita	+
ix	Ixiolite	Ixiolita	cf. rt (F13-14)
jac	Jacobsite	Jacobsita	+
jar	Jarosite	Jarosita	cf. lim

(continued)

(continued)

Abbreviation	Name (EN)	Nombre (ES)	Observations
jas	Jasper	Jaspe	gg
jd	Jordanite	Jordanita	+
jl	Jalpaite	Jalpaíta	cf. Sulf
jm	Jamesonite	Jamesonita	+
ker, (km)	Kermesite	Kermesita	+
Kfs	K Feldspar	Feldespato Potásico	gg
ko	Kobellite	Kobellita	
krn	Krennerite	Krennerita	+
ks	Kësterite	Kësterita	cf. stn
kut	Kutinaite	Kutinaita	
lcx	Leucoxeno	Leucoxeno	cf. tit, ana (gg)
ldw	Ludwigite	Ludwigita	cf. von
lgs	Langisite	Langisita	
lim, lm	Limonite	Limonita	+
ln	Linnaeite	Linneíta	+
lo	Löllingite	Löllingita	+
lor	Lorandite	Lorandita	
lpc	Lepidocrocite	Lepidocrocita	+
lph	Lithiophorite	Litioforita	
lrt	Laurite	Laurita	cf. erl
lz	Luzonite	Luzonita	+
M, min, mnl	Mineral (undifferentiated)	Mineral (indiferenciado)	
m1, m2…	M not identified	M no identificado	
mag	Magnesite	Magnesita	cf. Carb, calc
mau	Maucherite	Maucherita	+
mc	Marcasite	Marcasita	+
mch	Malachite	Malaquita	+
mcin	Metacinnabar	Metacinabrio	cf. cin
mck	Mackinawite	Mackinawita	+
mgh	Maghemite	Maghemita	+
Mica	Mica (Group)	Mica (Grupo)	gg
ml	Millerite	Millerita	+
mln	Melnicovite	Melnicovita	cf. py
mlt	Microlite	Microlita	cf. pyc
mng	Manganite	Manganita	+
mngn	Manganosite	Manganosite	
mobg	Moschellandsbergite	Moschellandsbergita	cf. am
mol	Molybdenite	Molibdenita	+
msbt	Metastibnite	Metastibnita	cf. sbt
mt, (mg)	Magnetite	Magnetita	+
mv, mus	Muscovite	Moscovita	gg
my	Miargyrite	Miargirita	+
mz	Monazite	Monazita	+
nau	Naumannite	Naumannita	+
nc, (nk)	Nickeline, Niccolite	Niquelita, Niquelina	+
Oanf, Oamp	Orthoamphibole (Group)	Ortoanfíbol (Grupo)	gg
ocm	Oxycalciomicrolite	Oxicalciomicrolita	cf. mlt, bim
Ocre	Ochre, Ocher	Ocres	cf. Óxidos
OM	Organic matter	Materia Orgánica	gg
OpM	Opaque minerals	Minerales Opacos	

(continued)

(continued)

Abbreviation	Name (EN)	Nombre (ES)	Observations
Opx	Orthopyroxene (Group)	Ortopiroxeno (Grupo)	gg
or	Orthoclase	Ortoclasa, ortosa	gg
orp, op	Orpiment	Oropimente	+
Os	Osmium	Osmio	+
ov	Olivin (Series)	Olivino (Serie)	gg
pc	Pearceite	Pearceíta	+
pch	Pitchblende	Pechblenda	cf. urn
Pd	Palladium	Paladio	cf. PGE
pdm	Polydymite	Polidimita	cf. ln
PGE	Platinum Group Elements	Elementos del Grupo del Platino	cf. Pt, Os, Ir
PGM	Platinum Group Minerals	Minerales del Grupo del Platino	cf. spy
Phos	Phosphate	Fosfato	gg
pie	Piemontite	Piemontita	cf. bra
Plag	Plagioclase (Series)	Plagioclasa (Grupo)	gg
plb	Polybasite	Polibasita	+
pn	Pentlandite	Pentlandita	+
po	Pyrrhotite	Pirrotita (Pirrotina)	+
poli	Polianite	Polianita	cf. prl
prg	Pyrargyrite	Pirargirita	+
prl	Pyrolusite	Pirolusita	+
prm	Pararammelsbergite	Pararammelsbergita	+
pru	Proustite	Proustita	+
psi	Psilomelane	Psilomelana	+
Pt	Platinum	Platino	+
pvk	Perovskite	Perovskita	gg
Px, Pyx	Pyroxene (Group)	Piroxeno (Grupo)	gg
py	Pyrite	Pirita	+
pyc (pyl)	Pyrochlore	Pirocloro	+
pymp	Pyromorphite	Piromorfita	gg
pyst	Pyrostilpnite	Pirostilpnita	cf. prg
q, (qz, qtz)	Quartz	Cuarzo	gg
rdl	Ramsdellite	Ramsdellita	+
REE, (ETR)	Rare Earth Elements	Tierras Raras (Elementos de)	
ren	Renierite	Renierita	+
rhc	Rhodochrosite	Rodocrosita	gg
rhn	Rhodonite	Rodonita	gg
rl	Realgar	Rejalgar	+
rm	Rammelsbergite	Rammelsbergita	+
rob	Robinsonite	Robinsonita	
rom	Romanechite	Romanechita	+
rt	Rutile	Rutilo	+
ruiro	Rutheniridosmine	Ruteniridosmina	cf. Os, Ir
ryn	Rynersonite	Rynersonita	cf. mlt, pyc
S	Sulfur	Azufre	gg
Sb	Antimony	Antimonio	+
sba	Stibarsen	Stibarsen	+
sbl	Stibioluzonite	Estibioluzonita	cf. fm
sbt	Stibnite	Estibina, Antimonita	+
sch	Scheelite	Scheelita	+
sco	Scorodite	Escorodita	gg (cf. asp)
Sel	Selenides	Seleniuros	cf. nau, um
sen	Senarmontite	Senarmontita	cf. ocres Sb
ser	Sericite	Sericita	gg

(continued)

(continued)

Abbreviation	Name (EN)	Nombre (ES)	Observations
sf	Safflorite	Safflorita	+
sg	Siegenite	Siegenita	cf. ln
Sic	Silica	Sílice	cf. q
sid, (sd)	Siderite	Siderita	cf. Carb
sk	Skutterudite	Skutterudita	+
sl	sensu lato	sensu lato	
Slc	Silicate (Class)	Silicato (Clase)	gg
smal	Smaltite (≈sk)	Esmaltina	cf. sk
sms	Smithsonite	Smithsonita	gg
smtt	Smithite	Smithita	cf. Ss
snr	Schachnerite	Schachnerita	cf. am, F stm8
sp	Sphalerite	Esfalerita	+
spk	Spionkopite (Blaubl. cv)	Spionkopita	cf. cv &F dg1-2
spn	Spinel	Espinela	+
spy	Sperrylite	Sperrylita	+
srp	Serpentine	Serpentina	gg
Ss	Sulfosalt	Sulfosal	
stb	Sternbergite	Sternbergita	+
stm	Stromeyerite	Stromeyerita	+
stn	Stannite	Estannita	+
stnd	Stannoidite	Estannoidita	
stp	Stephanite	Stephanita	+
Sulf	Sulfide	Sulfuro	
sx	Souxite	Souxita	cf. cst
syl (sv)	Sylvanite	Silvanita	cf. Tel
tan (tt)	Tantalite	Tantalita	cf. cbt
td	Tetrahedrite	Tetraedrita (Tetrahedrita)	+
tdy	Tedradymite	Tetradimita	cf. Tel
Te	Tellurium	Teluro	+
Tel (Tell)	Telluride (Group)	Telururo (Grupo)	
tit (sph)	Titanite	Titanita	+
tlk	Talnakhita	Talnakhita	cf. ccp
tn	Tennantite	Tennantita	cf. td
tno	Tenorite	Tenorita	
tod	Todorokite	Todorokita	
tp (tap)	Tapiolite	Tapiolita	+
tr (tro)	Troilite	Troilita	cf. po, Fe
tur	Tourmaline	Turmalina	gg
ul	Ullmannite	Ullmannita	+
ulv	Ulvöspinel (Ulvite)	Ulvöespinela (Ulvita)	+
um	Umangite	Umangita	+
urn	Uraninite	Uraninita	+
vae	Vaesite	Vaesita	cf. ctt
val	Valentinite	Valentinita	cf. ocres Sb
van	Vanadinite	Vanadinita	+
van	Vanadinite	Vanadinita	gg
vio	Violarite	Violarita	cf. ln
vll	Valleriite	Valleriita	+
von	Vonsenite	Vonsenita	gg
wf	Wolframite	Wolframita	+
wil	Willemite	Willemita	cf. Slc
wz	Wurtzite	Wurtzita	+
xan	Xanthoconite	Xantoconita	cf. pru

(continued)

(continued)

Abbreviation	Name (EN)	Nombre (ES)	Observations
xtm	Xenotime	Xenotimo	cf. Phos
Zeo	Zeolite (Group)	Zeolita (Grupo)	gg
zir	Zircon	Circón (Zircón)	gg
zk	Zinkenite	Zinkenita	cf. Ss
znc	Zincite	Zincita	gg (cf. fkl)
	Other names	Otros nombres	Observations
	Allemontite	Allemontita	cf. sba
	Antimonite	Antimonita	cf. sbt
	Blende	Blenda	cf. sp
	Chalcopyrrhotite	Calcopirrotita	cf. isc
	Hollandite	Holandita	cf. rom, psi
	Martite	Martita	cf. ht, mt
	Mispickel	Mispíquel	cf. asp
	Polianite	Polianita	cf. prl
	Schalenblende	–	cf. sp
	Sphene (see: Titanite)	Esfena (ver: Titanita)	cf. tit
	Steel	Acero	cf. Fe
	Strüverite	Struverita	cf. rt
	Ulvite	Ulvita	cf. ulv

Annex 1.1.m All Described Minerals and their Symbols or Abbreviations and Comments

Name (EN)	Abbreviation	Nombre (ES)	Observations
Acanthite	ac	Acantita	+
Adularia	adu	Adularia	gg
Aguilarite	agl	Aguilarita	+
Aikinite	aik	Aikinita	cf. bm
Alabandite	alb	Alabandita	+
Albite	ab	Albita	gg
Allargentum	aag	Allargentum	+
Alloclasite	alc	Aloclasita	cf. cob
Altaite	alt	Altaíta	
Amalgam	am (amal)	Amalgama	+
Amblygonite	amb	Ambligonita	cf. Phos
Amphibole (Group)	Anf, Amp	Anfíbol (Grupo)	gg
Anatase	ana	Anatasa	+
Anglesite	ang	Anglesita	gg
Anhydrite	anh	Anhidrita	gg
Anilite	anil	Anilita	cf. cc, dg
Ankerite	ank	Ankerita	gg
Annabergite	anbg	Annabergita	gg
Anorthite	an	Anortita	gg
Antimony	Sb	Antimonio	+
Apatite	ap	Apatito	gg
Argentite	arg	Argentita	+
Argentopyrite	agp	Argentopirita	+
Arsenic	As	Arsénico	+
Arsenopyrite	asp	Arsenopirita	+
Asbolane	asb	Asbolana	
Aurian silver Ag–Au	*ausi*	*Plata áurea Ag–Au*	cf. *el*

(continued)

(continued)

Name (EN)	Abbreviation	Nombre (ES)	Observations
Azurite	az	Azurita	+
Baddeleyite	bad	Baddeleyita	+
Barite	bar	Barita	gg
Bastnäsite	bst, (bas)	Bastnäsita	+
Bauxite	baux	Bauxita	
Berthierite	brt	Berthierita	+
Berzelianite	bzl	Berzelianita	cf. dg
Biotite	bio	Biotita	gg
Bismite	bim	Bismita	gg
Bismuth	Bi	Bismuto	+
Bismuthinite	bm	Bismutinita	+
Bismutite	bis	Bismutita	gg
Bornite	bn	Bornita	+
Boulangerite	bl	Boulangerita	+
Bournonite	bnn	Bournonita	+
Braggite	brg	Braggita	cf. erl
Braunite	bra (br)	Braunita	+
Bravoite	brv	Bravoíta	+
Breithauptite	bth	Breithauptita	+
Briartite	bri	Briartita	cf. ren
Brookite	bk	Brookita	gg
Bukovite	buk	Bukovita	
Calaverite	clv	Calaverita	+
Calciotantite	cta	Calciotantita	cf. mlt, pyc
Calcite	calc	Calcita	gg
Carbonate (Class)	Carb	Carbonato (Clase)	
Carrollite	car	Carrolita	cf. ln
Cassiterite	cst (cs)	Casiterita	+
Cattierite	ctt	Cattierita	cf. brv
Cerargyrite	crg	Cerargirita	cf. cag
Cerussite	crs	Cerusita	cf. gn
Cervantite	cerv	Cervantita	cf. ocres Sb
Chalcedony	cld	Calcedonia	cf. q
Chalcocite	cc	Calcocita = Calcosina	+
Chalcopyrite	ccp	Calcopirita	+
Chert	ch	Chert	gg
Chloanthite (disc. IMA)	cloa	Cloantita	cf. sk
Chlorargyrite	cag	Clorargirita	+
Chlorite (Series)	Chl	Clorita (Serie)	gg
Chromite	chr	Cromita	+
Chrysocolla	crc	Crisocola	+
Cinnabar	cin	Cinabrio	+
Clausthalite	clt (cla)	Clausthalita	+
Clay	Arc	Arcilla	gg
Clay minerals (Group)	CM, MArc	Minerales Arcillosos (Grupo)	gg
Coal—organic matter	COM	Carbón-Materia Orgánica	cf. gg & gra
Cobaltite	cob	Cobaltita	+
Cohenite	coh	Cohenita	cf. Fe
Coloradoite	colo (cld)	Coloradoíta	+
Columbite	cb	Columbita	cf. cbt
Columbotantalite	cbt	Columbotantalita	+

(continued)

(continued)

Name (EN)	Abbreviation	Nombre (ES)	Observations
Cooperite	cpr	Cooperita	cf. erl
Copper	Cu	Cobre	+
Coronadite	crn	Coronadita	cf. rom, psi
Corundum	cor	Corindón	gg
Cosalite	cos	Cosalita	+
Covellite	cv	Covellita	+
Cryptomelane	cry	Criptomelana	+
Cubanite	cbn	Cubanita	+
Cuprite	cup	Cuprita	+
Cylindrite	cyl	Cilindrita	cf. F 41.04
Digenite	dg	Digenita	+
Djurleite	dju (dj)	Djurleíta	cf. cc, dg
Dolomite	dol	Dolomita	gg
Domeykite	dom	Domeiquita	
Dyscrasite	dy	Discrasita	+
Electrum	el	Electrum	+
Enargite	en	Enargita	+
Epidote (Group)	Ep	Epidota (Grupo)	gg
Erlichmanite	erl	Erlichmanita	+
Erythrite	ery	Eritrina	gg
Eugenite	eug	Eugenita	cf. am, aag
Euxenite	eux	Euxenita	cf. REE
Falkmanite	flk	Falkmanita	cf. bl
Famatinite	fm	Famatinita	+
Feldspar (Group)	Fs, Felds	Feldespato (Grupo)	gg
Ferberite	fb	Ferberita	cf. wf
Ferroselite	fse	Ferroselita	cf. clt, um
Fletcherite	fle	Fletcherita	cf. ln
Fluorite	fl	Fluorita	gg
Franklinite	fkl (fk)	Franklinita	+
Freibergite	fbg	Freibergita	cf. td
Freieslebenite	flb	Freieslebenita	+
Galena	gn	Galena	+
Galenobismuthite	gnb	Galenobismutina	
Gallite	gal	Galita	
Gangue (undif.)	gg	Ganga (indiferenciada)	gg
Garnet (Group)	grt	Granate (Grupo)	gg
Geocronite	gc	Geocronita	+
Germanite	ger	Germanita	cf. ren
Gersdorffite	gf	Gersdorffita	+
Glaucodot	gld	Glaucodoto	+
Glauconite	gln	Glauconita, Glauconia	gg
Goethite	gth	Goethita	+
Gold	Au	Oro	+
Gossan	gos	Gossan	cf. lim
Graphite	gra	Grafito	+
Gratonite	gtt	Gratonita	+
Greenockite	gk	Greenockita	gg
Greigite	grg	Greigita	cf. ln
Gudmundite	gud (gd)	Gudmundita	+
Hausmannite	hs	Hausmanita	

(continued)

(continued)

Name (EN)	Abbreviation	Nombre (ES)	Observations
Hematite	ht	Hematites	+
Hessite	hss	Hessita	
Heterogenite	htg	Heterogenita	
Hollandite	hol	Hollandita	cf. psi
Hornblende	hb	Hornblenda	gg
Horobetsuite	hrb	Horobetsuita	cf. bm
Hübnerite	hub	Hübnerita	cf. wf
Illite	ili	Illita	gg
Ilmenite	Il	Ilmenita	+
Ilmenorutile	ilr	Ilmenorutilo	+
Iridium	Ir	Iridio	+
Iridosmine (discredited)	iros	Osmiridio (desacreditado)	cf. Os, Ir
Iron	Fe	Hierro	+
Isocubanite	isc	Isocubanita	+
Ixiolite	ix	Ixiolita	cf. rt (F13-14)
Jacobsite	jac	Jacobsita	+
Jalpaite	Jl	Jalpaíta	cf. Sulf
Jamesonite	jm	Jamesonita	+
Jarosite	jar	Jarosita	cf. lim
Jasper	jas	Jaspe	gg
Jordanite	jd	Jordanita	+
K Feldspar	Kfs	Feldespato Potásico	gg
Kermesite	ker, (km)	Kermesita	+
Kësterite	ks	Kësterita	cf. stn
Kobellite	ko	Kobellita	
Krennerite	krn	Krennerita	+
Kutinaite	kut	Kutinaita	
Langisite	lgs	Langisita	
Laurite	lrt	Laurita	cf. erl
Lepidocrocite	lpc	Lepidocrocita	+
Leucoxeno	lcx	Leucoxeno	cf. ana (gg)
Limonite	lim, lm	Limonita	+
Linnaeite	ln	Linneíta	+
Lithiophorite	lph	Litioforita	
Löllingite	lo	Löllingita	+
Lorandite	lor	Lorandita	
Ludwigite	ldw	Ludwigita	cf. von
Luzonite	lz	Luzonita	+
M not identified	m1, m2	M no identificado	
Mackinawite	mck	Mackinawita	+
Maghemite	mgh	Maghemita	+
Magnesite	mag	Magnesita	cf. Carb, calc
Magnetite	mt (mg)	Magnetita	+
Malachite	mch	Malaquita	+
Manganite	mng	Manganita	+
Manganosite	mngn	Manganosite	
Marcasite	mc	Marcasita	+
Maucherite	mau	Maucherita	+
Melnicovite	mln	Melnicovita	cf. py
Mercury (Quicksilver)	Hg	Mercurio	cf. cin
Metacinnabar	mcin	Metacinabrio	cf. cin

(continued)

(continued)

Name (EN)	Abbreviation	Nombre (ES)	Observations
Metastibnite	msbt	Metastibnita	cf. sbt
Miargyrite	my	Miargirita	+
Mica (Group)	Mica	Mica (Grupo)	gg
Microlite	mlt	Microlita	cf. pyc
Millerite	ml	Millerita	+
Mineral (undif.)	M, min, mnl	Mineral (indiferenciado)	
Molybdenite	mol	Molibdenita	+
Monazite	mz	Monazita	+
Moschellandsbergite	mobg	Moschellandsbergita	cf. am
Muscovite	mv, mus	Moscovita	gg
Naumannite	nau	Naumannita	+
Nickeline, Niccolite	nc, (nk)	Niquelita, Niquelina	+
Ochre, Ocher	Ocre	Ocres	cf. Óxidos
Olivin (Series)	ov	Olivino (Serie)	gg
Opaque minerals	OpM	Minerales Opacos	
Organic matter	OM	Materia Orgánica	gg
Orpiment	orp, op	Oropimente	+
Orthoamphibole (Group)	Oanf, Oamp	Ortoanfíbol (Grupo)	gg
Orthoclase	or	Ortoclasa, ortosa	gg
Orthopyroxene (Group)	Opx	Ortopiroxeno (Grupo)	gg
Osmium	Os	Osmio	+
Oxycalciomicrolite	ocm	Oxicalciomicrolita	cf. mlt, bim
Palladium	Pd	Paladio	cf. PGE
Pararammelsbergite	prm	Pararammelsbergita	+
Pearceite	pc	Pearceíta	+
Pentlandite	pn	Pentlandita	+
Perovskite	pvk	Perovskita	gg
Phosphate	Phos	Fosfato	gg
Phyllosilicates (Subclass)	FSlc, PSlc	Filosilicatos (Subclase)	gg
Piemontite	pie	Piemontita	cf. bra
Pitchblende	pch	Pechblenda	cf. urn
Plagioclase (Series)	Plag	Plagioclasa (Grupo)	gg
Platinum	Pt	Platino	+
Platinum Group Elements	PGE	Elem. Grupo del Platino	cf. Pt, Os, Ir
Platinum Group Minerals	PGM	Minerales Grupo del Platino	cf. spy
Polianite	poli	Polianita	cf. prl
Polybasite	plb	Polibasita	+
Polydymite	pdm	Polidimita	cf. ln
Proustite	pru	Proustita	+
Psilomelane	psi	Psilomelana	+
Pyrargyrite	prg	Pirargirita	+
Pyrite	py	Pirita	+
Pyrochlore	pyc (pyl)	Pirocloro	+
Pyrolusite	prl	Pirolusita	+
Pyromorphite	pymp	Piromorfita	gg
Pyrostilpnite	pyst	Pirostilpnita	cf. prg
Pyroxene (Group)	Px, Pyx	Piroxeno (Grupo)	gg
Pyrrhotite	po	Pirrotita (Pirrotina)	+
Quartz	q (qz, qtz)	Cuarzo	gg
Rammelsbergite	rm	Rammelsbergita	+
Ramsdellite	rdl	Ramsdellita	+

(continued)

(continued)

Name (EN)	Abbreviation	Nombre (ES)	Observations
Rare Earth Elements	REE, (ETR)	Tierras Raras (Elementos de)	
Realgar	rl	Rejalgar	+
Renierite	ren	Renierita	+
Rhodochrosite	rhc	Rodocrosita	gg
Rhodonite	rhn	Rodonita	gg
Robinsonite	rob	Robinsonita	
Romanechite	rom	Romanechita	+
Rutheniridosmine	ruiro	Ruteniridosmina	cf. Os, Ir
Rutile	rt	Rutilo	+
Rynersonite	ryn	Rynersonita	cf. mlt, pyc
Safflorite	sf	Safflorita	+
Schachnerite	snr	Schachnerita	cf. am, F stm8
Scheelite	sch	Scheelita	+
Scorodite	sco	Escorodita	gg (cf. asp)
Selenides	Sel	Seleniuros	cf. nau, um
Senarmontite	sen	Senarmontita	cf. ocres Sb
sensu lato	*sl*	sensu lato	
Sericite	ser	Sericita	gg
Serpentine	srp	Serpentina	gg
Siderite	sid, (sd)	Siderita	cf. Carb
Siegenite	sg	Siegenita	cf. ln
Silica	Sic	Sílice	cf. q
Silicate (Class)	Slc	Silicato (Clase)	gg
Silver	Ag	Plata	+
Skutterudite	sk	Skutterudita	+
Smaltite (≈sk)	smal	Esmaltina	cf. sk
Smithite	smtt	Smithita	cf. Ss
Smithsonite	sms	Smithsonita	gg
Souxite	sx	Souxita	cf. cst
Sperrylite	spy	Sperrylita	+
Sphalerite	sp	Esfalerita	+
Spinel	spn	Espinela	+
Spionkopite (Blaubl. cv)	spk	Spionkopita	cf. cv &F dg1-2
Stannite	stn	Estannita	+
Stannoidite	stnd	Estannoidita	
Stephanite	stp	Stephanita	+
Sternbergite	stb	Sternbergita	+
Stibarsen	sba	Stibarsen	+
Stibioluzonite	sbl	Estibioluzonita	cf. fm
Stibnite	sbt	Estibina, Antimonita	+
Stromeyerite	stm	Stromeyerita	+
Sulfide	Sulf	Sulfuro	
Sulfur	S	Azufre	gg
Sulfosalt	Ss	Sulfosal	
Sylvanite	syl (sv)	Silvanita	cf. Tel
Talnakhita	tlk	Talnakhita	cf. ccp
Tantalite	tan (tt)	Tantalita	cf. cbt
Tapiolite	tp, (tap)	Tapiolita	+
Tedradymite	tdy	Tetradimita	cf. Tel
Telluride (Group)	Tel (Tell)	Telururo (Grupo)	
Tellurium	Te	Teluro	+

(continued)

(continued)

Name (EN)	Abbreviation	Nombre (ES)	Observations
Tennantite	tn	Tennantita	cf. td
Tenorite	tno	Tenorita	
Tetrahedrite	td	Tetraedrita (Tetrahedrita)	+
Titanite	tit (sph)	Titanita	+
Todorokite	tod	Todorokita	
Tourmaline	tur	Turmalina	gg
Troilite	tr (tro)	Troilita	cf. po, Fe
Ullmannite	ul	Ullmannita	+
Ulvöspinel (Ulvite)	ulv	Ulvöespinela (Ulvita)	+
Umangite	um	Umangita	+
Uraninite	urn	Uraninita	+
Vaesite	vae	Vaesita	cf. ctt
Valentinite	val	Valentinita	cf. ocres Sb
Valleriite	vll	Valleriíta	+
Vanadinite	van	Vanadinita	+
Vanadinite	van	Vanadinita	gg
Violarite	vio	Violarita	cf. ln
Vonsenite	von	Vonsenita	gg
Willemite	wil	Willemita	cf. Slc
Wolframite	wf	Wolframita	+
Wurtzite	wz	Wurtzita	+
Xanthoconite	xan	Xantoconita	cf. pru
Xenotime	xtm	Xenotimo	cf. Phos
Zeolite (Group)	Zeo	Zeolita (Grupo)	gg
Zincite	znc	Zincita	gg (cf. fkl)
Zinkenite	zk	Zinkenita	cf. Ss
Zircon	zir	Circón (Zircón)	gg
Other names		Otros nombres	Observations
Allemontite		Allemontita	cf. sba
Antimonite		Antimonita	cf. sbt
Blende		Blenda	cf. sp
Chalcopyrrhotite		Calcopirrotita	cf. isc
Hollandite		Holandita	cf. rom, psi
Martite		Martita	cf. ht, mt
Mispickel		Mispíquel	cf. asp
Polianite		Polianita	cf. prl
Schalenblende		–	cf. sp
Sphene (see: Titanite)		Esfena (ver: Titanita)	cf. tit
Steel		Acero	cf. Fe
Strüverite		Struverita	cf. rt
Ulvite		Ulvita	cf. ulv

Annex 1.2.a Ore Mineral Symbols or Abbreviations and Comments

Abbreviation	Name	Nombre	Observations	Priority
aag	*Allargentum*	*Allargentum*	+	*2*
ac	**Acanthite**	**Acantita**	+	**1**
Ag	**Silver**	**Plata**	+	**1**
agl	Aguilarite	Aguilarita	+	
agp	*Argentopyrite*	*Argentopirita*	+	*2*
aik	Aikinite	Aikinita	cf. bm	
alb	*Alabandite*	*Alabandita*	+	*2*
alc	Alloclasite	Aloclasita	cf. cob	
alt	Altaite	Altaíta		
am (amal)	*Amalgam*	*Amalgama*	+	*2*
ana	Anatase	Anatasa	+	
anil	Anilite	Anilita	cf. cc, dg	
arg	Argentite	Argentita	+	
As	*Arsenic*	*Arsénico*	+	*2*
asb	Asbolane	Asbolana		
asp	**Arsenopyrite**	**Arsenopirita**	+	**1**
Au	**Gold**	**Oro**	+	**1**
ausi	*Aurian silver Ag–Au*	*Plata áurea Ag–Au*	cf. *el*	*2*
az	*Azurite*	*Azurita*	+	*2*
bad	Baddeleyite	Baddeleyita	+	
baux	Bauxite	Bauxita	Mezcla	
Bi	*Bismuth*	*Bismuto*	+	*2*
bl	*Boulangerite*	*Boulangerita*	+	*2*
bm	*Bismuthinite*	*Bismutinita*	+	*2*
bn	**Bornite**	**Bornita**	+	**1**
bnn	Bournonite	Bournonita	+	
bra (br)	*Braunite*	*Braunita*	+	*2*
brg	Braggite	Braggita	cf. erl F 1, 3	
bri	Briartite	Briartita	cf. ren	
brt	*Berthierite*	*Berthierita*	+	*2*
brv	*Bravoite*	*Bravoíta*	+	*2*
bst (bas)	**Bastnäsite**	**Bastnäsita**	+	**1**
bth	*Breithauptite*	*Breithauptita*	+	*2*
buk	Bukovite	Bukovita		
bzl	Berzelianite	Berzelianita	cf. dg	
cag	*Chlorargyrite*	*Clorargirita*	+	*2*
car	Carrollite	Carrolita	cf. ln	
cb	Columbite	Columbita	cf. cbt	
cbn	*Cubanite*	*Cubanita*	+	*2*
cbt	**Columbotantalite**	**Columbotantalita**	+	**1**
cc	**Chalcocite**	**Calcocita = Calcosina**	+	**1**
ccp	**Chalcopyrite**	**Calcopirita**	+	**1**
chr	**Chromite**	**Cromita**	+	**1**
cin	**Cinnabar**	**Cinabrio**	+	**1**
cloa	Chloanthite (disc. IMA)	Cloantita	cf. sk	
clt (cla)	Clausthalite	Clausthalita	+	
clv	*Calaverite*	*Calaverita*	+	*2*
cob	**Cobaltite**	**Cobaltita**	+	**1**
coh (chn)	*Cohenite*	*Cohenita*	cf. *Fe*	*2*

(continued)

(continued)

Abbreviation	Name	Nombre	Observations	Priority
colo (cld)	Coloradoite	Coloradoíta	+	
COM	Coal—organic matter	Carbón-Materia Orgánica	cf. gra & gg	
cos	Cosalite	Cosalita	+	*2*
cpr (coo)	Cooperite	Cooperita	cf. erl F 1, 3	
crc	Chrysocolla	Crisocola	+	
crg	Cerargyrite	Cerargirita	cf. cag	
crn	Coronadite	Coronadita	cf. rom, psi	
cry	*Cryptomelane*	*Criptomelana*	+	*2*
cst (cs)	**Cassiterite**	**Casiterita**	+	**1**
cta	Calciotantite	Calciotantita	cf. mlt, pyc	
ctt	Cattierite	Cattierita	cf. brv	
Cu	**Copper**	**Cobre**	+	**1**
cup	**Cuprite**	**Cuprita**	+	**1**
cv	**Covellite**	**Covellita**	+	**1**
cyl	Cylindrite	Cilindrita	cf. F 41.04	
dg	**Digenite**	**Digenita**	+	**1**
dju (dj)	Djurleite	Djurleíta	cf. cc, dg	
dom	Domeykite	Domeiquita		
dy	*Dyscrasite*	*Discrasita*	+	*2*
el	*Electrum*	*Electrum*	+	*2*
en	*Enargite*	*Enargita*	+	*2*
erl	Erlichmanite	Erlichmanita	+	
eug	Eugenite	Eugenita	cf. am, aag	
eux	Euxenite	Euxenita	cf. REE	
fb	Ferberite	Ferberita	cf. wf	
fbg	Freibergite	Freibergita	cf. td	
Fe	**Iron**	**Hierro**	+	**1**
fk (fkl)	*Franklinite*	*Franklinita*	+	*2*
flb	Freieslebenite	Freieslebenita	+	
fle	Fletcherite	Fletcherita	cf. ln	
flk	Falkmanite	Falkmanita	cf. bl	
fm	*Famatinite*	*Famatinita*	+	*2*
fse	Ferroselite	Ferroselita	cf. clt, um	
gal	Gallite	Galita		
gc	Geocronite	Geocronita	+	
ger	Germanite	Germanita	cf. ren	
gf	*Gersdorffite*	*Gersdorffita*	+	*2*
gg	Gangue (undiff.)	Ganga (indiferenciada)		
gk	Greenockite	Greenockita	cf. Sulf	
gld	*Glaucodot*	*Glaucodoto*	+	*2*
gn	**Galena**	**Galena**	+	**1**
gnb	Galenobismuthite	Galenobismutina		
gos	Gossan	Gossan	cf. lim	
gra	**Graphite**	**Grafito**	+	**1**
grg	Greigite	Greigita	cf. ln	
gth	**Goethite**	**Goethita**	+	**1**
gtt	Gratonite	Gratonita	+	
gud (gd)	*Gudmundite*	*Gudmundita*	+	*2*
Hg	Mercury (Quicksilver)	Mercurio	cf. cin	
hol	Hollandite	Hollandita	cf. psi	
hrb	Horobetsuite	Horobetsuita	cf. bm	
hs	Hausmannite	Hausmanita		

(continued)

(continued)

Abbreviation	Name	Nombre	Observations	Priority
hss	Hessite	Hessita		
ht	**Hematite**	**Hematites**	+	**1**
htg	Heterogenite	Heterogenita		
hub	Hübnerite	Hübnerita	cf. wf	
il	**Ilmenite**	**Ilmenita**	+	**1**
ilr	*Ilmenorutile*	*Ilmenorutilo*	+	*2*
Ir	*Iridium*	*Iridio*	+	*2*
iros	Iridosmine (discredited)	Osmiridio (desacreditado)	cf. Os, Ir	
isc	*Isocubanite*	*Isocubanita*	+	*2*
ix	Ixiolite	Ixiolita	cf. rt (F13-14)	
jac	Jacobsite	Jacobsita	+	
jd	Jordanite	Jordanita	+	
jl	Jalpaite	Jalpaíta		
jm	Jamesonite	Jamesonita	+	2
ker (km)	Kermesite	Kermesita	+	
ko	Kobellite	Kobellita		
krn	Krennerite	Krennerita	+	
ks	Kësterite (kösterite)	Kësterita (kesterita)	cf. stn	
kut	Kutinaite	Kutinaita		
lgs	Langisite	Langisita	Fpc7-8, pru5-6	
lim, lm	**Limonite**	**Limonita**	+	**1**
ln	**Linnaeite**	**Linneíta**	+	**1**
lo	*Löllingite*	*Löllingita*	+	*2*
lor	Lorandite	Lorandita		
lpc	*Lepidocrocite*	*Lepidocrocita*	+	*2*
lph	Lithiophorite	Litioforita		
lrt (lau)	Laurite	Laurita	cf. erl	
lz	*Luzonite*	*Luzonita*	+	*2*
M, min, mnl	Mineral (undiff.)	Mineral (indiferenc.)		
m1, m2	M not identified	M no identificado		
mau	*Maucherite*	*Maucherita*	+	*2*
mc	**Marcasite**	**Marcasita**	+	**1**
mch	*Malachite*	*Malaquita*	+	*2*
mcin	Metacinnabar	Metacinabrio	cf. cin	
mck	*Mackinawite*	*Mackinawita*	+	*2*
mgh	*Maghemite*	*Maghemita*	+	*2*
ml	**Millerite**	**Millerita**	+	**1**
mln	Melnicovite	Melnicovita	cf. py	
mlt	Microlite	Microlita	cf. pyc	
mng	**Manganite**	**Manganita**	+	**1**
mngn	Manganosite	Manganosite		
mobg	Moschellandsbergite	Moschellandsbergita	cf. am	
mol	**Molybdenite**	**Molibdenita**	+	**1**
msbt	Metastibnite	Metaestibina	cf. sbt	
mt (mg)	**Magnetite**	**Magnetita**	+	**1**
my	**Miargyrite**	**Miargirita**	+	**1**
mz	*Monazite*	*Monazita*	+	*2*
nau	Naumannite	Naumannita	+	
nc (nk)	**Nickeline, Niccolite**	**Niquelita, Niquelina**	+	**1**
ocm	Oxycalciomicrolite	Oxicalciomicrolita	cf. mlt, bim	

(continued)

(continued)

Abbreviation	Name	Nombre	Observations	Priority
OpM	Opaque Minerals	Minerales Opacos		
orp, op	*Orpiment*	*Oropimente*	+	2
Os	*Osmium*	*Osmio*	+	2
pc	*Pearcite*	*Pearceíta*	+	2
pch	Pitchblende	Pechblenda	cf. urn	
Pd	Palladium	Paladio	cf. PGE	
pdm	Polydymite	Polidimita	cf. ln	
PGE	Platinum Group Elements	Elementos Gr. Platino	cf. Pt, Os, Ir	
PGM	*Platinum Group Minerals*	*Minls del Grupo del Pt*	cf. *spy*	2
Phos	Phosphate	Fosfato		
plb	*Polybasite*	*Polibasita*	+	2
pn	**Pentlandite**	**Pentlandita**	+	**1**
po	**Pyrrhotite**	**Pirrotita (Pirrotina)**	+	**1**
poli	Polianite	Polianita	cf. prl	
prg	**Pyrargyrite**	**Pirargirita**	+	**1**
prl	**Pyrolusite**	**Pirolusita**	+	**1**
prm	*Pararammelsbergite*	*Pararamelsbergita*	+	2
pru	**Proustite**	**Proustita**	+	**1**
psi	**Psilomelane**	**Psilomelana**	+	**1**
Pt	**Platinum**	**Platino**	+	**1**
py	**Pyrite**	**Pirita**	+	**1**
pyc (pyl)	*Pyrochlore*	*Pirocloro*	+	2
pyst	Pyrostilpnite	Pirostilpnita	cf. prg	
rdl	Ramsdellite	Ramsdellita	+	
REE (ETR)	Rare Earth Elements	Tierras Raras (Elem)		
ren	*Renierite*	*Renierita*	+	2
rl	Realgar	Rejalgar	+	
rm	**Rammelsbergite**	**Rammelsbergita**	+	**1**
rob	Robinsonite	Robinsonita		
rom	**Romanechite**	**Romanechita**	+	**1**
rt	**Rutile**	**Rutilo**	+	**1**
ruiro	Rutheniridosmine	Ruteniridosmina	cf. Os, Ir	
ryn	Rynersonite	Rynersonita	cf. mlt, pyc	
Sb	**Antimony**	**Antimonio**	+	**1**
sba	*Stibarsen*	*Stibarsen*	+	2
sbl	Stibioluzonite	Estibioluzonita	cf. fm	
sbt	**Stibnite**	**Estibina, Antimonita**	+	**1**
sch	*Scheelite*	*Scheelita*	+	2
Sel	Selenides	Seleniuros	cf. nau, um	
sen	Senarmontite	Senarmontita	cf. F prg5-6	cf. ocres Sb
sf	**Safflorite**	**Safflorita**	+	**1**
sg	Siegenite	Siegenita	cf. ln	
sk	**Skutterudite**	**Skutterudita**	+	**1**
smal	Smaltite (≈sk)	Esmaltina	cf. sk	
smtt	Smithite	Smithita	cf. Ss	
snr	Schachnerite	Schachnerita	cf. F atm8	cf. am
sp	**Sphalerite**	**Esfalerita**	+	**1**
spk	Spionkopite (Blaub. cv)	Spionkopita	cf. cv &F dg1-2	
spn	*Spinel*	*Espinela*	+	2
spy	*Sperrylite*	*Sperrylita*	+	2

(continued)

(continued)

Abbreviation	Name	Nombre	Observations	Priority
Ss	Sulfosalt	Sulfosal		
stb	Sternbergite	Sternbergita	+	
stm	*Stromeyerite*	*Stromeyerita*	+	*2*
stn	*Stannite*	*Estannita*	+	*2*
stnd	Stannoidite	Estannoidita		
stp	*Stephanite*	*Stephanita*	+	*2*
Sulf	Sulfide	Sulfuro		
syl (sv)	Sylvanite	Silvanita	cf. Tel	
tan (tt)	Tantalite	Tantalita	cf. cbt	
td	**Tetrahedrite**	**Tetraedrita/Tetrahed**	+	**1**
tdy	Tedradymite	Tetradimita	cf. Tel	
Te	*Tellurium*	*Teluro*	+	*2*
Tel (Tell)	Telluride (Group)	Telururo (Grupo)		
tit (sph)	*Titanite*	*Titanita*	+	*2*
tlk	Talnakhita	Talnakhita	cf. ccp	
tn	Tennantite	Tennantita	cf. td	
tno	Tenorite	Tenorita		
tod	Todorokite	Todorokita		
tp (tap)	*Tapiolite*	*Tapiolita*	+	*2*
tr (tro)	Troilite	Troilita	cf. po, Fe	
ul	Ullmannite	Ullmannita	+	
ulv	*Ulvöspinel (Ulvite)*	*Ulvöespinela (Ulvita)*	+	
um	*Umangite*	*Umangita*	+	*2*
urn	**Uraninite**	**Uraninita**	+	**1**
vae	Vaesite	Vaesita	cf. brv	
vio	Violarite	Violarita	cf. ln	
vll	*Valleriite*	*Valleriíta*	+	*2*
wf	**Wolframite**	**Wolframita**	+	**1**
wz	*Wurtzite*	*Wurtzita*	+	*2*
xan	Xanthoconite	Xantoconita	cf. pru	
zk	Zinkenite	Zinkenita	cf. Ss	
	Other names	Otros nombres	Observations	Priority
	Allemontite	Allemontita	cf. sba	
	Antimonite	Antimonita	cf. sbt	
	Blende	Blenda	cf. sp	
	Chalcopyrrhotite	Calcopirrotita	cf. isc	
	Hollandite	Holandita	cf. rom, psi	
	Martite	**Martita**	cf. **ht, mt**	**1**
	Mispickel	**Mispíquel**	cf. **asp**	**1**
	Polianite	**Polianita**	cf. **prl**	**1**
	Schalenblende	**–**	cf. **sp**	**1**
	Sphene (*see* Titanite)	**Esfena (ver Titanita)**	cf. **tit**	**1**
	Steel	**Acero**	cf. **Fe**	**1**
	Strüverite	Struverita	cf. rt	
	Ulvite	Ulvita	cf. ulv	

Annex 1.2.m Ore Minerals and their Symbols or Abbreviations and Comments

Name (EN)	Abbreviation	Nombre (ES)	Observations	Priority
Acanthite	**ac**	**Acantita**	+	**1**
Aguilarite	agl	Aguilarita	+	
Aikinite	aik	Aikinita	cf. bm	
Alabandite	*alb*	*Alabandita*	+	2
Allargentum	*aag*	*Allargentum*	+	2
Alloclasite	alc	Aloclasita	cf. cob	
Altaite	alt	Altaíta		
Amalgam	*am (amal)*	*Amalgama*	+	2
Anatase	ana	Anatasa	+	
Anilite	anil	Anilita	cf. cc, dg	
Antimony	**Sb**	**Antimonio**	+	**1**
Argentite	arg	Argentita	+	
Argentopyrite	*agp*	*Argentopirita*	+	2
Arsenic	*As*	*Arsénico*	+	2
Arsenopyrite	**asp**	**Arsenopirita**	+	**1**
Asbolane	asb	Asbolana		
Aurian silver Ag–Au	*ausi*	*Plata áurea Ag–Au*	cf. *el*	2
Azurite	*az*	*Azurita*	+	2
Baddeleyite	bad	Baddeleyita	+	
Bastnäsite	**bst (bas)**	**Bastnäsita**	+	**1**
Bauxite	baux	Bauxita	Mixture	
Berthierite	*brt*	*Berthierita*	+	2
Berzelianite	bzl	Berzelianita	cf. dg	
Bismuth	*Bi*	*Bismuto*	+	2
Bismuthinite	*bm*	*Bismutinita*	+	2
Bornite	**bn**	**Bornita**	+	**1**
Boulangerite	*bl*	*Boulangerita*	+	2
Bournonite	bnn	Bournonita	+	
Braggite	brg	Braggita	cf. erl F 1, 3	
Braunite	*bra (br)*	*Braunita*	+	2
Bravoite	*brv*	*Bravoíta*	+	2
Breithauptite	*bth*	*Breithauptita*	+	2
Briartite	bri	Briartita	cf. ren	
Bukovite	buk	Bukovita		
Calaverite	*clv*	*Calaverita*	+	2
Calciotantite	cta	Calciotantita	cf. mlt, pyc	
Carrollite	car	Carrolita	cf. ln	
Cassiterite	**cst (cs)**	**Casiterita**	+	**1**
Cattierite	ctt	Cattierita	cf. brv	
Cerargyrite	crg	Cerargirita	cf. cag	
Chalcocite	**cc**	**Calcocita=Calcosina**	+	**1**
Chalcopyrite	**ccp**	**Calcopirita**	+	**1**
Chloanthite (discr.)	cloa	Cloantita	cf. sk	
Chlorargyrite	*cag*	*Clorargirita*	+	2
Chromite	**chr**	**Cromita**	+	**1**
Chrysocolla	crc	Crisocola	+	
Cinnabar	**cin**	**Cinabrio**	+	**1**
Clausthalite	*clt (cla)*	Clausthalita	+	
Coal—organic matter	COM	Carbón-Mat. Orgánica	cf. gra & gg	
Cobaltite	**cob**	**Cobaltita**	+	**1**

(continued)

(continued)

Name (EN)	Abbreviation	Nombre (ES)	Observations	Priority
Cohenite	*coh (chn)*	*Cohenita*	cf. *Fe*	*2*
Coloradoite	colo (cld)	Coloradoíta	+	
Columbite	cb	Columbita	cf. cbt	
Columbotantalite	**cbt**	**Columbotantalita**	+	**1**
Cooperite	cpr (coo)	Cooperita	cf. erl F 1, 3	
Copper	**Cu**	**Cobre**	+	**1**
Coronadite	crn	Coronadita	cf. rom, psi	
Cosalite	cos	Cosalita	+	*2*
Covellite	**cv**	**Covellita**	+	**1**
Cryptomelane	*cry*	*Criptomelana*	+	*2*
Cubanite	*cbn*	*Cubanita*	+	*2*
Cuprite	**cup**	**Cuprita**	+	**1**
Cylindrite	cyl	Cilindrita	cf. F 41.04	
Digenite	**dg**	**Digenita**	+	**1**
Djurleite	dju (dj)	Djurleíta	cf. cc, dg	
Domeykite	dom	Domeiquita		
Dyscrasite	*dy*	*Discrasita*	+	*2*
Electrum	*el*	*Electrum*	+	*2*
Enargite	*en*	*Enargita*	+	*2*
Erlichmanite	erl	Erlichmanita	+	
Eugenite	eug	Eugenita	cf. am, aag	
Euxenite	eux	Euxenita	cf. REE	
Falkmanite	flk	Falkmanita	cf. bl	
Famatinite	*fm*	*Famatinita*	+	*2*
Ferberite	fb	Ferberita	cf. wf	
Ferroselite	fse	Ferroselita	cf. clt, um	
Fletcherite	fle	Fletcherita	cf. ln	
Franklinite	*fk (fkl)*	*Franklinita*	+	*2*
Freibergite	fbg	Freibergita	cf. td	
Freieslebenite	flb	Freieslebenita	+	
Galena	**gn**	**Galena**	+	**1**
Galenobismuthite	gnb	Galenobismutina		
Gallite	gal	Galita		
Gangue (undiff.)	gg	Ganga (indif.)		
Geocronite	gc	Geocronita	+	
Germanite	ger	Germanita	cf. ren	
Gersdorffite	*gf*	*Gersdorffita*	+	*2*
Glaucodot	*gld*	*Glaucodoto*	+	*2*
Goethite	**gth**	**Goethita**	+	**1**
Gold	**Au**	**Oro**	+	**1**
Gossan	gos	Gossan	cf. lim	
Graphite	**gra**	**Grafito**	+	**1**
Gratonite	gtt	Gratonita	+	
Greenockite	gk	Greenockita	cf. Sulf	
Greigite	grg	Greigita	cf. ln	
Gudmundite	*gud (gd)*	*Gudmundita*	+	*2*
Hausmannite	hs	Hausmanita		
Hematite	**ht**	**Hematites**	+	**1**
Hessite	hss	Hessita		
Heterogenite	htg	Heterogenita		
Hollandite	hol	Hollandita	cf. psi	

(continued)

(continued)

Name (EN)	Abbreviation	Nombre (ES)	Observations	Priority
Horobetsuite	hrb	Horobetsuita	cf. bm	
Hübnerite	hub	Hübnerita	cf. wf	
Ilmenite	**il**	**Ilmenita**	+	**1**
Ilmenorutile	*ilr*	*Ilmenorutilo*	+	2
Iridium	*Ir*	*Iridio*	+	2
Iridosmine (discr.)	iros	Osmiridio (desacr.)	cf. Os, Ir	
Iron	**Fe**	**Hierro**	+	**1**
Isocubanite	*isc*	*Isocubanita*	+	2
Ixiolite	ix	Ixiolita	cf. rt (F13-14)	
Jacobsite	jac	Jacobsita	+	
Jalpaite	jl	Jalpaíta		
Jamesonite	jm	Jamesonita	+	2
Jordanite	jd	Jordanita	+	
Kermesite	ker (km)	Kermesita	+	
Kësterite (Kösterite)	ks	Kësterita (kesterita)	cf. stn	
Kobellite	ko	Kobellita		
Krennerite	krn	Krennerita	+	
Kutinaite	kut	Kutinaita		
Langisite	lgs	Langisita	Fpc7-8, pru5-6	
Laurite	lrt (lau)	Laurita	cf. erl	
Lepidocrocite	*lpc*	*Lepidocrocita*	+	2
Limonite	**lim, lm**	**Limonita**	+	**1**
Linnaeite	**ln**	**Linneíta**	+	**1**
Lithiophorite	lph	Litioforita		
Löllingite	*lo*	*Löllingita*	+	2
Lorandite	lor	Lorandita		
Luzonite	*lz*	*Luzonita*	+	2
M not identified	m1, m2…	M no identificado		
Mackinawite	*mck*	*Mackinawita*	+	2
Maghemite	*mgh*	*Maghemita*	+	2
Magnetite	**mt (mg)**	**Magnetita**	+	**1**
Malachite	*mch*	*Malaquita*	+	2
Manganite	**mng**	**Manganita**	+	**1**
Manganosite	mngn	Manganosite		
Marcasite	**mc**	**Marcasita**	+	**1**
Maucherite	*mau*	*Maucherita*	+	2
Melnicovite	mln	Melnicovita	cf. py	
Mercury (*Quicksilver*)	Hg	Mercurio	cf. cin	
Metacinnabar	mcin	Metacinabrio	cf. cin	
Metastibnite	msbt	Metaestibina	cf. sbt	
Miargyrite	**my**	**Miargirita**	+	**1**
Microlite	mlt	Microlita	cf. pyc	
Millerite	**ml**	**Millerita**	+	**1**
Mineral (undiff.)	M, min, mnl	Mineral (indiferenc.)		
Molybdenite	**mol**	**Molibdenita**	+	**1**
Monazite	*mz*	*Monazita*	+	2
Moschellandsbergite	mobg	Moschellandsbergita	cf. am	
Naumannite	nau	Naumannita	+	
Nickeline, Niccolite	**nc (nk)**	**Niquelita, Niquelina**	+	**1**
Opaque minerals	OpM	Minerales Opacos		
Orpiment	*orp, op*	*Oropimente*	+	2

(continued)

(continued)

Name (EN)	Abbreviation	Nombre (ES)	Observations	Priority
Osmium	*Os*	*Osmio*	+	*2*
Oxycalciomicrolite	ocm	Oxicalciomicrolita	cf. mlt, bim	
Palladium	Pd	Paladio	cf. PGE	
Pararammelsbergite	*prm*	*Pararamelsbergita*	+	*2*
Pearcite	*pc*	*Pearceíta*	+	*2*
Pentlandite	**pn**	**Pentlandita**	+	**1**
Phosphate	Phos	Fosfato		
Pitchblende	pch	Pechblenda	cf. urn	
Platinum	**Pt**	**Platino**	+	**1**
Platinum Group Elements	PGE	Elem. Grupo Platino	cf. Pt, Os, Ir	
Platinum Group Minerals	*PGM*	*Minls. Grupo del Pt*	cf. *spy*	*2*
Polianite	poli	Polianita	cf. prl	
Polybasite	*plb*	*Polibasita*	+	*2*
Polydymite	pdm	Polidimita	cf. ln	
Proustite	**pru**	**Proustita**	+	**1**
Psilomelane	**psi**	**Psilomelana**	+	**1**
Pyrargyrite	**prg**	**Pirargirita**	+	**1**
Pyrite	**py**	**Pirita**	+	**1**
Pyrochlore	*pyc (pyl)*	*Pirocloro*	+	*2*
Pyrolusite	**prl**	**Pirolusita**	+	**1**
Pyrostilpnite	pyst	Pirostilpnita	cf. prg	
Pyrrhotite	**po**	**Pirrotita (Pirrotina)**	+	**1**
Rammelsbergite	**rm**	**Rammelsbergita**	+	**1**
Ramsdellite	rdl	Ramsdellita	+	
Rare Earth Elements	REE (ETR)	Tierras Raras (Elem)		
Realgar	rl	Rejalgar	+	
Renierite	*ren*	*Renierita*	+	*2*
Robinsonite	rob	Robinsonita		
Romanechite	**rom**	**Romanechita**	+	**1**
Rutheniridosmine	ruiro	Ruteniridosmina	cf. Os, Ir	
Rutile	**rt**	**Rutilo**	+	**1**
Rynersonite	ryn	Rynersonita	cf. mlt, pyc	
Safflorite	**sf**	**Safflorita**	+	**1**
Schachnerite	snr	Schachnerita	cf. F atm8	cf. am
Scheelite	*sch*	*Scheelita*	+	*2*
Selenides	Sel	Seleniuros	cf. nau, um	
Senarmontite	sen	Senarmontita	cf. F prg5-6	cf. Sb ocr
Siegenite	sg	Siegenita	cf. ln	
Silver	**Ag**	**Plata**	+	**1**
Skutterudite	**sk**	**Skutterudita**	+	**1**
Smaltite ($\approx$sk)	smal	Esmaltina	cf. sk	
Smithite	smtt	Smithita	cf. Ss	
Sperrylite	*spy*	*Sperrylita*	+	*2*
Sphalerite	**sp**	**Esfalerita**	+	**1**
Spinel	*spn*	*Espinela*	+	*2*
Spionkopite (Blub. cv)	spk	Spionkopita	cf. cv, F dg1-2	
Stannite	*stn*	*Estannita*	+	*2*
Stannoidite	stnd	Estannoidita		
Stephanite	*stp*	*Stephanita*	+	*2*

(continued)

(continued)

Name (EN)	Abbreviation	Nombre (ES)	Observations	Priority
Sternbergite	stb	Sternbergita	+	
Stibarsen	*sba*	*Stibarsen*	+	2
Stibioluzonite	sbl	Estibioluzonita	cf. fm	
Stibnite	**sbt**	**Estibina, Antimonita**	+	**1**
Stromeyerite	*stm*	*Stromeyerita*	+	2
Sulfide	Sulf	Sulfuro		
Sulfosalt	Ss	Sulfosal		
Sylvanite	syl (sv)	Silvanita	cf. Tel	
Talnakhita	tlk	Talnakhita	cf. ccp	
Tantalite	tan (tt)	Tantalita	cf. cbt	
Tapiolite	*tp (tap)*	*Tapiolita*	+	2
Tedradymite	tdy	Tetradimita	cf. Tel	
Telluride (Group)	Tel (Tell)	Telururo (Grupo)		
Tellurium	*Te*	*Teluro*	+	2
Tennantite	tn	Tennantita	cf. td	
Tenorite	tno	Tenorita		
Tetrahedrite	**td**	**Tetraedrita o Tetrah**	+	**1**
Titanite	*tit (sph)*	*Titanita*	+	2
Todorokite	tod	Todorokita		
Troilite	tr (tro)	Troilita	cf. po, Fe	
Ullmannite	ul	Ullmannita	+	
Ulvöspinel (Ulvite)	*ulv*	*Ulvöespinela (Ulvita)*	+	
Umangite	*um*	*Umangita*	+	2
Uraninite	**urn**	**Uraninita**	+	**1**
Vaesite	vae	Vaesita	cf. brv	
Valleriite	*vll*	*Valleríta*	+	2
Violarite	vio	Violarita	cf. ln	
Wolframite	**wf**	**Wolframita**	+	**1**
Wurtzite	*wz*	*Wurtzita*	+	2
Xanthoconite	xan	Xantoconita	cf. pru	
Zinkenite	zk	Zinkenita	cf. Ss	
Other names		Otros nombres	Observations	Priority
Allemontite		Allemontita	cf. sba	
Antimonite		Antimonita	cf. sbt	
Blende		Blenda	cf. sp	
Chalcopyrrhotite		Calcopirrotita	cf. isc	
Hollandite		Holandita	cf. rom, psi	
Martite		**Martita**	cf. **ht, mt**	**1**
Mispickel		**Mispíquel**	cf. **asp**	**1**
Polianite		**Polianita**	cf. **prl**	**1**
Schalenblende		–	cf. **sp**	**1**
Sphene (*see* Titanite)		**Esfena (ver Titanita)**	cf. **tit**	**1**
Steel		**Acero**	cf. **Fe**	**1**
Strüverite		Struverita	cf. rt	
Ulvite		Ulvita	cf. ulv	

Annex 1.3.a Gangue Mineral Abbreviations and Comments

Abbreviation	Name (EN)	Nombre (ES)	Observations
ab	Albite	Albita	cf. Slc, Fs
adu	Adularia	Adularia	cf. Slc, Fs
amb	Amblygonite	Ambligonita	cf. Phos
an	Anorthite	Anortita	cf. Slc, Fs
anbg	Annabergite	Annabergita	Arseniatos
Anf, Amp	Amphibole (Group)	Anfíbol (Grupo)	cf. Slc
ang	Anglesite	Anglesita	cf. Sulfatos
anh	Anhydrite	Anhidrita	cf. Sulfatos
ank	Ankerite	Ankerita	cf. Carb
ap	Apatite	Apatito	cf. Phos
Arc	Clay	Arcilla	cf. FSlc
bar	Barite	Barita	cf. Sulfatos
bim	Bismite	Bismita	cf. ocres Bi
bio	Biotite	Biotita	cf. Slc, FSlc
bis	Bismutite	Bismutita	cf. ocres Bi
bk	Brookite	Brookita	Óxidos, cf. lcx
calc	Calcite	Calcita	cf. Carb
Carb	Carbonate (Class)	Carbonato (Clase)	cf. Carb, gg
cerv	Cervantite	Cervantita	cf. ocres Sb
ch	Chert	Chert	cf. Slc, Sic
Chl	Chlorite (Series)	Clorita (Serie)	cf. Slc, FSlc
cld	Chalcedony	calcedonia	cf. q
CM, MArc	Clay minerals	Mins Arcillosos	cf. Slc, FSlc
COM	Coal—organic matter	Carbón-Materia Orgán	cf. Organ. Mat
cor	Corundum	Corindón	
crc	Chrysocolla	Crisocola	cf. Slc
crs	Cerussite	Cerusita	cf. Carbonatos
dol	Dolomite	Dolomita	cf. Carb
Ep	Epidote (Group)	Epidota (Grupo)	cf. Slc
ery	Erythrite	Eritrina	Arseniatos
fl	Fluorite	Fluorita	cf. Haluros
Fs, Felds	Feldspar (Group)	Feldespato (Grupo)	cf. Slc
FSlc, PSlc	Phyllosilicates (Subclass)	Filosilicatos (Subclase)	cf. Slc
gg	Gangue (undifferenciated)	Ganga (indiferenciada)	
gln	Glauconite	Glauconita, Glauconia	cf. Slc, FSlc
grt	Garnet (Group)	Granate (Grupo)	cf. Slc
hb	Hornblende	Hornblenda	cf. Slc, Inoslc
ili	Illite	Illita	cf. Slc, FSlc
jar	Jarosite	Jarosita	cf. lim
jas	Jasper	Jaspe	cf. q
Kfs	K Feldspar	Feldespato Potásico	cf. Slc, Fs
lcx	Leucoxene	Leucoxeno	cf. tit, ana
ldw	Ludwigite	Ludwigita	cf. von
mag	Magnesite	Magnesita	cf. Carb
Mica	Mica (Group)	Mica (Grupo)	cf. Slc, FSlc
mv, mus	Muscovite	Moscovita	cf. Mica
Ocre	Ochre, Ocher	Ocres	cf. Óxidos
OM	Organic matter	Materia Orgánica	
or	Orthoclase	Ortoclasa, ortosa	cf. Slc, Fs
ov	Olivine (Series)	Olivino (Serie)	cf. Slc

(continued)

(continued)

Abbreviation	Name (EN)	Nombre (ES)	Observations
Phos	Phosphate	Fosfato	gg
pie	Piemontite	Piemontita	cf. bra, Ep
Plag	Plagioclase (Series)	Plagioclasa (Serie)	cf. Slc, Fs
pvk	Perovskite	Perovskita	Óxidos
Px, Pyx	Pyroxene (Group)	Piroxeno (Grupo)	cf. Slc, Inoslc
pymp	Pyromorphite	Piromorfita	cf. Phos
q (qz, qtz)	Quartz	Cuarzo	cf. Slc, Sic
rhc	Rhodochrosite	Rodocrosita	cf. Carb
rhn	Rhodonite	Rodonita	cf. rdl F 1-4
S	Sulfur	Azufre	gg
sco	Scorodite	Escorodita	cf. asp; Arseniato
sen	Senarmontite	Senarmontita	cf. ocres Sb
ser	Sericite	Sericita	cf. Slc, FSlc
Sic	Silica	Sílice	cf. Slc, q
sid (sd)	Siderite	Siderita	cf. Carb
Slc	Silicate (Class)	Silicato (Clase)	gg
sms	Smithsonite	Smithsonita	cf. Carb
srp	Serpentine	Serpentina	cf. Slc, FSlc
sx	Souxite	Souxita	cf. cst
tur	Tourmaline	Turmalina	cf. Slc
val	Valentinite	Valentinita	cf. ocres Sb
van	Vanadinite	Vanadinita	cf. Vanadatos
von	Vonsenite	Vonsenita	cf. Carb-Boratos
wil	Willemite	Willemita	cf. Slc
xtm	Xenotime	Xenotimo	cf. Phos
Zeo	Zeolite (Group)	Zeolita (Grupo)	cf. Slc
zir	Zircon	Circón (Zircón)	cf. Slc
znc	Zincite	Zincita	Óxidos
	Other names	Otros nombres	Observaciones
	Sphene	Esfena	cf. tit

Annex 1.3.m Gangue Minerals and their Abbreviations and Comments

Name (EN)	Abbreviation	Nombre (ES)	Observations
Adularia	adu	Adularia	cf. Slc, Fs
Albite	ab	Albita	cf. Slc, Fs
Amblygonite	amb	Ambligonita	cf. Phos
Amphibole (Group)	Anf, Amp	Anfíbol (Grupo)	cf. Slc
Anglesite	ang	Anglesita	cf. Sulfates
Anhydrite	anh	Anhidrita	cf. Sulfates
Ankerite	ank	Ankerita	cf. Carb
Annabergite	anbg	Annabergita	Arsenates
Anorthite	an	Anortita	cf. Slc, Fs
Apatite	ap	Apatito	cf. Phos
Barite	bar	Barita	cf. Sulfates
Biotite	bio	Biotita	cf. Slc, FSlc
Bismite	bim	Bismita	cf. ocres Bi
Bismutite	bis	Bismutita	cf. ocres Bi
Brookite	bk	Brookita	Oxides, cf. lcx

(continued)

(continued)

Name (EN)	Abbreviation	Nombre (ES)	Observations
Calcite	calc	Calcita	cf. Carb
Carbonate (Class)	Carb	Carbonate (Clase)	cf. Carb, gg
Cerussite	crs	Cerusita	cf. Carbonates
Cervantite	cerv	Cervantita	cf. ocres Sb
Chalcedony	cld	calcedonia	cf. q
Chert	ch	Chert	cf. Slc, Sic
Chlorite (Series)	Chl	Clorita (Serie)	cf. Slc, FSlc
Chrysocolla	crc	Crisocola	cf. Slc
Clay	Arc	Arcilla	cf. FSlc
Clay minerals	CM, MArc	Mins Arcillosos	cf. Slc, FSlc
Coal—organic matter	COM	Carbón-Mat. Orgánica	cf. Mat Organ
Corundum	cor	Corindón	
Dolomite	dol	Dolomita	cf. Carb
Epidote (Group)	Ep	Epidota (Grupo)	cf. Slc
Erythrite	ery	Eritrina	Arsenates
Feldspar (Group)	Fs, Felds	Feldespato (Grupo)	cf. Slc
Fluorite	fl	Fluorita	cf. Halides
Gangue (undiff.)	gg	Ganga (indiferenciada)	
Garnet (Group)	grt	Granate (Grupo)	cf. Slc
Glauconite	gln	Glauconita, Glauconia	cf. Slc, FSlc
Hornblende	hb	Hornblenda	cf. Slc, Inoslc
Illite	ili	Illita	cf. Slc, FSlc
Jarosite	jar	Jarosita	cf. lim
Jasper	jas	Jaspe	cf. q
K Feldspar	Kfs	Feldespato Potásico	cf. Slc, Fs
Leucoxene	lcx	Leucoxeno	cf. tit, ana…
Ludwigite	ldw	Ludwigita	cf. von
Magnesite	mag	Magnesita	cf. Carb
Mica (Group)	Mica	Mica (Grupo)	cf. Slc, FSlc
Muscovite	mv, mus	Moscovita	cf. Mica
Ochre, Ocher	Ocre	Ocres	cf. Oxides
Olivine (Series)	ov	Olivino (Serie)	cf. Slc
Organic matter	OM	Materia Orgánica	
Orthoclase	or	Ortoclasa, ortosa	cf. Slc, Fs
Perovskite	pvk	Perovskita	Oxides
Phosphate	Phos	Fosfato	gg
Phyllosilicates (Subclass)	FSlc, PSlc	Filosilicatos (Subclase)	cf. Slc
Piemontite	pie	Piemontita	cf. bra, Ep
Plagioclase (Series)	Plag	Plagioclasa (Serie)	cf. Slc, Fs
Pyromorphite	pymp	Piromorfita	cf. Phos
Pyroxene (Group)	Px, Pyx	Piroxeno (Grupo)	cf. Slc, Inoslc
Quartz	q (qz, qtz)	Cuarzo	cf. Slc, Sic
Rhodochrosite	rhc	Rodocrosita	cf. Carb
Rhodonite	rhn	Rodonita	cf. rdl F 1-4
Scorodite	sco	Escorodita	cf. asp; Arsenate
Senarmontite	sen	Senarmontita	cf. Sb ochers
Sericite	ser	Sericita	cf. Slc, FSlc
Serpentine	srp	Serpentina	cf. Slc, FSlc
Siderite	sid (sd)	Siderita	cf. Carb
Silica	Sic	Sílice	cf. Slc, q

(continued)

(continued)

Name (EN)	Abbreviation	Nombre (ES)	Observations
Silicate (Class)	Slc	Silicato (Clase)	gg
Smithsonite	sms	Smithsonita	cf. Carb
Souxite	sx	Souxita	cf. cst
Sulfur	S	Azufre	gg
Tourmaline	tur	Turmalina	cf. Slc
Valentinite	val	Valentinita	cf. Sb ochers
Vanadinite	van	Vanadinita	cf. Vanadates
Vonsenite	von	Vonsenita	cf. Borates
Willemite	wil	Willemita	cf. Slc
Xenotime	xtm	Xenotimo	cf. Phos
Zeolite (Group)	Zeo	Zeolita (Grupo)	cf. Slc
Zincite	znc	Zincita	Oxides
Zircon	zir	Circón (Zircón)	cf. Slc
Other names		Otros nombres	Observations
Sphene		Esfena	cf. tit

Annex 1.4 Glossary MAAS Meanings of Acronyms, Abbreviations and Symbols

A.1.4.1 Acronyms and Abbreviations

+nic	microscope setting + p
+p	setting with double polarization, i.e. polarizer and analyzer crossed at 90° (previously called "crossed nicols")
+p.w.	setting with polarizer, analyzer and interference contrast (i.c.) inserting Wollanston prism
Abrv	Abbreviations
AMCO	Automated Microscopic Characterization of Ores. Automated ore mineral characterization system developed at the LMA (Applied Microscopy Laboratory, ETSI Minas y Energia, UPM)
Anis + P	Description of polarization colors or anisotropism (+p setting)
Bir	bireflectance, or variation of R with orientation
Bk	black color (no colour, actually)
Blu	blue color
BP	Block (ore or rock) Polished
Br	brown color
BSE	Backscattered-Electron Imaging
C illuminant	type of illumination similar to daylight
ca (*circa*)	approximately
CAMEVA	VNIR database
cf. (confer)	view, compare or consult
cf. ibid.	consult the same place
CIE	Commission Internationale de l'Éclairage, or International Commission on Illumination
COM-IMA	Commission on Ore Mineralogy of the International Mineralogical Association
CRIRSCO	*Committee for Mineral Reserves International Reporting Standards*, organization that controls the various bodies that certify quality standards
cx	cryptocrystalline
d = dry	observation with dry lens (in air)
d+p	observation in air with polarizer and analyzer crossed at 90°
d+pw	observation in air with polarizer and analyzer crossed and with i.c. or interference contrast
discr	*discredited* (term, name or mineral rejected by IMA)
DPTS	double polished thin section, the usual one for the study of fluid inclusions
dsp	observation with dry lens (in air) with only the polarizer (*single polarizer*)
dxp	observations with dry objective (in air) and with polarizer and analyzer crossed a little less than 90° (about 87°)
EIT Raw Materials	EU organization promoting technological innovation in raw materials and mineral resources
EMP (EMPA)	Electron Microprobe (Analyzer or Analysis)

ETSI	Escuela Tecnica Superior de Ingenieros (Mines and Energy, UPM) de Minas y Energía, Universidad Politecnica de Madrid
EU	European Union
Gn	green color
Gy	grey colour
H	hardness
i.c.	interference contrast, obtained by inserting the Wollanston prism and rotating the analyzer until the optimum contrast is obtained
IMA	International Mineralogical Association
IMA/COM	Commission on Ore Mineralogy of the IMA: Commission on Ore Mineralogy of the International Mineralogical Association
IMEB	organization that certifies quality standards in Spain and Portugal
IR	Internal Reflection, also described by some authors as diffuse reflectance or diffuse light: light reflected by reticular planes, microcracks or defects from inside the crystal. Perceived by the eye as additional luminosity, showing the mineral's own color, which interferes with the polarization colors and can even mask them
JORC	Joint Ore Reserves Committee, an Australian organization that certifies quality standards
l.t. (*locus typicus*)	place of first definition of a species, mineral or rock, considered typical
LAS	Leica Application System v3.8, image acquisition software
LED	Light Emitting Diode
LLE	Life-Long-Education
LLL	Life-Long-Learning
m, min, mnl	mineral (undifferentiated)
m1, m2, etc.	unidentified minerals
MAN	Mean Atomic Number or average atomic number
MicroH	Microhardness
MLA	commercial system for automated mineralogy
Mohs	quantitative Mohs hardness
MS-DOS	MicroSoft Disk Operating System
mx	microcrystalline
N	refractive index of the medium
n	refractive index of the mineral
N// (or parallel nicols)	obsolete term for sp setting (polarizer only)
NI 43 101	National Instrument: Canadian organization that certifies quality standards
NX (or crossed nicols)	obsolete designation for +p (polarizer and analyzer crossed at 90°)
o (= oil)	observation with oil immersion lens
o+p	observation with oil and polarizer and analyzer crossed at 90°
o+pw	observation in oil with crossed polarizer and analyzer and with i.c. or interference contrast
Og	orange colour
Ol	olive color
OM	Optical Microscopy
osp	observation with oil and with *single polarizer*
oxp	observations with objective immersed in oil and with polarizer and analyzer crossed a little less than 90° (about 87°)
P_e	excitation purity
PERC	European organization that certifies quality standards
PGE	Platinum Group Elements
PGM	Platinum Group Minerals
Pi:	pink color
PolH	Qualitative polishing hardness
PPL	Plane Polarized Light, sp setting (polarizer only)
PS	Polished Section
PTS	Polished Thin Section
Pu	purple color
QDF3	*Quantitative Data File*, for reflectance values in the visible range (Criddle and Stanley 1993)
QemScan	commercial automated mineralogy system
R	specular Reflectance
Rd	red color
REE	Rare earth elements
REM	Rare earth minerals
resp	respectively
RL	reflected light
rltv	relative
ROM (Run-of-Mine)	raw ore, as extracted from the mine and prior to any kind of selection or processing
RRUFF	project to create a catalogue of spectral data of well-characterized minerals
s.l. (sensu lato)	in a general, approximate sense
s.str. (sensu stricto)	in the strict, precise sense
SAMREC	South African organization that certifies quality standards

SEM	Scanning Electron Microscope
sp	single polarizer setting (no analyzer, formerly referred to as *parallel nichols*, now obsolete), equivalent to the PPL (Plain Polariser Light) setting
sRGB	color space, based on sRGB color definition
T, or TL	transmitted light
T+p	transmitted light setting, with polarizer and analyzer crossed at 90°
Talm	Talmage Scratch hardness
TESCAN TIMA	Commercial automated mineralogy system
TL	transmitted light
Tsp	transmitted light setting, with only polarizer
undiff	undifferentiated
UPM	Universidad Politécnica de Madrid
UV	Ultra Violet wavelength range
VHN	Vickers Microhardness Number
Vi	violet color
VNIR	wavelength range between 370 and 1000 nm (visible and near infrared)
Wh	white color
x/xx	crystal/crystals
xp	observations with polarizer and analyzer crossed a little less than 90º (about 87º)
XPL (*Crossed Polars*)	+p, microscope setting with polarizer and analyzer crossed at 90º
XRD	X-Ray Diffraction
Y (%)	luminance
Yl	yellow

A.1.4.2 Symbols

§	chapter, paragraph or subparagraph of text.
ø	grain size, diameter measurement.
±	with or without (a mineral or object).
≈	approximate value, approximately.
>	greater than (*used to compare measurements, e.g. ø, or properties of minerals*)
≥	equal to or greater than.
>>	much larger than.
<	less than (*used to compare measurements, e.g. ø, or properties of minerals*)
≤	equal to or less than.
<<	much smaller than.
~	similar to, nearly.
α, β, γ, n_α, n_β, $n\gamma$	values of n (or *refractive index*) in the crystallographic directions α, β, γ
ε, ω = n_ε, n_ω or n_e, n_o	values of n (or *refractive index*) in the directions of the extraordinary and ordinary rays (uniaxial minerals)
λ	wavelength

Annex 2: Common Mineral Associations

«A mineral most commonly occurs in nature as a part of an assemblage, and this association itself is of considerable value in identification, because there are numerous examples of characteristic assemblages». (Vaughan 1990, p. 85)

The many types of deposits comprise a variety of mineral associations, knowledge of which is an important aid to identifying an ore. To facilitate this help, from a simplified typology of metal deposits limited to the most common classes (**Annex 2.1**) a table of common mineral associations (**Annex 2.2**) has been compiled. Both are based on my own experience and on published data (Castroviejo 2004; Castroviejo and Berrezueta 2009; Edwards and Atkinson 1986; Laznicka 2006; Ridley 2013; Robb 2005; Schneiderhöhn 1962; Vaughan 1990).

References

Castroviejo R (2004) Gold in ophiolites. Ophiolithic Complexes in Iberoamerica. YTED, Madrid, pp 25–70

Castroviejo R, Berrezueta E (2009) Automated recognition of metallic ores by digital image analysis: asupport to the mineralogical process. II: discriminating metallogenic criteria. Revista de Metalurgia 45(6):439–456, CSIC, Madrid (ISSN: OO34-8570)

Edwards R, Atkinson K (1986) Ore deposit geology. Chapman, London

Laznicka P (2006) Giant metallic deposits—future sources of industrial metals. Springer, Berlin

Ridley J (2013) Ore deposit geology. Cambridge University Press

Robb L (2005) Introduction to ore-forming processes. Blackwell, London

Schneiderhöhn H (1962) Erzlagersttaetten. Fischer-Verlag, 4e. Auflage, Stuttgart

Vaughan D (1990) Advanced microscopic studies of ore minerals. in: Jambor JL, Vaughan DJ (eds) Short course handbook 17. Mineral Association of Canada, Ottawa, p 85

Annex 2.1 Main Classes of Metal Deposits: Simplified Typology

Magmatic environment		
1	In ultramafic rocks and ophiolitic complexes (Cr, PGE, Ni, Cu, Fe-Ti)	
	1.1	Chromitite/Podiform Chromitite (ophiolitic)
	1.2	PGE (Platinum Group Elements)
	1.3	Ni-Cu sulfides (cf. Mafic R, *class 2.1*). Ophiolitic VMS (cf. *cl. 8.1*)
2	In mafic rocks (Ni, Cu, Fe, Ti, PGE)	
	2.1	Ni-Cu sulfides (±PGE)
	2.2	Fe-Ti oxides ± V
3	In granitic rocks and alkaline complexes	
	3.1	Pegmatitic—Pneumatolytic (Sn, W, Nb–Ta, Au…)
	3.2	Alkali complexes and carbonatites (REE)
Hydrothermal and magmatic-hydrothermal environments		
4	Hydrothermal veins and *mantos*	
	4.1	Pb–Zn-Ag and other veins
5	Porphyries (Cu, Mo, Au)	
	5.1	Cu and Cu-Au deposits
	5.2	Mo and Cu-Mo deposits
6	Epithermal (Ag, Au, Pb, Zn, Cu, Hg…)	
	6.1	High sulfidation
	6.2	Low sulfidation
7	Disseminated Carlin type (Au)	
Sedimentary environment		
8	Volcanic-sedimentary deposits. Fe/Mn banded formations	
	8.1	Volcanogenic massive sulfides (VMS: Cu, Pb, Zn, Ag, Au)
	8.2	Banded Iron Formations (BIF) and Mn Formations (BMF)
9	Cu-Pb–Zn deposits in sedimentary rocks	
	9.1	SEDEX and Mississippi Valley types (Pb, Zn, Ag)
	9.2	Kupferschiefer/Copperbelt types (Cu and polymetallic)
10	Detrital and residual types	
	10.1	Placers, paleoplacers, meta-conglomerate (Au, Sn, W, U, etc.)
	10.2	Laterites (Ni, Al, Au, etc.)
Metamorphic and complex environments		
11	Skarn and metasomatic processes (Cu, Au, Zn, Ag, U, REE, etc.)	
	11.1	Skarn and contact metasomatic deposits (Cu, Zn, Au, Sn, W, Ag, etc.)
	11.2	IOCG: Iron Oxide–Copper–Gold deposits (Cu, Au, Ag, U, REE)
12	In greenstone belts and metamorphic terrains (Au, Cu, Zn, Fe, Mn, etc.)	
	12.1	Orogenic gold deposits (Au)
	12.2	Metamorphosed deposits: VMS, SEDEX, BIF/Itabirites (§ 8, 9)

R. Castroviejo, *A Practical Guide to Ore Microscopy—Volume 2*,
https://doi.org/10.1007/978-3-031-18954-8

Annex 2.2 Mineral Associations by Deposit Class (cf. Annex 2.1)

Type of deposit	Class (Ax2.1)	Mineral associations: ores*		Examples	Remarks
		Principal ores	Minor ores		
Chromitite	1.1	chr (±mt/spn, il)	PGM, sulfides	Bushveld, RSA; Moa, Cuba	Very restricted PGM
Platinum or PGE	1.2	Pt, spy, Ir, Os, etc	Sulf, erl, chr…	Merensky Reef (RSA)	Exceptional mineralogy
Ni-Cu sulfides associated with ultramafites	1.3 a	po, pn, py, ccp, cbn, brv	ml, chr, vll, mt, il, PGM	Kambalda, Australia	Sulf. subclass Ni-Cu
VMS (Cu–Zn type)	1.3 b	py, ccp, Au	sp-po-mc-cc-cv-mt	Cyprus	Subcl. *Ophiolitic VMS* ± Au
Ni-Cu sulfides associated with gabbros	2.1	po, pn, py, mt, ccp	cbn, PGM, nc, ml	Sudbury; Noril'sk (Russia)	Occasional PGM
Titano-magnetite	2.2	mt, il, ht, rt	po, ccp, py	Allard Lake (Quebec, Can.)	Gabro association
Pegmatitic and pneum-hydrothermal	3.1	cst, wf, py, asp	mol, po, ccp, stn, REM, etc	Kalima, R.D. Congo; Cornwall UK; Finland; Erzgebirge, Germany	Very varied mineralogy (pegmatites)
Carbonatite complexes	3.2	RE Minerals: very varied mineralogy		Tsumeb, Namibia	Strategic interest REE
Pb–Zn-Ag veins and *mantos*	4.1	py, sp, gn, ccp, td	Ag, bn, cc, cv, Pb-Ag Ss	Coeur d'Alène (USA); Linares (Spain); Peru	Mineralogy can be varied; Sb-As and Ag sulfosalts
Copper porphyries (Cu-Au and Cu-Mo)	5.1/5.2	py, ccp, mol, bn, cc, cv, dg, Ss	mt, ht, en, cbn, il, rt, Au–Ag, gth, mch	Chuquicamata (Chile); Toquepala (Peru)	Hypo- and superg. minerals; potential secondary enrichment
High sulfidation epithermal	6.1	Au ± Ag, py, ccp, en	cv, sp, gn, gth	Yanacocha (Peru) Rodalquilar (Almeria)	≈ Quartz-alunitic or acid-sulfate type
Low sulfidation epithermal	6.2	py, gn, sp, Au/Ag/el, td, ac	ccp, ht, Sel, Ss Ag	Creede, Colorado, USA Cabo de Gata (Almeria)	Formerly known as the adularia-sericite type
Disseminated Au	7	Au, py, asp, Ss	sbt, rl, orp, gth	Carlin Trend (Nevada, USA)	*Invisible Au* (tiny particles)
Volcanogenic massive sulfides (VMS)	8.1	py, ccp, sp, po, gn, td	cc, cv, bn, dg, mc, cst, bm, el, asp, Ss, gth, ht	Faja Pirítica Iberica (Río Tinto, Neves Corvo, etc.)	Varied subclasses. Potential supergene enrichment (Au/Ag), Cu
Banded Iron Formations (*BIF*)	8.2	mt, ht, chert	sid, py, martite, gth	Guyana; Liberia; Superior, USA; Hammersley B, Australia; India	Large tonnages in Precambrian shields. Possible Au metallotect
Pb, Zn, Cu stratabound deposits (various types: Sedex/Carbonate Rocks/Kupferschiefer/Copperbelt)	9.1/9.2	ccp, bn, cc, cv, gn, sp, dg	td/Ss, ln, py	• Sedex: Rammelsberg, Germany; • Carb R: Miss. Valley, USA; • Kupferschiefer: Lubin, Pol.; Mansfeld, Germany; • Copperbelt. Mufulira (Zambia), Shaba (DR Congo)	Dominant metals: • Cu/Cu-Co (Copperbelt: Shaba, DR Congo) • Pb–Zn (Sedex/Miss. Vall.) • Cu-Ag (Pt) (Kupferschiefer)

(continued)

(continued)

Type of deposit	Class (Ax2.1)	Mineral associations: ores*		Examples	Remarks
		Principal ores	Minor ores		
Detrital, placer and paleoplacer deposits	10.1	Au, PGE, urn, cst, il, rt, mz, wf, diamonds, etc	ht, mt, cbt, zir, silicates, etc	Malaysia (cst placers), Witwatersrand, RSA (Au, U)	Witwatersrand: the world's largest Au concentration (meta-conglomerates)
Lateritic	10.2	gth, lim, jar, residual Al-Ni minerals: garnierite, gibbs., boehmite, Ni-nontronite, etc	Traces of PGE, Au	Laterites, bauxites and nickeliferous laterites: • New Caledonia • Moa (Cuba), etc	Mineralogy unsuitable for reflected light microscopy techniques
Skarn	11.1	mt, ccp, wf, sch, Au, asp, mol, sp, cst, etc	po, ht, Ss, gth; Ca-Mg-Mn Silicates and Carbonates, etc	• Tintaya (Cu-Au), Antamina (Zn), Peru • Cala, Badajoz (Fe, Au) • Carlés, Asturias (Au-Cu)	Very varied and complex deposits and mineralogy
IOCG	11.2	py, bn, ccp, Au, cc	mt, ht, fl, sid	Olympic Dam (Australia)	Large tonnage, complex gen
Orogenic gold deposits	12.1	Au, Tel, el, py, asp, po, etc	gn, sbt, sp, ccp, gra, rt, mol, Ss	Val d'Or, Hemlo, etc. Quebec, Canada	Synonyms: *greenstone, archaean, shear zone, gold-only, vein type, metamorphic, mesothermal,* etc., *gold deposits,* etc
Metamorphosed and polymetamorphic deposits	12.2	Generally analogous mineralogy to Sedex, SMV, BIF, etc., protoliths, although re-equilibrated under metamorphic conditions. Catazonal metamorphic changes —Broken Hill, Arinteiro—may mask protolith		Pb: Broken Hill (Australia) Fe: Itabira, Quadrilatero Ferrífero (Brazil) VMS: Arinteiro (Santiago de Compostela, Spain); Norway	Metamorphism recognizable by modified textures

*ORES: due to space constraints, the ores are represented by their corresponding symbols or abbreviations (cf. **Annex 1.1**)

Annex 3: Preparation of Polished Sections

Ursula Grunwald-Romera, *MSc Geological Eng.*
LMA_UPM (Technical University of Madrid).

1. **Preliminary Considerations**

The preparation of samples, in this case polished sections, is an important yet often neglected preliminary step in mineralogical analysis. However, we must bear in mind that the quality of an analysis will inevitably depend on the quality of the polished section. Indeed, whether the object to be worked on makes visible or loses the information that determines the success of the final results depends on it.

Although the procedures to follow to prepare a polished section are simple, the reality is that the quality of polish ultimately depends on the skill and responsibility of the technician, whether the polishing is manual or automatic.

It is important that the polishing is adjusted to the hardness of the various minerals present in the sample. In particular, special attention must be paid to the intergrowths between hard and soft minerals: poor polishing can lead to the creation of pronounced reliefs around the harder minerals. Experience is undoubtedly what helps the expert to choose the best polishing cloths and techniques according to the equipment, which means that the preparation techniques for polished sections differ according to the laboratory preparing them and the degree of automation of the process.

For this reason, in what follows, rather than systematizing the method through a series of fixed steps (difficult to adapt to the specific needs of each user), essential guidelines are given on the preparation of samples that, in any case, must be adjusted to the conditions of each. Following the basic steps to be taken into account to prepare a polished section, the most common problems derived from polishing that affect mineralogical analysis are explained, trying to anticipate a practical solution. These indications are especially relevant when the objective is to perform digital image analysis, either by means of electron microscopy equipment or by using innovative automatic characterization systems based on optical microscopy, as is the case of the AMCO system.

The main difference between doing an analysis manually with an expert in mineralogy and doing it with an automated system is that the human eye is able to discriminate certain flaws—and even benefit from them—and reconstruct the ideal aspect of the grains or mineral particles that are observed. A good example can be found in the acicular aggregates of wolframite, with totally irregular relief due to its difficult polishing; or in the shadow or bevelling produced by a simple edge around a mineral particle, irregular in appearance due to differential polishing. However, what is relatively easy for a person to understand (sometimes even unconsciously) can be an insurmountable barrier for an automated system, and the only way to overcome it is to take the utmost care in sample preparation.

The sample preparation process can be carried out either completely manually, by polishing and grinding with carborundum and glass, or fully automated, with purpose-designed machinery. A basic automation stage is considered here, for which a vacuum hood, a precision saw and a semi-automated polishing machine are essential. Whenever possible, the highest degree of automation available is recommended, to ensure, above all, that the polished surface is flat and remains parallel to the base, without the sometimes unavoidable surface undulations due to unevenly applied pressure at different points on the sample by manual polishing.

The basic process of preparing a polished section, either from a hand sample or from a mineral concentrate, can be summarized in four phases:

I. *Sample preparation*
II. *Embedding*
III. *Cutting, grinding and pre-polishing*
IV. *Polishing.*

As a first step, however, it is necessary to make some general considerations on the use of abrasives and polishing cloths. These apply both to roughening and polishing, and also in the previous phases of the embedding, when a *polished thin section* (**PTS**) is to be produced.

R. Castroviejo, *A Practical Guide to Ore Microscopy—Volume 2*,
https://doi.org/10.1007/978-3-031-18954-8

- Nature of the polished section: before starting, the predominant mineral composition and hardness must be taken into account. As it is normal to find mixed typologies in the same sample (hard minerals such as pyrite and soft minerals such as galena), it is recommended to make careful polishing at low revolutions and use hard polishing surfaces.
- Type of abrasive: it must be selected according to the result to be obtained and the nature of the mineral sample. During the grinding and polishing process, the type of abrasive can be changed, if necessary. The most commonly used are sandpapers impregnated with carborundum and those marketed in paste or grain form, to be mixed with a lubricating liquid.
- Grain size of the abrasive: in each phase of preparation it is necessary to use the correct grain size (whatever the type used), always taking care to move from the coarser to the finer sizes. The sample and the supports used must be washed thoroughly before moving on to the next grain size, to avoid contamination, which can scratch the polished surface or even cause material to be removed. It must be ensured that in each polishing phase all the scratches produced in the previous phase are removed, otherwise they will remain in all subsequent phases and the polishing will be defective.

These considerations are documented with additional information (see Appendix at the end) about the materials currently available on the market.

2. **Sample Preparation Process**

In the following, the many cases that may arise when dealing with each phase of the process are developed in detail.

I. *Sample Preparation*

If the polished section is made from a hand specimen, simply cut a section with the disc to the desired shape and size, clean it well to remove all traces of abrasive or oils that may have been left in the pores and dry it. The cleaning can be done with acetone (which will facilitate the drying process), immersing the sample in an ultrasonic bath. Finally, all moisture is removed in a 60 °C oven. It is important not to exceed this temperature so as not to compromise the stability of the minerals present in the sample. Above 75 °C, transformation processes can occur, which must be avoided at all costs so as not to alter it and thus be able to study the sample with its original ore assemblages.

This is the first step in preparing a polished block (**PB**) from a rock sample, and the process is somewhat more complex if a polished thin section (**PTS**) is to be prepared from a sample of milled ore, as required for the usual analyses needed to support processing plants. Therefore, the techniques described below will focus on this type of preparation although they generally cover both cases, as detailed below.

When the objective is to characterize the metallic ore fed to a concentration plant, or even to study the mineral content of the effluents, it is normal to start from a sample of mineral concentrate, previously crushed or milled, to make a PTS.

The first thing to consider when dealing with a milled sample is the particle size. Although it is true that the study of the particle size is one of the aspects to be studied through a mineralogical analysis, it is easy to answer some of the questions with a careful observation of the sample: to recognize if it is a milled concentrate (powder) or only a crushed or ground sample (visible fragments), or a mixture without previous sieving; that is, both options at the same time. Something as simple as anticipating the basic characteristics of the concentrate in advance can determine the success or failure of the final result. Whenever possible, it is recommended to do a grain-size screening to make the sample size as homogeneous as possible. However, we are often faced with heterogeneous samples, without prior screening. In these cases, the presence of fines and their quantity will be key in the preparation of the sample.

The fines tend to agglomerate, forming lumps that hinder the application of digital image analysis programs, as they produce a large number of "unclassified" pixels. It is also important that the particles do not touch each other (Fig. A.1) otherwise the particles in contact might be processed as a single particle during digital segmentation. This is a fairly common problem in electron microscopy systems. It introduces errors in determining the real particle size; moreover, and perhaps even more importantly, it gives inaccurate or erroneous information about the degree of liberation of the sample, an essential aspect in the study of a concentrate for processing in a mineral concentration plant.

Then there is the problem of segregation. Almost inevitably, after mixing with the resin the particles will stratify according to size and density. This stratification means that the particle size results obtained by studying the surface of the polished specimen may be biased, and not true to reality. In fact, at the time of setting this surface corresponded to the bottom of the mold, where a certain segregation by gravity is to be expected, so there will be a greater presence of large (or denser) particles and a smaller presence of fines (or lighter particles) deposited on the larger particles (i.e. when the specimen is inverted to proceed to polishing, they will remain below the surface to be studied or, at best, in the interstices only). They can thus give the false impression of

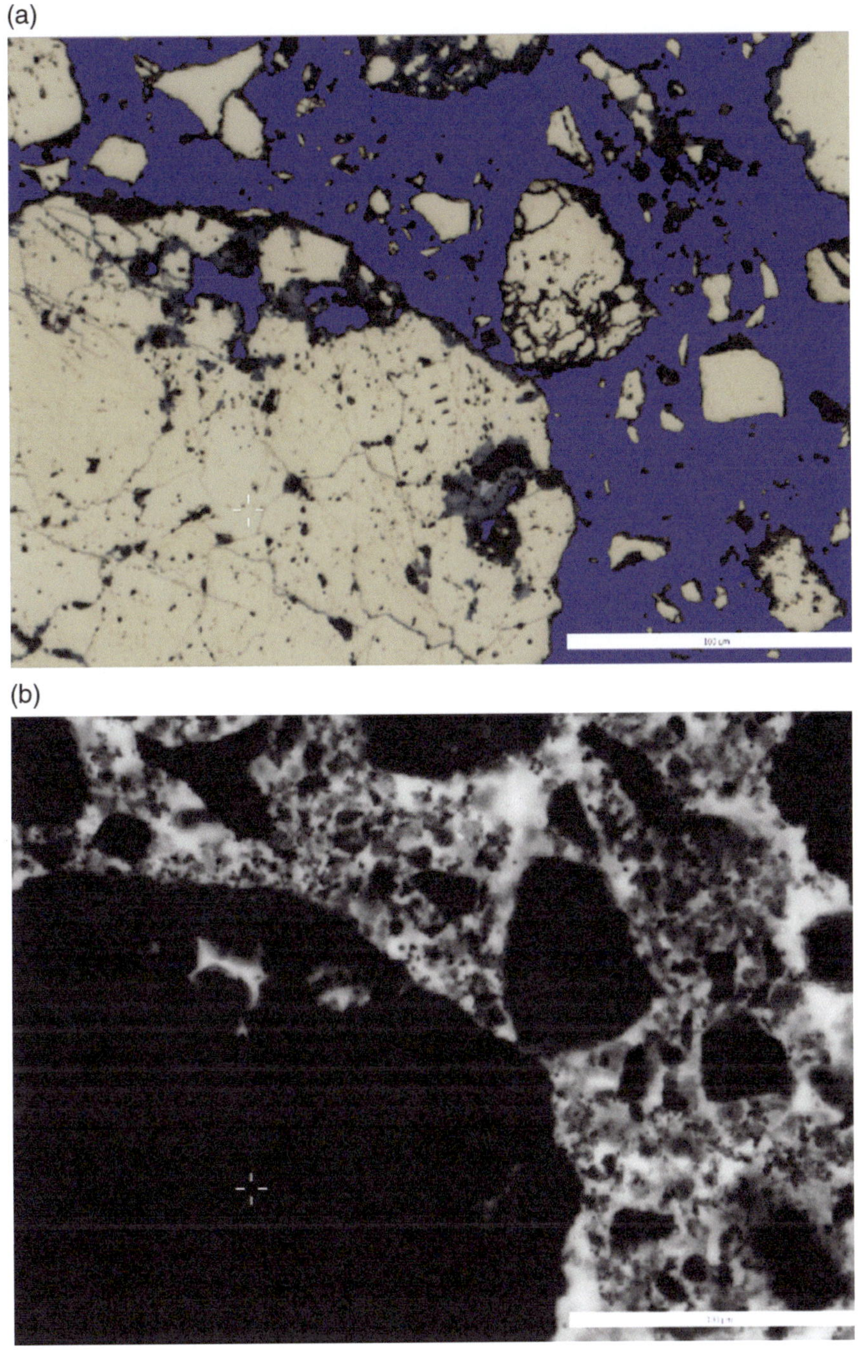

Fig. A.1 Two independent particles touching each other. On the Top, false color image generated with the AMCO system with a blue mask on the resin. On the bottom, a fluorescence image. It can be seen that these are two independent pyrite particles (a large one in lower left corner and a smaller one beside its upper-right border) but, when touching each other, if the image is not properly treated digitally, the automatic analysis systems could consider it as a single particle

having an average particle size that is coarser (or with a higher proportion of dense components) than the actual size.

Although there is no definitive and completely effective solution to this problem, some actions can be performed in this first phase to facilitate the process in subsequent phases. First, it is recommended to wash the sample, preferably with acetone, to remove possible traces of oils or abrasives and to facilitate drying. The objective is to disintegrate the agglomerations of material as much as possible (Fig. A.2), aided by a spatula. Once the acetone has evaporated, the sample should be dried in an oven at 60 °C so that it loses all moisture. As a general rule, the sample should be left to dry for two hours, but

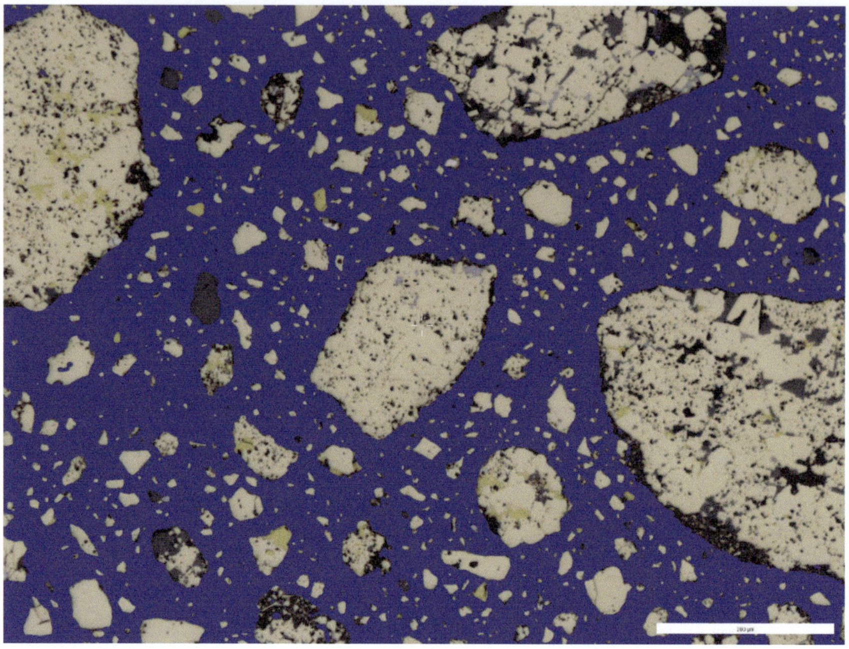

Fig. A.2 Example of well-distributed particles of different sizes. Blue mask on the resin generated with the AMCO system

this can be extended depending on the initial state of the concentrate. In general, the greater the amount of fines, the longer the drying time required.

For some years now, moreover, the addition of pure graphite particles to the mineral concentrate has been spreading in the preparation of samples for electron microscopy studies Although it has not been shown to have a great effect in reducing particle segregation, there is evidence, both empirical and theoretical, of its effectiveness in preventing the formation of particle agglomerates (Røisi and Aasly 2018), which is directly reflected in a lower contact ratio between them. Graphite, as it is not detected by EDS sensors, does not affect the final results of electron microscopy analysis.

Regardless of the system used, lumps are a widespread problem in the study of metal ore concentrates through automatic image analysis programs because, above all, they make it difficult or even impossible to carry out particle size and mineral liberation studies. This is because the contours of each individual particle cannot be accurately determined.

These problems disappear if the object is to make a preparation from a rock fragment (to make a PB), because the only thing to take care of during the preparation is that no bubbles are left on the surface to be polished and, later on, to avoid differential pull-outs and polishing.

II. *Embedding*

For embedding samples, currently the most widespread option is to use silicone molds that allow the sample to be easily extracted once the resin has set (Fig. A.3). If it is a concentrate, it is essential to use a fluid epoxy resin, although it is advisable to use this type of resin for blocks as well.

Although there is a wide range of resins on the market, a widespread example is the Epofix kit by Struers, which consists of two parts: part A is the epoxy resin itself; and part B is the hardener.

To increase the fluidity of the resins, it is important to work them hot. Ideally (if working with a resin with hardener such as Epofix) it is best to put aside in a smaller pot the amount of resin required for the work. In this way, this pot can be kept warm in an oven at 40 to 50 °C during the sample preparation process.

Before mixing the resin with the hardener, 2–3 g of the ground sample to be analysed are put in each mold. The sample is spread at the bottom of the molds as evenly as possible, without piling up at one point. Next, a small amount of resin (parts A and B already mixed according to the manufacturer's instructions) is poured gradually to cover the concentrate (Fig. A.5). The molds are placed in a vacuum hood to remove the bubbles by slowly bringing the resin to effervescence. The decrease in surface tension will wet all the particles, thus avoiding agglomeration of the material. Special care must be taken to ensure that it does not overflow, reducing the vacuum by allowing air to enter, if necessary. Then, the test molds are removed and filled with the resin, making the vacuum again, if necessary. In order to concentrate the particles at the bottom, the samples can be centrifuged (Castroviejo 1980).

In this step, to make a PB it is necessary only to cut the rock fragment to a suitable size to fit the mold and

Fig. A.3 Cylindrical silicon molds. This format allows the same sample to be studied both in the AMCO system and in an electron microscopy system or by traditional methods

cover it with the resin, making sure that at the base of the cylinder the face of the fragment is as flat and as large as possible.

The samples should then be left in the oven at 60 °C for at least three hours to accelerate the setting of the resin. After this time, they are left at room temperature for half an hour before being removed from the mold, to allow them to finish setting. Once out of the mold, the test piece is sufficiently hardened to allow preliminary roughening to even out the surface.

Nowadays, resins are available that, with the help of an ultraviolet lamp, can harden in a few minutes. This type of technique makes it possible to have a polished section ready to be analysed on the same day that the sample is taken.

Box Annex 3.1. Automated Gangue-Resin Discrimination (Sample Preparation for the Amco System) The AMCO system for automated ore characterisation, based on RL (reflected light) microscopy, requires polished sections, either PB (from rocks/run-of-mine-ores) or PTS (from milled ore, plant feed or ore concentrates of effluents). To take full advantage of the system and perform particle size and particle analysis, the latter requires the addition of a fluorescent dye to the resin to ensure its automated distinction from the gangue.

To discriminate between gangue and resin, a procedure was developed in LMAAI, because both resin and gangue have very similar reflectances and cannot be distinguished with RL. The goal is for the image analysis program (AMCO, in this case) to be able to create a mask over the resin so that it can be discounted from the image classification process. This involves altering some aspect of the resin so that AMCO can detect it and distinguish it. While this problem is negligible in whole-rock samples directly embedded in resin to form polished blocks, ground mineral samples have to be mixed with epoxy resin to which a fluorescent additive (e.g. EpoDye, also from Struers) is added. Subsequently, they need to be cut and polished as thin transparent slides (about 30 µm thick), so that no fluorescent resin remains under the transparent gangue minerals; otherwise, it will shine from below and interfere with distinguishing resin (Grunwald et al. 2019) Fig. A.4a and A.4b.

This sample preparation method allows effective discrimination between the fluorescent resin matrix and mineral particles, including gangue minerals. The solution is applied automatically by the program. First, an additional fluorescence band is acquired from the polished transparent thin section (Fig. A.4a), which highlights all fluorescein-stained resin regions. Processing with AmcoAnalysis generates a mask on the resin, removing those pixels from the final modal analysis (Fig. A.4b).

If, in the preparation of the PTS, it is suspected that the segregation of the material has been significant, it is advised to make two samples from the same concentrate: one as explained above; and another by cutting the specimen vertically. In this way it is possible to study the segregation of particles according to their size and density, and to make a complementary or more reliable particle-size study.

(a)
(b)

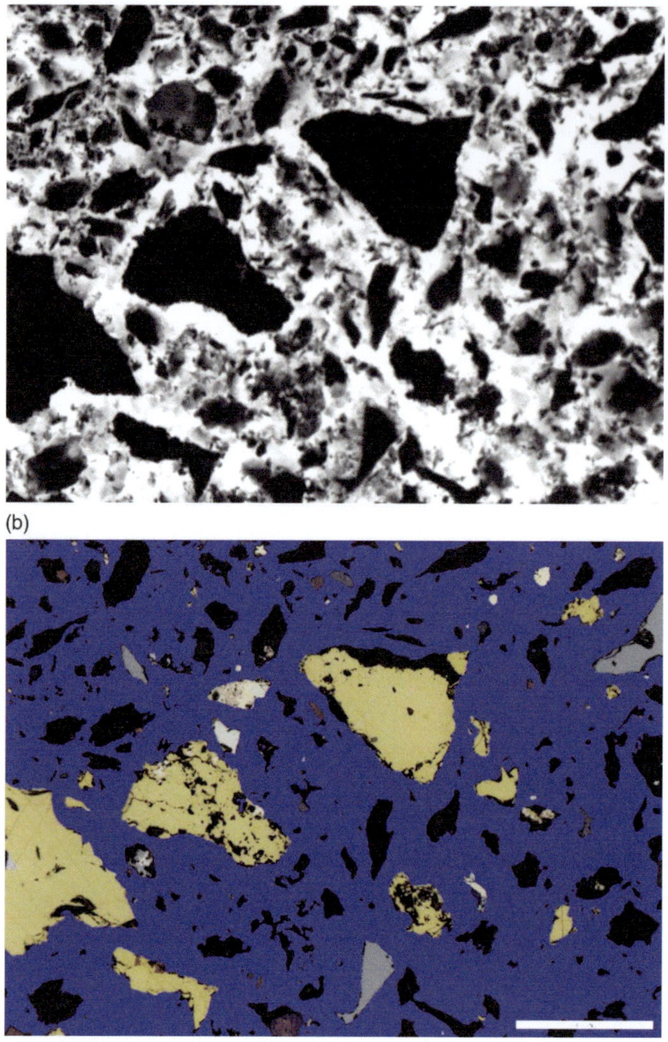

Fig. A.4 **a** Fluorescence image of a polished thin section. **b** Blue mask on the resin in the final digital image. Images acquired with AmcoCapture and processed with AmcoAnalysis

Fig. A.5 Embedding of ground samples in fluorescent resin

Fig. A.6 Preparation of the specimen surface to be glued to the object holder glass. A light roughening is done to even the surface, and to facilitate gluing the edges are rounded with the same polishing machine

So far the process for making a polished block or a PTS is the same. Now, to make PTS, it is necessary to glue the bottom surface of the block that has just been prepared onto a glass slide, after an initial roughening of the block's surface. If you wish to continue making a PB, go directly to the explanation of the **pre-polishing** process (§ III).

One of the possible roughening sequences to prepare the surface to be bonded is as follows:

1. Manual grinding is performed on a polishing machine with medium-grit sandpaper (P600 to P1200, depending on the hardness of the predominant material in the sample), using water as a lubricant. The aim is to achieve a flat surface and to round the edges of the block to facilitate gluing to a slide (Fig. A.6). Before proceeding to the next step, it is important to clean the samples to remove the abrasive residues.
2. The second grinding is done on an automatic polisher, using 6–7 μm carborundum for six minutes, or fine grit sandpaper (P2400).
3. Finally, the blocks are cleaned with the help of an ultrasonic bath and dried with a compressed air nozzle.

To glue the samples to the slide, the easiest way is to use the same epoxy resin used during embedding, in this case without the fluorescent additive. It is important to glue with an undyed epoxy resin in order not to modify the behaviour of the transparent gangue grains under fluorescent light.

For the glass to fit the polishers optimally, it is recommended to polish the bottom side of the glass (the side opposite to where the sample will be glued), as well as rounding the corners and edges. In addition, in order to permanently mark the name of each sample and not to confuse them, it is advisable to scratch the glass with the corresponding code with an awl or similar tool.

To prepare the glue, parts A and B of the resin mentioned above are mixed again. One gram of the final product is sufficient to glue 12 samples. To avoid the production of new bubbles, it is very important to pay attention to the process of gluing the block to the glass. Perfect technique requires practice but, as general advice, place a few drops of the resin both on the slide and on the surface of the sample and then centre the sample. Carefully put them together while making small circular movements, pressing the centre of the preparation to remove all the bubbles. Check that there are no air bubbles by carefully lifting and turning the sample over. If necessary, peel off the specimen and start again (Fig. A.7). If the surface is extremely porous, use the vacuum hood.

To finish using the resin, place the glass on a hot plate at 45° (a higher temperature could create bubbles) with weights in the centre of the preparation (Fig. A.8). Leave to dry for one to two hours.

III. *Slicing, grinding and pre-polishing*

To speed up the process and keep the sample surface as flat as possible, it is essential to cut the specimen using a precision slicer (Fig. A.9). Ideally, if the machine allows it, it should be cut to a thickness of approximately 1 mm (Fig. A.10).

Fig. A.7 Example of ineffective bonding of the specimen to the glass. It can be seen that, while in half of the specimen the epoxy resin is adhering the surfaces correctly, in the other half it is missing. The name of the sample is marked on the glass with a chisel

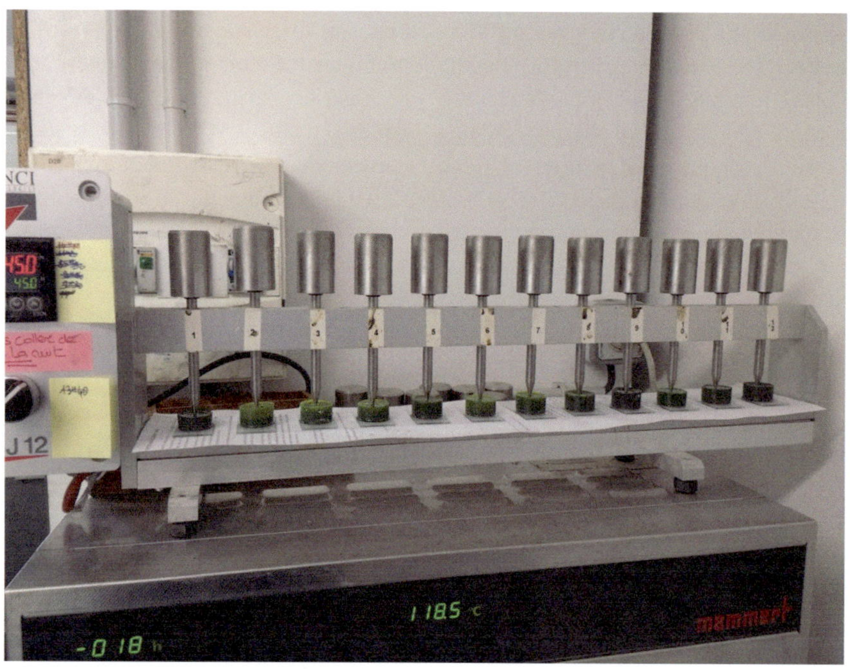

Fig. A.8 Drying test specimens glued to slides

From this point on, there is a wide range of machinery to facilitate the process. The most complete is the multiplate grinder type (Fig. A.11), a grinding machine that allows automatic thinning of the section adhered to a slide, down to a thickness determined by the technician (normally 100 μm, if it is a PTS, to prevent the preparation from disappearing during the polishing process). This type of technology ensures that the surface to be polished remains parallel to the base of the sample in order to guarantee its parallelism to the stage when, once polished, it is studied under the microscope. This is an indispensable condition for the validity of reflectance measurements, which require the orthogonality of the incident beam on the sample. For the reader's convenience, a commercial example of this type of machine is the multiplate grinder 1.03.12 G, by BROTLAB, although there is a wide variety on the market.

Fig. A.9 Precision slicer to reduce the specimen thickness to 1 mm

Fig. A.10 Cutting result with the precision-slicing blade

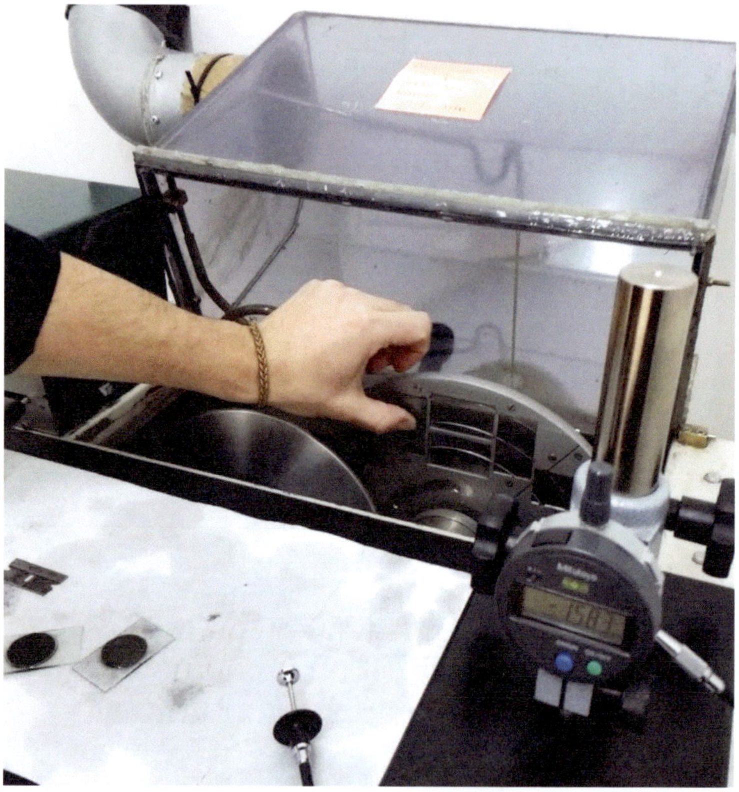

Fig. A.11 Multiplate grinder

Table A.1 Equivalence between abrasive grit sizes in μm and FEPA P

Grain size (μm)	200	125	82	68	46	30	22	18	15	7	4
FEPA P (Europe)	P80	P120	P180	P220	P320	P500	P800	P1000	P1200	P2000	P4000

Source Struers

If only a classic *lapping-machine* is available, the most common procedure is to use carborundum mixed with water on the specific metal plate on the polishing rotor; discs with abrasive; or sandpaper glued on magnetic surfaces that adhere to the polishing rotor. The advice given below is for a general case, using sandpaper on magnetic support.

The general sequence of **roughening** or **pre-polishing** is done with water or a commercial preparation as a lubricant, progressing from larger carborundum diameters to finer and finer grit sizes of abrasive: P120, P240, P320, P400, P600, P800, P1000, P1200 and P1500. These sequences correspond to the European nomenclature of particle size equivalents FEPA P. Their particle size equivalents in μm are listed in Table A.1. This pattern is only an example. Depending on the manufacturer, the variety of abrasive grit sizes may change. As a guideline, Buehler offers a larger number of grit sizes in its catalog than Struers.

When working with a PTS, it is recommended to start directly with P800 or P1000, depending on the starting thickness. This sequence is not rigid and can be adapted as needed, taking care always to go from coarser to finer grain size and cleaning the sample well with water and an ultrasonic bath before changing abrasive size. Several short cycles will be carried out, especially with the coarser grits, some five to ten minutes each, taking care that there is no lack of lubrication or abrasive (in the case of using carborundum powder). In each cycle the result should be checked with an auxiliary microscope to decide whether to repeat or to move on to the next particle size.

The boundary between roughening or pre-polishing and polishing is very thin, and will depend more on the grain size of the abrasive, the revolutions and pressure exerted by the machine, and the duration of each cycle than by the materials used. In the first stages of grinding it is normal to use between 120 to 200 rpm (less if the minerals are soft) and to decrease the speed as the polishing and final finishing progresses. As an empirical orientation, the grinding ends when the thin section starts to become transparent, starting the more delicate phases of polishing and finishing. In the case of a polished specimen, polishing begins when, with the sample completely dry (as an observation for a quick intuitive check; it is always better to check with a microscope) the surface begins to shine when placed under the light at a certain inclination.

IV. *Polishing*

This is the most delicate step. If a PTS is being prepared, it is recommended to start polishing when the section is between 60 and 40 μm thick. Before starting, it is necessary to check the thickness and parallelism with a precision digital micrometer. The indications given here are valid for PB and PTS, with finishing using cloths and abrasive suspension with diamond powder of 6, 3 and 1 μm, but the general features are easily exportable to other materials.

Depending on the thickness and the scratches present, start the process with 6 or 3 μm abrasive. The pressure of the sample on the cloth should be minimal, and the abrasive fluid should form a constant film between the preparation and the cloth. It is recommended not to exceed 100 rpm. If, due to the number and depth of the scratches, it is necessary to start with 6 μm abrasive, it is recommended that the section be about 50 μm thick to avoid excessive wear and to prevent it going below 30 μm. Short cycles of five minutes (shorter if there are many contacts between soft and hard minerals) will be carried out until the deepest scratches have disappeared, taking care and observing that relief is not created.

The objective is to start with the abrasive of 1 μm when the sample is practically polished, so that this last step serves only as finishing, at 50 rpm, for the shortest time possible, making cycles of no more than ten minutes. Of course, the cloth must be changed with each grain size. The ideal is to have three cloths already mounted on a magnetized support, marked for which grain size that each is intended, to avoid cross-contamination.

If the pre-polishing has left deep marks or relief, it is recommended to continue polishing on a rigid support with alumina, carborundum or very fine grit paper, between P2000 and P4000 (equivalent to a grit size between 7 and 4 μm). However, if the damage is deep and long-lasting, it is necessary to return to pre-polishing.

The final polishing step, if necessary, can be done with diamond paste, applied with a lubricant. The diameter of the diamond monocrystals in the pastes offered by the commercial companies ranges from 15 to 0.25 μm. However, for polishing or finishing, it is recommended to start with 6 or 3 μm and gradually decrease the particle size of the diamond paste towards finer and finer sizes, ending with 0.25 μm. The finish should always be fine-grained (≤ 1 μm).

3. **Polishing Defects and Artefacts**

When sample preparation or polishing is faulty or insufficient, various artefacts may appear. Below are some examples of the various problems caused by polishing defects that mineralogists may encounter during the study of polished sections.

I. *Lumps or aggregates of material*

As indicated, lumping is the main problem faced by automated image analysis systems on finely ground samples (Fig. A.12).

II. *Poor resin setting*

In Fig. A.13, two comparative images of a polished section are shown, the top image taken with polarized light and the lower image with interference contrast, using the Wollaston prism. The prism highlights the differences in relief. In this case, an apparently well-polished sample reveals a rough surface due to poor resin set, which contrasts with the smooth, well-polished surface of ore and gangue.

III. *Loss of material, relief and pores*

If care is not taken, the presence of minerals of different hardnesses leads to the appearance of reliefs on the edges of the grains. It is also very common for particle pits or pull-outs to occur. Generally, as shown in Figs. A.14, A.15 and A.16, these problems are associated with insufficient polishing.

IV. *Exudate and patina*

Figures A.17 and A.18 show the same particle in air (Fig. A.17) and with oil immersion objective (Fig. A.18). Figure A.17 shows a zonation that does not disappear with crossed polars (d + p setting), which could be mistakenly understood as a secondary marginal zonation. However, after immersion in oil (Fig. A.18), the zonation disappears completely, both with single polarizer and with crossed polars, because the oil exudates from the polishing have dissolved in the immersion oil. This is precisely why the exudations appear on edges, pores and microfissures.

A similar type of zoning may occur due to the presence of a patina that is removed after cleaning with hexane, so it is very important to clean the samples whenever in doubt (Fig. A.19).

(a)

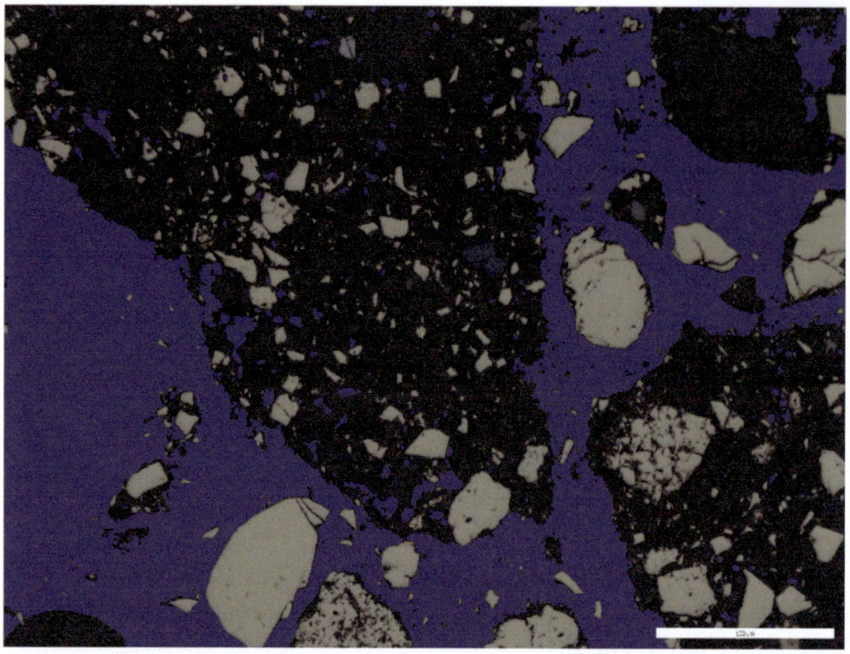

(b)

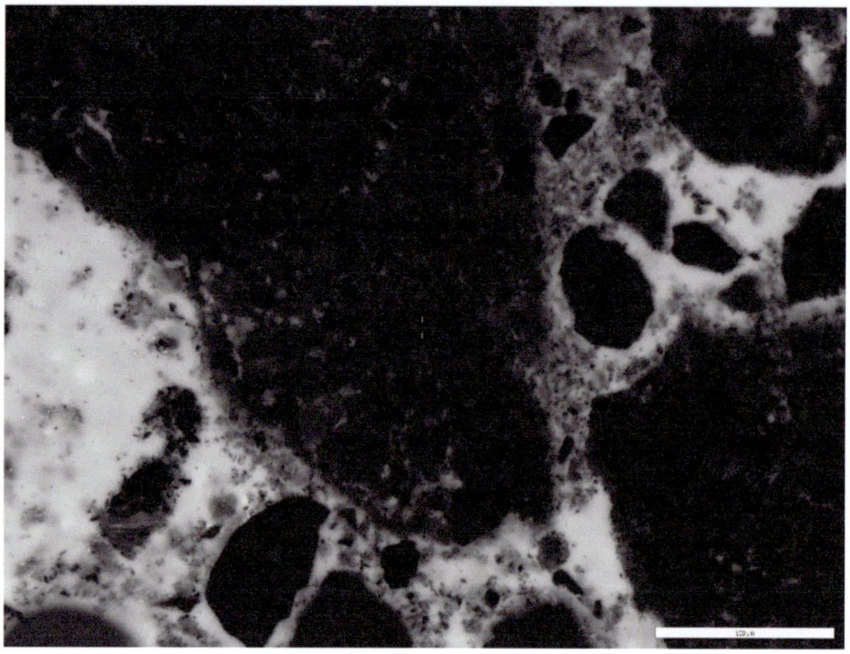

Fig. A.12 A field of an automated sample scan processed with amcoAnalysis. Top, image with the blue mask covering the area assigned to the resin. Below, image acquired in fluorescence observation mode to identify the fluorescent resin areas. It can be seen that large lumps have formed, composed of many particles stuck together, because there are interior areas where the fluorescent resin has entered

V. *Final finishing*

Finally, Fig. A.20 presents, as an example, good-quality polishing of a composite ore (pyrite, pyrrhotite, chalcopyrite, sphalerite, galena and quartz) with intergrown species of different hardnesses and no artefacts.

4. Final Considerations

As we have seen in the previous pages, there is no single method or formula for making good-quality polished sections. In fact, only some of the methods and products available on the market have been discussed here. If well chosen, they can facilitate and improve the process.

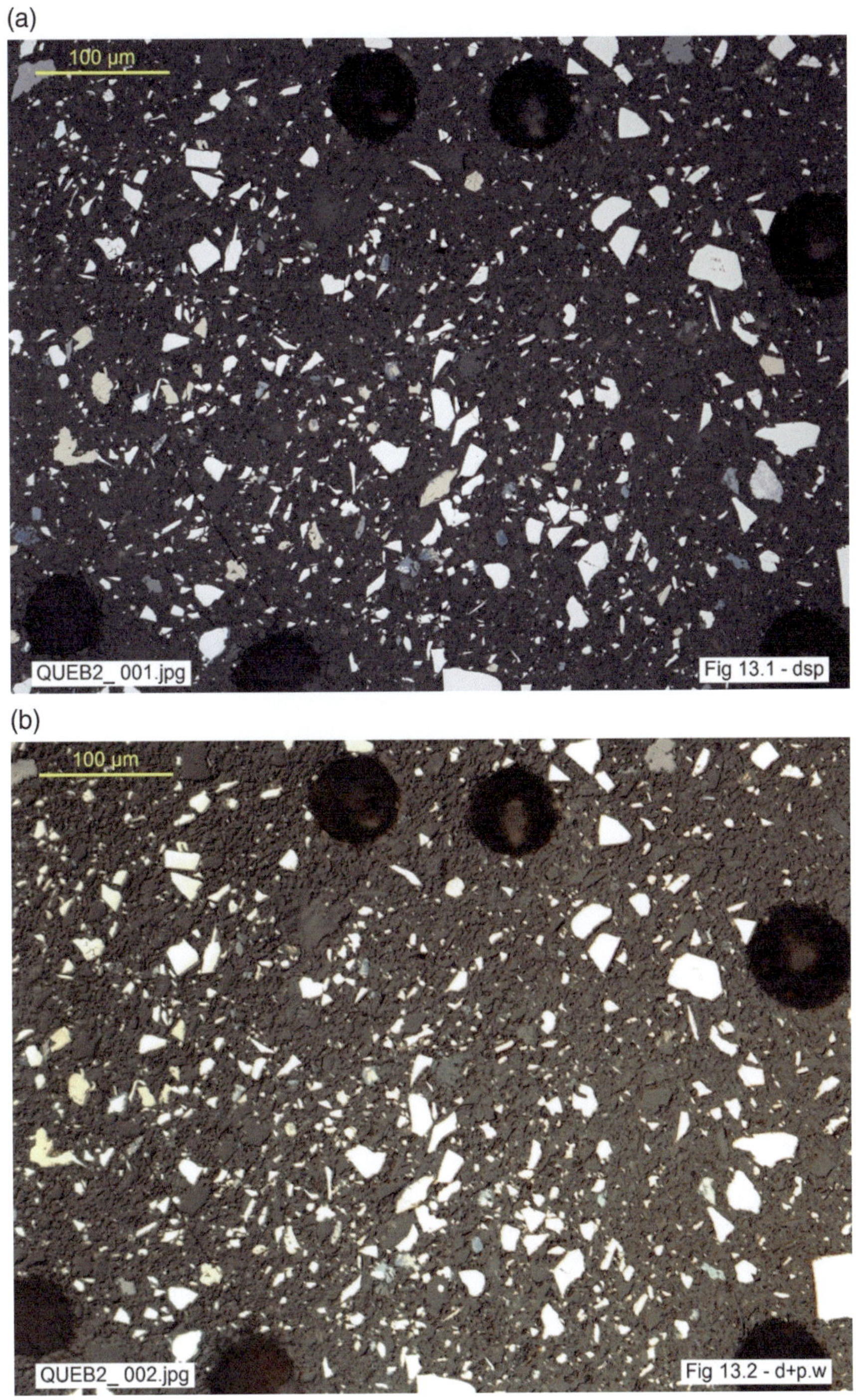

Fig. A.13 Above, image of a crushed mineral sample taken with the 10 × objective and polarizer. Below, same sample with the Wollaston prism, where the relief can be seen

Keep in mind that once the polished section is finished, tarnishing starts, and darkening or an oxidation patina may begin to develop. The speed depends on the minerals present (for example, bornite can create patina in a few hours; arsenic, in minutes). Dust or even dirt and grease are also deposited. Because of this, it is important to check the quality of the polish and, if necessary, clean the sample properly before studying it. It is recommended to study it freshly polished.

As an example of a general cleaning sequence giving very good results in removing the patina from diverse mineralogies, the following is recommended:

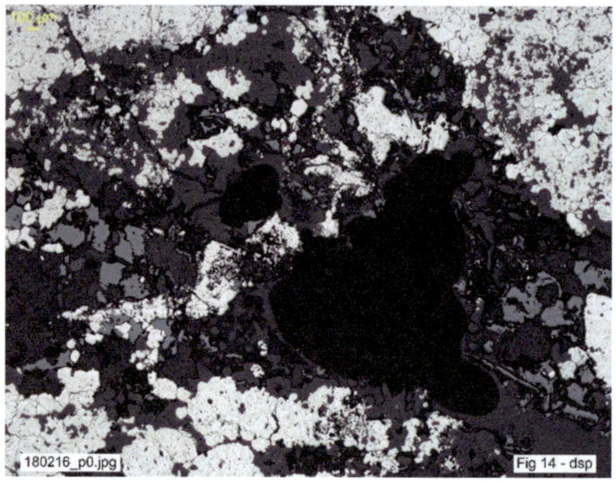

Fig. A.14 Presence of pores and relief in the polishing

Fig. A.15 Excessive relief on the edge of the particles

1. After washing hands, place a small amount (such as a grain of rice or a lentil) of ¼ micron (0.25 μm) diamond paste on the polished surface of the sample to be cleaned. Gently spread it loosely, with the fingertip, over the entire preparation, paying particular attention to the corners. Once it is well distributed, with a bit of good quality toilet paper or cloth that does not leave lint, gently rub the surface, pressing as evenly as possible with your thumb to avoid creating relief. If necessary, apply another drop of diamond paste and repeat the operation. It is important to make sure that you are getting to the contour of the sample, as this is usually the most difficult area to clean and the most neglected. Exceptionally, if the patina is very marked and widespread and cannot be removed by hand, a manual polisher can be used with a clean billiard cloth and a small amount of diamond paste. At the lowest possible speed, hold the sample firmly by pressing gently and evenly in the centre over the diamond paste, and polish for a maximum of one to two minutes in small circles.
2. Using an eyedropper, pour a small amount of hexane onto the surface to remove the diamond paste. Hexane evaporates very quickly, so just leave it for a few seconds to work well before removing it, carefully and without rubbing, with a piece of soft toilet paper. Make small circles with uniform pressure to reach the entire surface.
3. Finally, it is necessary to remove any residues that may remain on the surface. To do this, use pure alcohol, following the same procedure as with the hexane. Once they are removed, dry the surface with another piece of clean toilet paper to remove any marks that may remain.
4. Blow the surface of the preparation with an air pump or compressed air canister to remove any dust, lint or paper fibres that may have adhered to the preparation.

Once the process is completed, it is important not to touch the surface of the sample again and to handle it carefully. It is necessary to always check the cleaning result under the microscope. If drops, fingerprints or small drag marks are observed, go back to step 2. On the other hand, if a patina is still visible on the mineral, return to step 1.

Fig. A.16 Irregularities on the surface of the preparation due to different hardnesses in the material: loss of mineral grains, pores and relief

Depending on the initial condition of the polished section, sometimes it is only necessary to clean from step 2 or 3 (if the marks are slight fingerprints).

Cleaning is an essential process that should be carried out whenever the previous examination indicates the need (which is often the case when dealing with bornite, chalcopyrite or silver ores). Therefore, it should always be done with great care, because, although ¼ micron diamond paste will not thin the surface, it can create relief at the contact between hard and soft grains. If the samples are stored correctly, certain mineralogies that do not develop a great deal of patina may not need the first pass with diamond paste, which keeps their surface in good condition for longer.

(a)

(b)

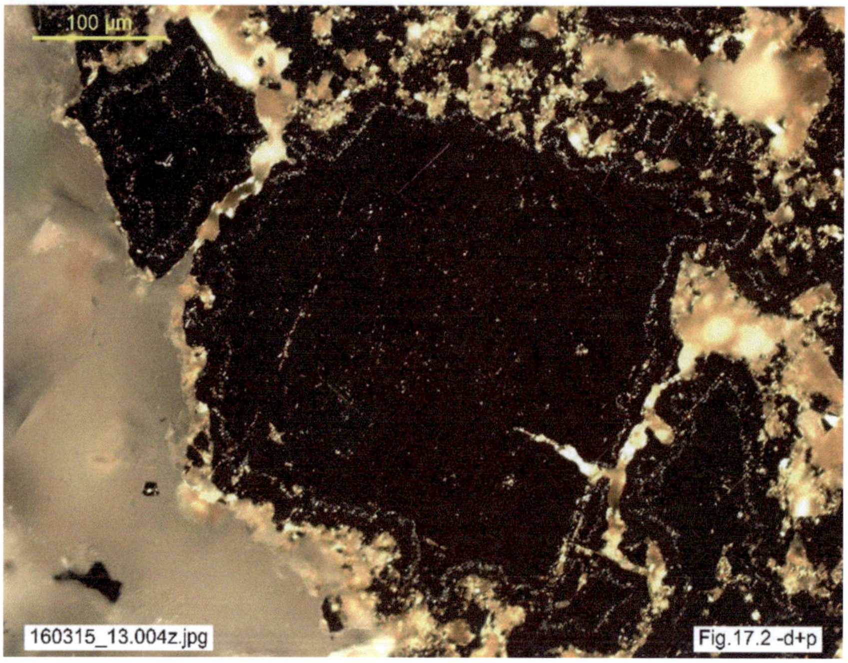

Fig. A.17 Apparent marginal zoning at the edge of the grains. The effect is removed by cleaning the sample (Fig. A.19). Image a with single polar (dsp), and b with crossed polars (d+p)

5. **Appendix: Introduction to Abrasive Types and Surfaces**

Three fundamental aspects must be taken into account: the type of abrasive; the size of the abrasive; and the nature of the minerals present in the polished section.

1. *Type of abrasive*: depending on the time and materials available and the nature of the sample, one type of abrasive or another is more suitable. According to their characteristics and finishes they are divided into five groups, ordered from most to least effective, explaining the advantages and disadvantages of each choice:

(a)

(b)

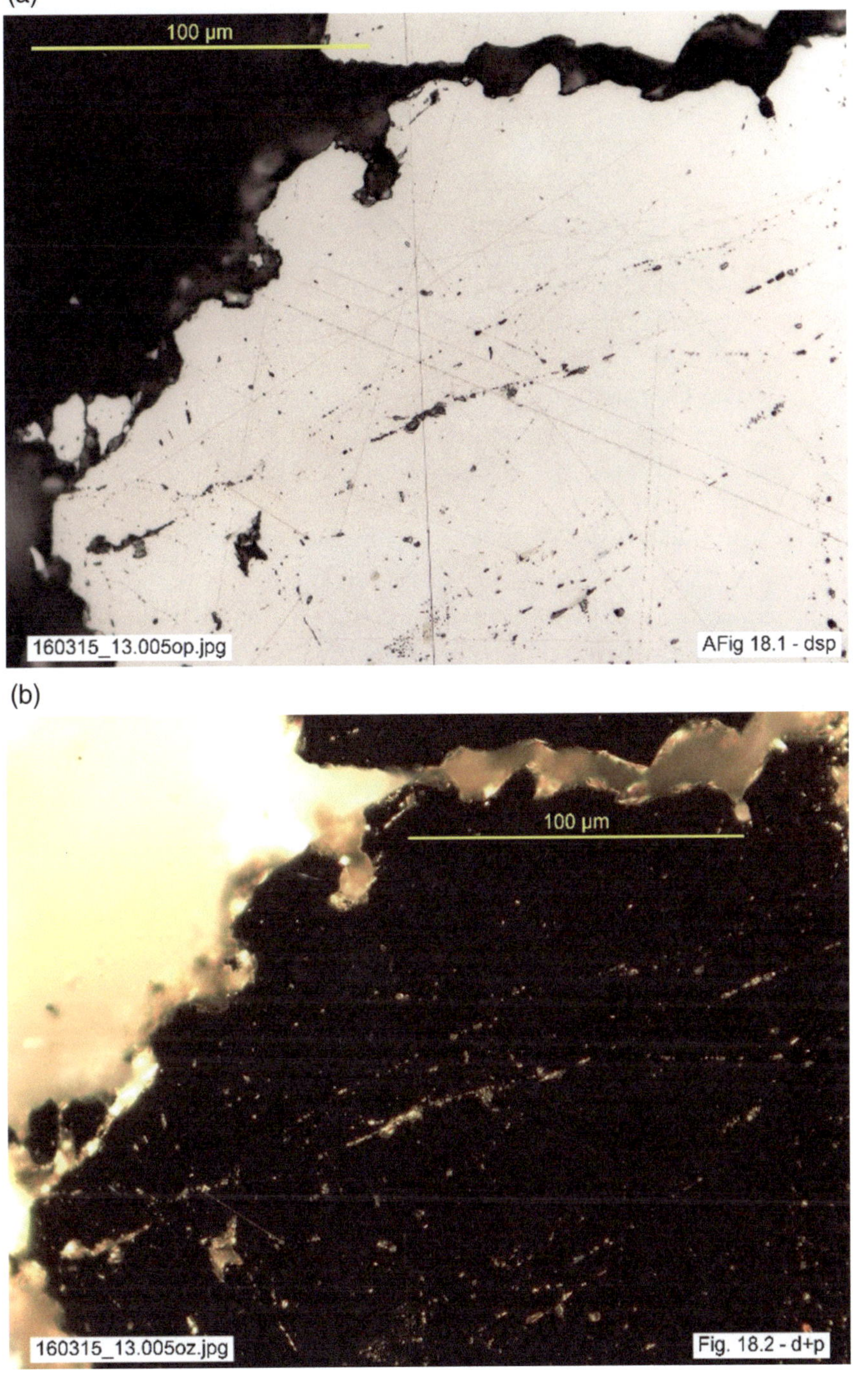

Fig. A.18 Detail of the field in Fig.. A.17 after immersion in oil. With the oil, the zoning disappears both without the analyser a and with it crossed b

1st. In the form of grains—whether carborundum, alumina or diamond—which are mixed with lubricating liquid (water or oil) to form a more or less fluid paste on the hard metal plate prepared for this purpose on the polishing machine rotor, or used directly on glass for manual grinding or polishing.

This is the method with which the most perfect polishing is obtained, especially when minerals of different hardness are treated. As the abrasive is on a hard, flat surface, this does not deform during the work, so that the surface of the section remains horizontal and the contacts between grains are not

(a)

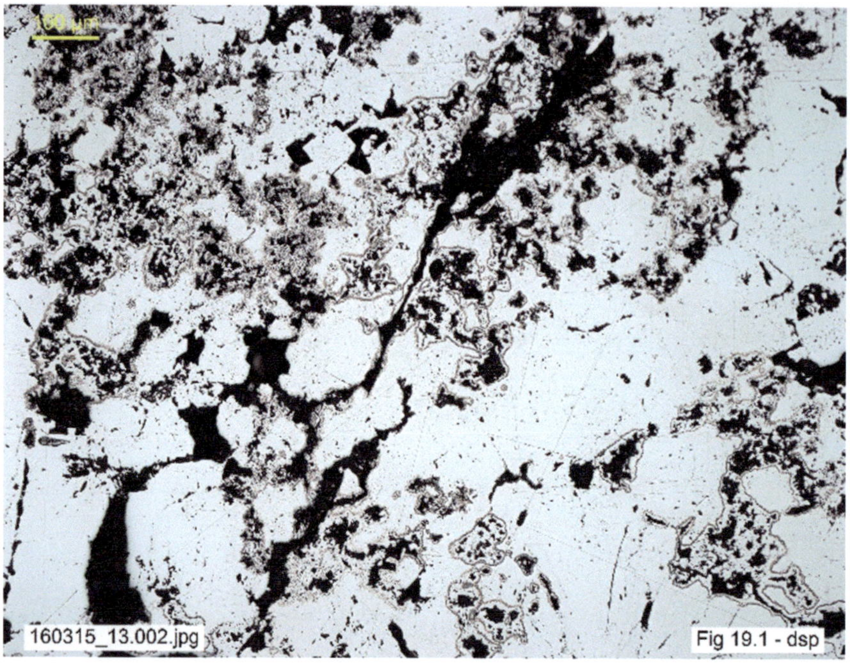

(b)

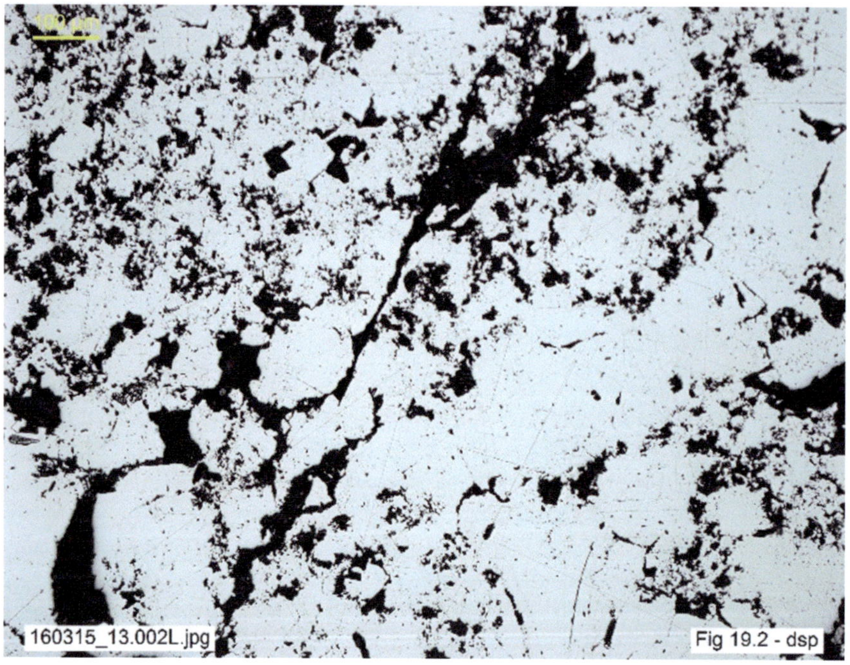

Fig. A.19 Zoning artefact a that disappears after cleaning with hexane when the patina is removed b

rounded, as they are all ground equally. If the abrasive were spread on a soft surface, it would tend to wear when polishing hard minerals. Therefore, irregularities would be created on the surface, causing the abrasive to sink deeper into the soft minerals and leaving deep scratches, and accentuating the relief.

Despite being the method giving the best results, it is a very slow process lasting up to 20 hours divided into four phases, the third being the longest (14–16 hours) with a very fine grain size of abrasive (0.5 µm). In addition, it is important to ensure that the abrasive forms a thin, continuous film on the disc and on the

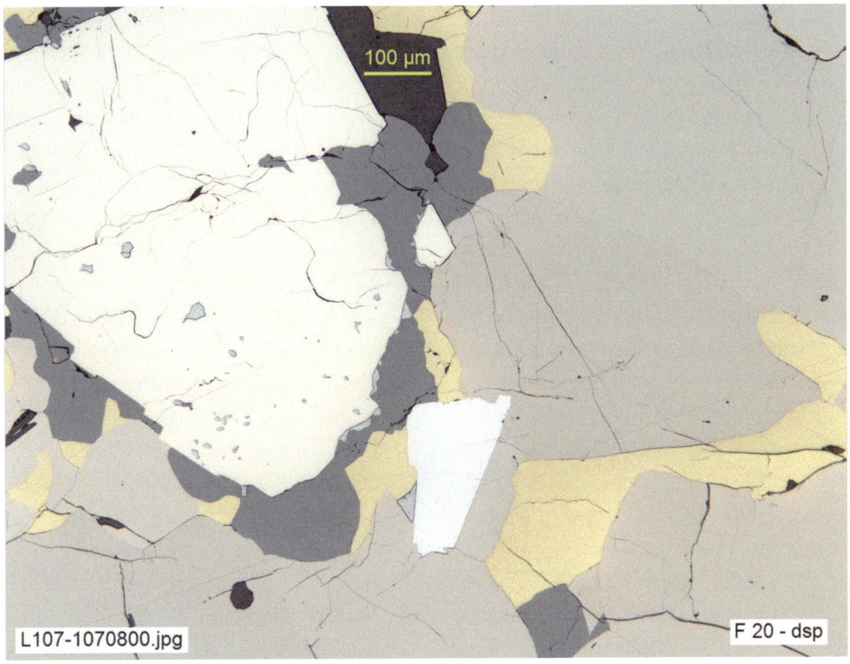

Fig. A.20 Example of quality finished polishing (without artefacts)

surface of the sections during all phases. Although the method is very time-consuming for manual polishing, performance increases considerably with polishing machines that work on several samples simultaneously. For example, the classic Rehwald-Depiereux polishes six specimens at a time, with excellent results (cf. Castroviejo 1980), reaching a polishing quality that is almost without equal (vbgr. that of the Ramdohr-Rehwald collection, used in Heidelberg for the fundamental descriptions by Ramdohr 1980).

2nd. Rather than a category in their own right, these are a variant of the first option: ready-made commercial abrasive liquids, such as Struers DiaDuo-2, a water-based solution with monocrystalline diamond and lubricant. Like the powder abrasives, they can be used on a rigid metal surface or on cloth. The implications of each choice are discussed later.
In our experience, commercial abrasive suspensions are most useful in the final stages of cloth polishing.

3rd. Discs formed from a special alloy of metals (usually lead and antimony) that embed the grains of the abrasive material. These types are very common and give good results with samples containing minerals of varying hardness, but it should be borne in mind that their optimal use involves rigorous maintenance. This involves periodic roughening and levelling of the surface so that it remains horizontal and without protrusions that can create relief in the polished section.

Usually, sample preparation laboratories have several grinders and polishers and have a standardized preparation process, because the constant change of discs in each grain size is slower than working with sandpaper on a magnetic backing. Instead of changing the abrasive, they change the preparations to another machine previously prepared with a disc of a lower grain size. This method is useful and saves time in production lines, but it is not very flexible when it comes to adjusting to the particularities of each sample or batch of samples.

The roughening process is faster, and two or three phases of about 15 minutes per cycle is sufficient. The number of cycles required depends on the initial state of the sample.

4th. Sanding papers (usually impregnated with carborundum) on magnetic backing that are easily fixed on the polisher rotor. The results are somewhere between the first and the third option, provided they are used in good condition. In addition, they are quicker and easier to maintain. When the surface starts to wear (usually after five hours of use), simply peel the paper off the magnetic support and replace with a new one.

They are very useful for use in small laboratories with little infrastructure and to perform the initial coarser roughening (especially when the objective is to make a PB), due to their versatility and ease of changing from one to another. It is only necessary to take care that the sandpaper is well stuck to the support, without wrinkles, and that the machine is free of the previous abrasive, cleaning it with a damp cloth. It is a fairly fast system, but it tends to leave numerous polishing streaks. It is normal to carry out three or four phases, in a similar way to the previous method, taking care not to make cycles of more than one or two minutes when using the coarser grits in the first stages of roughening (P120-600-800). From P1000 onwards, longer cycles are possible. For polished thin sections, it is recommended to dispense with grits coarser than P800 in order to prevent the sample from disappearing during roughening. In preparations with soft minerals or minerals of different hardness, our own experience advises to start directly with P1000-1200 paper to avoid creating coarse grooves in the softer minerals and excessive relief.

5th. Exempt cloths and sanding papers. Cloth or billiard cloth is an alternative to the use of hard metal discs with abrasives in powder or suspension. In general, with certain exceptions that are discussed in the next point, it is recommended to avoid generalized use of this method. Personally, I reserve it for finishing polishing or even for cleaning samples with oxidation patinas, at low revolutions and in short cycles to avoid creating relief as much as possible. This is because it eliminates a large number of the scratches that may remain from the last phase of polishing with abrasive on a rigid support.

Finally, although not recommended, there is the option of using a carborundum-embedded but exempt sandpaper; that is, it is not placed on a rigid backing, rather the preparation is rubbed directly onto it. Sometimes it can be attached to the polisher rotor with a metal ring that fits into the rotor. However, whenever possible, use of this option is discouraged because it is easy for the paper to wrinkle or move around, resulting in grain-edge relief and even particle pull-outs. It is a technique that wears quickly. In return, it produces much streaking and embossing. Its use is only recommended in situations demanding quick action, where the minerals present are of high and homogeneous hardness.

2. *Abrasive size*: Using the correct abrasive grit size at each stage of both roughening and polishing is the basis of any well-polished preparation. Always start with the largest grit size required by the sample, always wash it thoroughly before moving on to the next abrasive and always check with an optical reflection microscope that each phase has been carried out correctly, paying particular attention to the horizontality of the sample (i.e. the parallelism of the surface with the base) and the appearance of scratches. In each phase, the streaks left by the previous one must be removed. If the diameter of the abrasive is reduced too quickly or the correct abrasive is not used, the process may take too long, may not progress or may even spoil the preparation by creating relief or causing particles to jump off the surface of the sample. Keep in mind that if, for example, we are grinding with carborundum with a P800-P1000-P1500 sequence, the thicker polishing scratches created with P800 that were not removed by grinding with P1000 will not be removed with P1500. This is why it is important to periodically check the grinding, and not to change phases until you are sure that the previous step is complete. It is not recommended to go back to coarser grits once the abrasive type has been changed.
3. *Nature of the polished section*: Each preparation has to be treated according to the predominant mineral composition. However, it is very common to find preparations with high contents of hard and soft minerals, such as a sample composed mainly of pyrite, bornite, galena and gangue. This heterogeneity usually makes it impossible to choose a polishing method for exclusively hard or soft minerals, so a compromise has to be reached. This usually involves softer polishing with finer grits, increasing the duration of cycles at low revolutions (maximum 50 revolutions per minute if soft and difficult to polish minerals are present).

Although it is generally advisable to polish on hard surfaces, if we know in advance that the sample contains minerals that are difficult because of their tendency to generate pull-outs or their low hardness, it may be interesting to use a cloth with a suspension of abrasive in water and lubricant, as long as there are no significantly harder minerals present (for the reasons explained above). Also some hard minerals, such as pyrite, can benefit from this type of material during the finishing phase (Galopin and Henry 1972).

References

Castroviejo R (1980) Sobre el desarrollo de técnicas auxiliares para la investigación de menas metálicas: preparación de secciones pulidas para su estudio microscópico con luz reflejada. Jornadas Minerometalúrgicas (VI National and IV International). Huelva, pp 529–559

Galopin R, Henry N (1972) Microscopic study of Opaque minerals. W. Heffer & Sons, Cambridge

Grunwald Romera U, Catalina J, Alarcón D, López Benito A, Castroviejo R (2019) A reliable method for the automated distinction of quartz gangue and epoxy resin with reflected light microscopy for geometallurgical characterization. 15 Biennial SGA meeting—life with ore deposits on Earth, vol 4. Society for Geology Applied to Mineral Deposits, Glasgow, pp 1528–1531

Ramdorh P (1980) *The Ore Minerals and Their Intergrowths.* Pergamon Press

Røisi I, Aasly K (2018) The effect of graphite filler in sample preparation for automated mineralogy—a preliminary study. Mineralproduksjon 8:A1–A23

Annex 4: Compared Properties of Ore Minerals

Annex 4.1 Specular Reflectance of Common Ores(*) (Estimated Values in Air, White Light)

Group	Abrv	Mineral	R %
Very high R (R ≥ 60%)	el	Electrum	≈90–88
	Ag	Silver	85
	Cu	Copper	81
	Sb	Antimony	74
	Au	Gold	72
	Bi	Bismuth	63.5
	aag	Allargentum	70.9–69.5
	Te	Tellurium	67
	Ir	Iridium	65
	sba	Stibarsen	64
	dy	Discrasita	62.7–59.9
	Pt	Platinum	63
	krn	Krennerite	62
	clv	Calaverite	62
	prm	Pararammelsbergite	60
High R (60 > R ≥ 40)	Os	Osmio	59
	Fe	Iron	59
	rm	Rammelsbergite	58
	sk (Ni)	Skutterudite (Ni)	57
	gud	Gudmundite	54
	sk (Co)	Skutterudite (Co)	54
	lo	Löllingite	53.5
	sf	Safflorite	53
	ml	Millerite	53 (48)
	asp	Arsenopyrite	52
	mc	Marcasite	52 (46)
	As	Arsenic	52
	spy	Sperrylite	52
	py	Pyrite	51
	nc	Nickeline	50.5 (45)
	gld	Glaucodot	50, 1

(continued)

R. Castroviejo, *A Practical Guide to Ore Microscopy—Volume 2*,
https://doi.org/10.1007/978-3-031-18954-8

(continued)

Group	Abrv	Mineral	R %
	cob	Cobaltite	50
	mau	Maucherite	49.5 (39)
	bth	Breithauptite	49
	clt	Clausthalite	49
	pn	Pentlandite	49
	bm	Bismuthinite	49 (37)
	ln	Linneite	47.5
	sbt	Stibnite	47 (30)
	ul	Ullmannite	46.5
	gf	Gersdorffite	46
	ccp	Chalcopyrite	45
	gn	Galena	43
	brv	Bravoite	43–35
	erl	Erlichmanite	≈42.7–41.8
	bl	Boulangerite	42
	mck	Mackinawita	42 (24)
	mol	Molybdenite	42 (19)
	brt	Berthierite	41.5 (30)
	hss	Hessite	40.5
Medium R ($40 > R \geq 25$)	gc	Geochronite	39.5
	flb	Freieslebenite	39.5
	po	Pyrrhotite	39
	cbn	Cubanite	39
	colo	Coloradoite	36.5
	bnn	Bournonite	36
	my	Miargyrite	36
	nau	Naumannite	36
	prl	Pyrolusite	35
	agp	Argentopyrite	35
	hol	Hollandite	33
	agl	Aguilarite	33
	ac	Acanthite	32
	cc	Chalcocite	31.5
	stm	Stromeyerite	31
	plb	Polybasite	31
	prg	Pyrargyrite	31
	td	Tetrahedrite	31
	isc	Isocubanite	35.5
	pc	Pearceite	30.5
	pru	Proustite	30
	stp	Stephanite	30
	tn	Tennantite	29
	ht	Hematite	29
	stn	Stannite	28
	en	Enargite	28
	cin	Cinnabar	28
	lz	Luzonite	27.5
	cup	Cuprite	27
	fm	Famatinite	27
	psi	Psilomelane	27

(continued)

(continued)

Group	Abrv	Mineral	R %
	cry	Cryptomelane	≈26.7
	orp	Orpiment	26
	rom	Romanechite	25.5
Low R (25 > R ≥ 12)	mgh	Maghemite	24
	rt	Rutile	23.5
	orp	Orpiment	23.1
	cv	Covellite	23 (8)
	alb	Alabandite	23
	bn	Bornite	23
	dg	Digenite	22
	ilr	Ilmenorutile	≈22? (≤ rt)
	rl	Realgar	21
	mt	Magnetite	21
	ger	Germanite	21
	mng	Manganite	20 (14)
	il	Ilmenite	19.5
	bra	Braunita	19.5
	lpc	Lepidocrocite	19 (9)
	fk	Franklinite	18
	tp	Tapiolite	18
	wf	Wolframite	18.5–16
	gth	Goethite	18–15
	sp	Sphalerite	17
	wz	Wurtzite	17
	van	Vanadinite	≈17–16
	urn	Uraninite	16.8
	cbt	Columbotantalite	16.5
	vll	Valleriite	16 (9)
	hub	Hübnerite	16
	um	Umangite	16
	gra	Graphite	16 (6)
	ulv	Ulvöspinel (Ulvite)	15.5
	lim	Limonite	≈15–10
	bad	Baddeleyite	14.8–13.4
	pyc	Pyrochlore	13.5
	cst	Cassiterite	13
	chr	Chromite	12.5
Very low R (R < 12)	sch	Scheelite	10
	az	Azurite	9.5 (6)
	mch	Malachite	9.5 (6)
	mz	Monazite	≈9–8
	cag	Chlorargyrite	9
	rdl	Ramsdellite	9–41
	bst	Bastnäsite	≈8.5–7.2
	spn	Spinel	8
	crc	Chrysocolla	≈5.5–5
	q	Quartz	5

[(*)]The ores are listed in order of decreasing R%. Based on data from QDF3-Criddle and Stanley 1993; Picot and Johan 1982; Galopin and Henry 1972; Hagel 1979; Uytenbogaardt and Burke 1971; Schneiderhöhn 1952 and author's observations or measurements

Annex 4.2 Polishing Hardness and Scratch Hardness

Annex 4.2.1 characterizes common ores by their **Polishing Hardness** (***PolH***): in Table 4.2.1.1 they are sorted by direct comparison, and listed according to decreasing *PolH*. If there is no accurate data(*) for a precise position, they are sorted by *PolH* groups (Table 4.2.1.2).

Annex 4.2.2 defines the **Talmage** scratch hardness types (**Talm**) and arranges the documented ores into groups in relation to these types (listed according to Schneiderhöhn 1952).

Table 4.2.1.1 Compared polishing hardnesses (PolH decreasing) of common ores(*)

PolH	Abrv	Mineral
PolH Group 5 (very high)	C	Diamond
	SiC	Carborundum
	cor	Corundum
	spn	Spinel
	cst	Cassiterite
	prl	Polianite (xx-prl)
	q	Quartz
	tp	Tapiolite
	cbt	Columbotantalite
	spy	Sperrylite
PolH Group 4 (high)	py	Pyrite
	mc	Marcasite
	ht	Hematite
	rt	Rutile
	urn	Uraninite
	cob	Cobaltite
	asp	Arsenopyrite
	gld	Glaucodot
	sch	Scheelite
	bra	Braunita
	chr	Chromite
	il	Ilmenite
	wf	Wolframite
	ulv	Ulvöspinel (Ulvite)
	mgh	Maghemite
	mt	Magnetite
	lo	Löllingite
	ul	Ullmannite
	gf	Gersdorffite
	po	Pyrrhotite
	Ir	Iridium
PolH Group 3 (medium)	Pt	Platinum
	pn	Pentlandite
	sf	Safflorite
	rm	Rammelsbergite
	sk (Co)	Skutterudite (Co)
	sk (Ni)	Skutterudite (Ni)
	prm	Pararammelsbergite
	mau	Maucherite
	bth	Breithauptite
	nc	Nickeline
	ln	Linneite

(continued)

Table 4.2.1.1 (continued)

PolH	Abrv	Mineral
	gud	Gudmundite
	Fe	Iron
	sp	Sphalerite
	wf	Wurtzite
	mch	Malachite
	stn	Stannite
	az	Azurite
	cup	Cuprite
	lim	Limonite(**)
	ml	Millerite
	Cu	Copper
	td	Tetrahedite
	Ag	Silver
	el	Electrum
	Au	Gold
	en	Enargite
	lz	Luzonite
	gra	Graphite
	mol	Molybdenite
	cbn	Cubanite
	ccp	Chalcopyrite
	bn	Bornite
	bnn	Bournonite
	bm	Bismuthinite
	dg	Digenite
	gn	Galena
PolH Group 2 (low)	cc	Chalcocite
	bl	Boulangerite
	clv	Calaverite
	dy	Dyscrasite
	cin	Cinnabar
	clt	Clausthalite
	um	Umangite
	my	Miargyrite
	As	Arsenic
	sba	Stibarsen
	Sb	Antimony
	stm	Stromeyerite
	cv	Covellite
	Bi	Bismuth
	brt	Berthierite
	sbt	Stibnite
PolH Group 1 (very low)	*orp*	Orpiment
	stp	Stephanite
	rl	Rejalgar
	prg	Pyrargyrite
	pru	Proustite
	agp	Argentopyrite
	plb	Polybasite
	pc	Pearceite
	ac	Acanthite
	cag	Chlorargyrite

(*)The ores are listed in order of decreasing PolH. Compiled after Schneiderhöhn (1952), Ramdohr (1980) and author's observations

(**)**Goethite** should be listed in PolH Group 3 (Medium), but actually it is assumed to be comprised in **lim**: after Schneiderhöhn's German tradition goethite (=Nadeleisenerz) is one of the mixed components of **limonite**

Table 4.2.1.2 PolH of other ores (in Groups)*

Abv	Mineral	Group
agl	Aguilarite	1
alb	Alabandite	3
bst	Bastnäsite	3
colo (cld)	Coloradoite	2
cry	Cryptomelane	4
erl	Erlichmanite	5
fm	Famatinite	3
fk	Franklinite	4
flb	Freieslebenite	2
gth	Goethite	3
ilr	Ilmenorutile	4
isc	Isocubanite	3
krn	Krennerite	2
mck	Mackinawite	3
mng	Manganite	4
mz	Monazite	3
nau	Naumannite	2
Os	Osmium	5
pyc	Pyrochlore	3
psi	Psilomelane	4
rdl	Ramsdellite	3
rom	Romanechite	3–4
Te	Tellurium	2
tit	Titanite	4
vll	Valleriite	1

*The ores are ordered alphabetically by their abbreviations (**Annex 1**), and their PolH is reported as an estimated value, defined by the numbers of the corresponding groups in Table 4.2.1.1.

Annex 4.2.2 Talmage Hardness (*Talm H*)[(*)]

Group	Type mineral	Load (g)
A	Acanthite	0.105
B	Galena	0.201
C	Chalcopyrite	0.365
D	Tetrahedrite	0.520
E	Nickeline	0.775
F	Magnetite	1.035
G	Ilmenite	1.550

Common ores sorted by *Talm H*: increasing downwards and to the right[(**)]

Group A Acanthite type	Group B Galena type	Group C Chalcopyrite type	Group D Stibnite type	Group E Nickeline type	Group F Magnetite type	Group G Ilmenite type
ac	cin–	prg–	znc–	tno–	gf–	psi–
agl	Te–	colo–	brt–	lim–	mc–	cob–
hss+	mol–	dy–	in–	ml	py–	urn
clt+	gn	syl–	po–	mau	sk–	il
Bi+	cc	plb–	pn–	rm	lo	rt+
	Au	bnn–	wz–	nc	asp–	cst+
	ker	As	cup–	bth	mt	chr+
	nau	aik	stn–	sf	gld	
	rl	pc	lz	ul	fk+	
	Ag	ccp	td	ln+		
	stp	zk	tn	hm+		
	stm	alb				
	Cu	clv				
	my	cbn				
	um	jd				
	pru+	krn				
	bl+	fm+				
	sbt+	sp+				
	cv+					
	orp+					
	bn+					
	jm+					
	bm+					
	tdy+					

[(*)]Scale (A–G) defined by mineral type according to the load (g) of the microsclerometer which produces a standard scratch.
[(**)]The scratch hardness of ores marked +/– may vary, resulting in: DTalm $\geq / \leq$ the type, respectively (after Schneiderhöhn 1952). Mineral abbreviations: cf. **Annex 1**.

Annex 4.3 Minerals with Distinct Internal Reflections (I.R.)

IR colour(*)	Mineral (Abrv)	Remarks(**)
Blue	Azurite (az)	m
	Chrysocolla (crc)	m. Turquoise blue
Green	Malachite (mch)	m
	Annabergite (anbg)	m
	Spinel (spn)	m/d
	Alabandite (alb)	d
Pink	Erythrine (ery)	m
Yellow	Orpiment (orp)	m
	Sulfur (S)	m
Yellow to reddish brown	Sphalerite (sp)(***)	m/d
	Wurtzite (wz)	Colorless to *RdBr*
	Cassiterite (cst)	m. Colorless to YlBr
Yellowish red to reddish brown	Realgar (rl)	m
	Lepidocrocite (lpc)	m
	Goethite (gth)	m
	Titanite (tit)	m
	Pyrochlore (pyc)	m
	Rutile (rt)	m/d
	Wolframite (wf) (Hübnerite, hub)	d (m, like sp)
	Chromite (chr)	d
	Tetrahedrite (td)-Tenantite (tn)	o/d
	Boulangerite (bl)	o
	Columbotantalite (cbt)	o
	Uraninite (urn)	o
	Ilmenite (il)	o
Red	Cinnabar (cin)	m
	Pyrargyrite (prg)	m
	Proustite (pru)	m
	Kermesite (ker)	m
	Zincite (znc)	m
	Miargyrite (my)	m
	Pearceite (pc)—Polybasite (plb)	d/m
	Hematite (ht)	d
	Cuprite (cup)	d
	Manganite (mng)	d
	Franklinite (fk)	d
Orange	Vanadinite (van)	m
Varied	Fluorite (fl)	m
Colorless	Anatase (ana)—Leucoxene (lcx)	m
	Bismite (bim)	m
	Cervantite (cerv)	m
	Senarmontite (sen)	m
	Scheelite (sch)	m
	Zircon (zir)	m
	Gangue (gg, undif.)	m/d

(*)*Yl*: yellow; *Br*: brown; *RdBr*: reddish brown; *YlBr*: yellowish brown
(**)Abudance of IR: **m** = massive to abundant; **d** = discrete to frequent; **o** = scarce to occasional
(***)I.R. of sphalerite vary with composition from: colorless or yellowish (m, *blenda acaramelada:* pure ZnS) to *RdBr* and blood red (d, *marmatite*: Fe-rich sp); even green (Ga-bearing varieties)

Annex 4.4 Distinctly Bireflective or Pleochroic Ores

Very strong to distinct bireflectance versus pleochroism (*)			
Mineral (abrv)	B	P	Remarks(**)
Graphite (gra)	+++	++	**B & P** very strong. R minimum ≈ R_{gg}
Covellite (cv)	+++	+++	Dramatic color change: Blu/Vi (d/o)
Molybdenite (mol)	+++	++	Strong birrefl. White–gray, Rmin > R_{gg}
Umangite (um)	++	+++	Strongly pleochroic: *OgViRd-YlGyBr*
Valeriite (vll)	+++	++	**B & P** both strong, *YlBr-Gy*. Rmin > R_{gra}
Stibnite (sbt)	+++	+	Strongly birreflectant. Stays gray
Breithauptite (bth)	+	++	*ViPi to ViGy*
Nickeline (nc)	+	++	*PiWh-YlBrPi*
Kermesite (ker)	+	++	*BluGy-GyBr-OlGr*
Berthierite (brt)	+	++	*WhGy-BrPiGy*
Orpiment (orp)	++	+	Yellowish gray: **B >> P**
Stibnite (sbt)	++	–	White to gray: **B >>>> P**
Pyrolusite (prl)	++	++	Yellowish white to gray
Stephanite (stp)	+	+	*Gy-BrGy or PiGy*
Bismuthine (bm)	+	+	*YlWh-GyWh*
Stromeyerite (stm)	+	+	*PiGy-ViG/BluGr*
Millerite (ml)	+	+	*Yl-YlWh*
Cubanite (cbn)	+	+	*PiBr-GyBr*
Enargite (en)	+	+	*PiBr-ViGy*, dark
Gudmundite (gud)	+	+	*Wh-PiWh*
Jamesonite (jm)	+	+	Gray to greenish gray
Pyrrhotite (po)	+	+	*PiBr-YlBr*
Luzonite (lz)	+	+	*OgYl-PiYl*
Marcasite (mc)	–	+	*GnWh-PiWh (oil: osp)*
Manganite (mng)	+	–	Shades of gray
Miargirite (my)	+	–	White–gray
Boulangerite (blg)	+	–	Shades of gray
Calaverite (clv)	+	–	Yellowish white
Goethite (gth)	+	–	Shades of gray
Proustite (pru)	+	–	Shades *BluGy*
Pyrargyrite (prg)	+	–	Shades *BluGy*
Pearceite (pc)	+	–	*GnGy* shades
Rutile (rt)	+	–	Light shades of gray
Tellurium (Tea)	+	–	White (**B** < , **R** > > >)
Cinnabar (cin)	+	–	*GyWh*
Malachite (mch)	+	–	Carbonates: under reflected light they are gray colored ≈ gg, and show: • **R** low, but > most silicates • **B** >> silicate gg
Cerussite (crs)	+	–	
Siderite (sid)	+	–	
Carbonates (Carb)	+	–	

(*)Bireflectance (**B**) or Pleochroism (**P**) **intensity**, in air (*dsp*): +++ = very strong; ++ = strong; + = distinct; – = imperceptible
Visible **effects** (*dsp/osp*), according to B/P intensity: **P+** : color changes clearly with stage rotation. **B+/P–**: no color change, but *brightness* varies (lighter–darker) with stage rotation
(**)Color **abbreviations**. *Wh:* white—*Gy:* gray—*Bk:* black—*Blu:* blue—*Gn*: green—*Yl*: yellow—*Og*: orange—*Rd: red*—*Pu:* purple—*Vi:* violet—*Pi:* pink—*Ol:* olive—*Br:* brown. Color **combinations**: *BrRd*: brownish red—*GnGy*: greenish gray, etc.

Annex 4.5 Isotropic or Very Weakly Anisotropic Minerals(*)

Color	Mineral	Abrv	Remarks
White, gray or faintly colored ore minerals (**)			
White	Native Platinum	Pt	Very bright
	Iridium	Ir	Very bright
	Pentlandite	pn	Faint brownish-light yellow tone
	Galena	gn	Triangular pits (*T-pits*), soft
	Clausthalite	clt	T-pits; ≈ gn, but brighter
	Laurite	lrt	PGM, white with slightly bluish tinge, very hard
	Erlichmannite	erl	
	Sperrylite	spy	PGM, very hard
	Gersdorfite	gf	Frequently autom., zoned, T-pits
	Skutterudite	sk	Frequent zoning
	Ullmannite	ul	T-pits
Yellowish white	Pyrite	py	Occas. anomalous anisotropism
	Native Silver	Ag	Very shiny but tarnishes quickly
Pink white	Cobaltite	cob	Frequently automorphic
Grayish white	Coloradoite	colo	Grayish white, slightly brownish
	Maghemite	mgh	Bluish shade
	Tetrahedrite (*s.l.*)	td	Broad group. Olive-gray shades
Gray	Uraninite	urn	Internal reflections
	Spinel	spn	Internal reflections
	Chromite	chr	Internal reflections
	Pyrochlore	pyc	Internal reflections
	Franklinite	fk	Internal reflections
	Alabandite	alb	Internal reflections
	Sphalerite	sp	Internal reflections
Brownish gray	Magnetite	mt	Occas. anomalous anisotropism
	Jacobsite	jac	Occasional IR
Distinctly colored ore minerals(**)			
Yellow	Native Gold	Au	Very bright, imperfect extinction
	Electrum	el	Very bright, imperfect extinction
	Amalgam	am	Very bright (≈ Eugenite, Schachnerite)
Red	Copper	Cu	Quick tarnish (darker patina)
Blue	Digenite	dg	Pale blue color next to cv
Violaceous	Germanite	ger	GrPi hue ≈ bn (often associated)
Pink to yellowish-brown	Linneite (*s.l.*)	ln	Large and varied group
	Bravoite	brv	Zoned variety of py (Ni…)
	Melnicovite	mln	Pyritic gel

(*) Cubic (isotropic, but sometimes with imperfect extinction) or amorphous and other weakly anisotropic ores

(**) These are the natural colors of the mineral, seen **without analyzer** (i.e. ***dsp/osp***). They are the only colors relevant in isotropic ores, which with analyzer ($d + p$ / $o + p$) are always in extinction

They are described by *13 hues* or their combinations: *Wh: white—Gy: gray—Bk: black—Blu: blue—Gn: green—Yl: yellow—Og: orange—Rd: red—Pu: purple—Vi: violet—Pi: pink—Ol: olive—Br: brown*

Annex 4.6 Polarization Colors of Distinctly Anisotropic Ores(*)

Dominant colors (+P)	Mineral	Abrv	Remarks(**)
Yellow	Millerite	ml	Yl-GnBlueGr, light shades
	Graphite	gra	Anistr. extreme, very bright
	Valleriite	vll	Anistr. extreme, very bright
	Pyrolusite	prl	
	Native Arsenic	As	Very fast and dark tarnishing
	Famatinite	fm	GnYl vivid shades, twins
Orange (to OgBr-RdBr)	Osmium	Os	Very bright anisotropism
	Antimony	Sb	
	Stibarsen	sba	
	Pararammelsbergite	prm	
	Covellite	cv	Very bright golden tones
	Chalcocite	cc	Og-shaded but dull tones
	Gudmundite	gud	
	Ferroselite	fse	Bright anisotropism
	Vonsenite	von	GrViGn vivid, eye-catching tones
	Enargite	en	Strong, multicolored anisotr
Pink	Molybdenite	mol	Anistr. strong, very eye-catching
	Alloclasite	alc	
	Sternbergite	stb	Pi, RdBr, and GnGy shades
	Stibnite	sbt	Pi and bluish shades
Red	Umangite	um	OgRd shades
Brown	Pyrrhotite	po	
	Gratonite	gtt	
	Dyscrasite	dy	
	Maucherite	mau	
	Wolframite	wf	IR dominant in *hubnerite* (Mn)
	Argentopyrite	agp	Tones darker than stb
	Renierite	ren	
	Calaverite	clv	
	Stephanite	stp	Br + Yl and Gn shades
	Freieslebenite	flb	Br + Yl and BluGy shades
	Cosalite	cos	YlBr shades
Blue	Stromeyerite	stm	Luminous shades
	Cubanite	cbn	
	Nickeline	nc	Bright and vivid GnBlu shades
	Goetithe	gth	BluWh to Gy. IR abundant in *lim*
	Polybasite	plb	IR
	Bournonite	bnn	GnBlu anisotropism
	Arsenopyrite	asp	Assorted colors: BluPuVi…
	Löllingite		Blu-YlBr
	Rammelsbergite	rm	ViBlu shades
	Safflorite	sf	Vivid colors, twinning stars
	Miargyrite	my	BluWh, IR
	Jordanite	jd	GnBlu-Vi
	Geochronite	gc	GnBlu
Green	Marcasite	mc	Strong and very bright
	Pearceite	pc	Anistr. stronger than plb, IR
	Berthierite	brt	

(continued)

(continued)

Dominant colors (+P)	Mineral	Abrv	Remarks(**)
	Bismuthinite	bm	
	Native Bismuth	Bi	Glossy, but polishing scratches
	Breithauptite	bth	
	Luzonite	lz	
	Cinnabar	cin	Very abundant IR
	Cuprite	cup	Frequent/abundant I.R
	Jamesonite	jm	GnGr
	Boulangerite	blg	GnGr
Violet	Kermesite	ker	Violet to blue, eye-catching
	Stannite	stn	
Violaceous gray to greenish gray	Acanthite	ac	ViGy to BluGn dominant
	Ilmenite	il	GnGr
	Hematite	ht	GnGr, IR
Gray	Manganite	mng	
	Pyrargyrite	prg	IR masks strong anisotropism
	Sylvanite	syl	BrGy
	Krennerite	krn	
	Tellurium	Te	Gy-BrGy
	Romanechite	rom	
White (very bright)	Mackinawite	mck	Extreme, bright anisotropism
	Lepidocrocite	lpc	Very abundant IR

(*)Only clearly anisotropic ores. Cubic ores are excluded, even if showing anomalous anisotropism (such as pyrite or magnetite) or imperfect extinction (such as native metals: Ag, Au, etc.)

(**)The colors mentioned in this table are not the natural colors of the mineral, i.e. those seen without analyzer (*dsp/osp*), but those characterizing the *effects of anisotropism* (with analyzer crossed at 90°: ***d + p/o + p***): they refer exclusively to the **polarization colors**, especially those dominant in stage positions of maximum illumination

Color Abbreviations (13 *hues* used): *Wh*: White—*Gy*: gray—*Bk*: black—*Blu*: blue—*Gn*: green—*Yl:* yellow—*Og*: orange—*Rd*: red—*Pu*: purple—*Vi*: violet—*Pi*: pink—*Ol*: olive—*Br*: brown. Their **Combinations**: *GnGy*: greenish gray, etc.

Annex 4.7 Common Ores with Typical Natural Colors

Common Colored Ores (*)								
Color	Intense		Distinct		Faint		Just a nuance	
	Mineral	Abrv	Mineral	Abrv	Mineral	Abrv	Mineral	Abrv
Blue	Covellite	cv	Digenite	dg	Pyrargyrite	prg	Cuprite	cup (BluGy)
					Proustite	pru		
Yellow	Native Gold	Au	Millerite	ml	Pyrite	py	Native Silver	Ag (YlWh)
	Chalcopyrite	ccp			Electrum	el		
Red	Copper	Cu	–		–		–	
Pink	Bornite (fresh)	bn	–		–		Cobaltite	cob
							Maucherita	mau (PiWh)
Brown	–		Pyrrhotite	po	Magnetite	mt	–	
			Enargite	en				
Violet	Germanite	ger	–		–		–	
	Bornite (tarnished)	Bn_T						
OgYl	–		Renierite	ren	–		–	
			Luzonite	lz				
			Nickeline	nc				
BrYl	–		–		Melnicovite	mln	–	
PiBr	–		Ilmenite	il	Linneite	ln	–	
					Bravoite	brv		
ViPi	–		Breithauptite	bth	–		–	
			Famatinite	fm				
OlGn-GyBr	–		–		Tetrahedrite	td	–	
					Estannita	stn		

(*)Natural colors of the ore, without polarization. Microscope setting: *dsp/osp* (no analyzer). *White–Gray-Black "colors" excluded*
Color Symbols (or ***abbreviations***, § **Annex 1** or § **1-Box 1.1**). *Wh:* white—*Gy:* gray—*Bk:* black—*Blu:* blue—*Gn*: green—*Yl*: yellow—*Og*: orange—*Rd: red*—*Pu:* purple—*Vi:* violet—*Pi:* pink—*Ol:* olive—*Br:* brown. Color **combinations**: *BrRd*: brownish red—*GnGy*: greenish gray—*YlWh*: yellowish white, etc.

Annex 5: Determinative Tables

Annex 5.1 Background and Methodology

Since the initial determinative tables published by Schneiderhöhn and Ramdohr (1931), there were several proposals of tables intended to help the microscopist. The aim was to establish effective criteria to achieve an accurate diagnosis of the identity of the ore in practice. They comprised versions in both book (Schouten 1962; Uytenbogaardt and Burke 1971) and magazine format (Tarkian 1974; Spry and Gedlinske 1987; Tarkian and Liessmann 1991).

The first (*Erzmikroskopische Bestimmungstafeln*, Schneiderhöhn and Ramdohr 1931) responds in 47 pages to the objective of systematizing and guiding the observations aimed at identifying ores with the aid of the reflection microscope. They comprised the laboratory tests available then, including physical measurements, structural etching and microchemical tests, among others.

The means available have evolved greatly since then, so this early and valuable contribution is no longer applied in its original form. However, it opened the way to modern approaches. It prefigures and combines the two main determinative lines followed until now: qualitative (observation) and quantitative (determination of physical properties or chemical composition).

The quantitative pathway has diversified enormously, first with the application of precise instrumental techniques for the

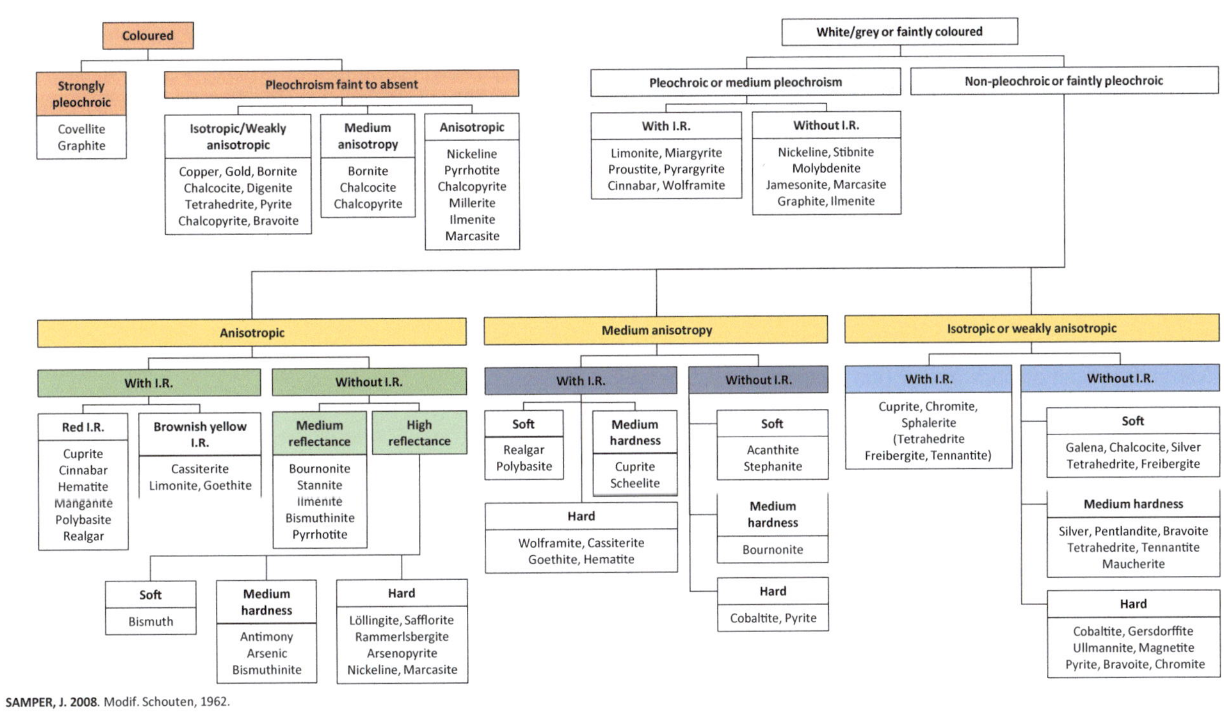

Fig. A.5.1 Determinative table (Samper 2008, after Schouten 1962)

R. Castroviejo, *A Practical Guide to Ore Microscopy—Volume 2*,
https://doi.org/10.1007/978-3-031-18954-8

measurement of properties such as hardness or specular reflectance. They had a great impact through their combination in *key diagrams* or **VHN-R% tables** (Bowie and Taylor 1958; McLeod and Chamberlain 1968; Tarkian 1974; Tarkian and Bernhardt 1984; Tarkian and Liessmann 1991). Then, with the current development of microanalytical techniques (EMPA, SEM, etc.), electronics and computer methods, automation was made possible (cf. **§ Ch. 1**, vol. 1, and **Part II**, vol. 2).

With regard to the tables following the qualitative line, in the most modern versions (e.g. Spry and Gedlinske 1987, after Schouten 1962) an attempt has been made to simplify their use, presenting the information in the form of systematic schemes or *flowcharts,* with decision nodes based on "yes/no" or "±" responses to opposing options. As a result, the number of ores selected for comparison decreases and the amount of information processed is smaller, yet for the beginner they can be a useful support (*vbgr.* Price 2013; Samper 2008: Fig. A.5.1), as they help to interpret, compare and critically assess the observations, and to work them out systematically, overcoming the insecurity of the first attempts.

However, as Schouten (1962) already noted, simplifying an identification process based on inevitably varied and complex microscopic observations, to the point of reducing it to a brief outline, limits the rigour of the method and, it has been shown, the accuracy of the results. In comparison, the methodology of Galopin and Henry (1972) is very enlightening. This combines explanatory text with a re-elaborated and abbreviated version of Schneiderhöhn and Ramdohr' tables (1931), reduced to nine items combined as follows:

Ores	Colorless	Slightly colored	Clearly colored
Isotropic	Table I	Table I B	Table I C
Weakly anisotropic	Table II A	Table II B	Table II C
Strongly anisotropic	Table III A	Table III B	Table III C

The use of these tables is related to the follow-up of the text itself (Galopin and Henry 1972) and of the practical exercises that it proposes to boost the beginner's progress.

Methodological approaches of this type are probably the most appropriate strategy. It is also the choice of the present text. It is illusory to think that an inexperienced microscopist can correctly interpret and apply an advanced determinative scheme without having developed the necessary observational skills and mastered the basic concepts. If the practitioner is an experienced microscopist, they will most likely no longer need a simple scheme. Therefore, it does not seem practical to insist on the elaboration of simple schemes of general application and questionable practical results. As a solution, the whole text is conceived with recognition criteria and as a determinative method. Instead of a simple scheme, synthesis or support tables are offered (**Annexes 4** and **5**), which help to progressively reduce the indeterminacy until the focus is on identification.

Annex 5.2 Determinative Strategy and Use of the Tables

The alternative planned in the text is to integrate the determinative keys or diagnostic criteria with the description of each ore, and to provide independently the necessary orientations so that the microscopist can progress in their observations or measurements until able to determine alone and with certainty the identity of the mineral. This cannot be achieved without the observer having acquired the necessary experience; to omit this step, raising false expectations of easy and immediate success, is to lead the beginner to failure. This is why we insist on a two-step strategy:

1. If you are a beginner, you should first of all practice with the appropriate instrumentation and specimens, always with the support of an expert and a conceptual background that you should not neglect, until you acquire basic confidence in your own observation skills with the reflected light microscope (cf. **Box Annex 5.1**).
2. If you are an experienced professional or have successfully passed step 1, you may find the following guidelines helpful. In any case, you should first make sure that you recognize the fundamental properties of at least the elementary selection of priority minerals with which you should already be familiar from step 1 (cf. **Box Annex 5.2**).

The **strategy** for microscopic ore recognition has this in common with the strategy for mineral exploration: the need to eliminate uncertainty as soon as possible and at minimum cost. In the case of exploration, it is a matter of discarding barren ground in order to focus attention (and investment) on the land potentially endowed with a deposit (*target*). In microscopic identification, it is a matter of discarding false options and focusing on diagnostic data to determine the identity as soon as possible. For this reason, attention should be paid first and foremost to **general properties such as luster**, **color**, **hardness**, **anisotropism,** and so on.

In addition, if the usual information is available on the origin of the ore, it is wise to pay attention to it from the beginning. Actually, each deposit type has a **mineral association** or **mineral assemblage** (**Annexes 2.1** and **2.2**) related to the ores' parageneses and is therefore meaningful for the sample to be studied. These data provide a powerful tool that can save the beginner false steps and much time, as seen in the examples that follow.

The general properties mentioned above have been synthesized for their application in tables (**Annex 4**), serving to support the identification process. Before discussing them, it is useful to show the method to be followed with an elementary example: the application of the specular reflectance, **R** (in this case in *white light* or, if collected as multispectral data from the ore descriptions, its *proxy* at $\lambda = 546\ nm$: R_{546}).

The estimation of the **brightness**—the *proxy* of the **specular reflectance *R***—is a good start to examining a polished section, since it can be appreciated at a simple glance (always taking the precaution of working with a constant intensity of illumination—in principle, the nominal one of the microscope—otherwise the appreciation can be much biased).

It is easy to approximate the *Reflectance Group, R* (**Annex 4.1**, values always referred here to white light) if one starts with the darkest components or those which appear gray in reflected light (*Group: very low R*, **Annex 4.1**);

✓ If the gray level (brightness) is similar to that of the resin (R ≈ 5%), the most likely option is gangue minerals, among which the most frequent are silicates and carbonates:

- o Carbonates are distinguished (*d + p*) by their remarkable bireflectance (**Annex 4.4**) and, once this has been checked, can be characterized according to Chapter 136, vol. 1; otherwise, there is a possibility of silicate or some other type of gg (cf. Part III, vol. 1).
- o The IR should also be checked, the absence of which suggests organic or carbonaceous matter (COM: cf. Chapter 138, vol. 1).

✓ Attention should be paid if the shade of gray is a bit lighter, because in the upper range of the same group there are certain possible ores like bastnäsite or monazite (both important potential sources of REE), malachite (green IR), azurite (blue IR) or scheelite (colorless IR), which should be contrasted with their descriptions (cf. bst, mz, mch, az, sch).

The *low R Group* starts with a slightly lighter gray color (R ≥ 12.5%) (**Annex 4.1**), typically beginning with:

- o Oxides, such as chromite or cassiterite, both gray, dark and very hard, yet easy to distinguish by their parageneses and by their symmetry (the former cubic and therefore isotropic, the latter tetragonal and anisotropic).
- o With increasing R, in addition to oxides, the group includes *graphite* (immediately recognizable by its very strong bireflectance, from R ≈ 6% almost like gg to R ≈ 27%, and extreme anisotropism) and a few *sulfides*, in general easy to identify, such as:
  - Valleriite (very similar to graphite, but with higher R_{min})
  - Digenite (delicate blue color, isotropic)
  - Bornite (pink color, fast tarnishing)
  - Covellite (extreme and unmistakable pleochroism and anisotropism)
  - Realgar and orpiment (IR abundant and colored).

For the remaining ores with progressively higher R values, a similar analysis is applied, but with a greater variety of species.

Knowledge of the mineral association, in correspondence with the typology, is again a valuable aid as it allows one to anticipate the predictable R-ranges in the sample, starting from the typical paragenesis in each class of deposits, as shown in **Box Annex 5.3**. The beginner soon discovers how useful it is to return again and again to the typological associations.

It should be stressed that the term of reference here is ***mineral association***, although it might seem at first more appropriate to refer to *paragenesis*; that is, the set of minerals formed simultaneously, in the same place and by the same genetic process. To establish the paragenesis it is indispensable to explain the genesis. For the moment that is not the problem: what the observer has to establish at first is the identity of the minerals present, whatever their genesis.

There is no doubt that, in a single sample, several episodes may have taken place from which as many parageneses may result. The assemblage of them all, that is to say the totality of the minerals present, is the mineral association. Thus, for example, if a massive sulfide ore composed of pyrite, chalcopyrite and galena (*primary minerals*) undergoes a process of supergene enrichment with partial replacement of pyrite and chalcopyrite by chalcocite, covellite and digenite (*secondary minerals*), when studying the samples these six minerals may be found together, and are the ones that the microscopist has to identify. These define the real existing association, although later it will be possible —and it will be very convenient—to establish by applying textural interpretation criteria that this association comprises two superimposed parageneses: the primary (py, ccp, gn); and the supergene (or secondary: cc, cv, dg).

The *medium R* (25 to 40%) and *high R* (40–60%) groups are the most numerous. Common ores such as hematite (R ≈ 29%) or tetrahedrite (R ≈ 31%) for the former, and galena (R ≈ 43%) or pyrite (R ≈ 51%) for the latter can be remembered as references points on the scale of R values.

The *very high* R group ($\geq$ 60%) gathers essentially native elements, mainly metals and precious metal alloys or compounds (Ag, Au, EGP). It is not a very numerous group and the various species can be diagnosed, generally quickly, by readily observable properties (provided that the abundance or grain size allows it, since they are usually very scarce minerals and almost always micro to cryptocrystalline).

For the **common cases** of application of the comparative tables (**Annex 4**), the *reflectance* (**R**: **Annex 4.1**) and *polishing hardness* (***PolH***: **Annex 4.2.1**) tables provide a general framework for contrasting the options obtained from other more directly diagnostic properties, such as bireflectance, color, anisotropism, and so on. However, the R and *PolH* tables may also be almost diagnostic in the extreme value ranges (e.g., very high or very low R), as we have just seen when dealing with brightness (native metals, R>>; gangue and some oxides, R<<). Nonetheless, as far as hardness is concerned, although it is undoubtedly an important set of properties, some reservations are unavoidable. The determination of VHN is now rare, due to the lack of instrumentation. As for the polishing, *PolH*, and scratch hardnesses, apart from the fact that they may vary with composition or, in anisotropic ores, with orientation, they are not always observable with the precision and rigour required and, consequently, the determinations vary according to the authors.[2] Although the maximum consistency of the data has been sought in the compilation presented (**Annex 4.2.1**) there remains a certain margin of uncertainty; this is normally insignificant, but it may make it advisable to compare to the Talmage hardness (**Annex 4.2.2**), even if it is seldom used today.

The other properties listed in the tables allow for discretionary use suited to the sample under study. As a general guideline, it is recommended to check by a quick initial examination if there are noticeable *internal reflections*, ***IR*** (**Annex 4.3**) or if there are *colored ores* (**Annex 4.7**), because finding any of these properties can eliminate many options and considerably focus the determination. The same is true for the *Bireflectance-B vs. Pleochroism-P* tables (**Annex 4.4**). Properties that are very easily checked can greatly reduce the range of options to be considered: for example, the very strong B/P combination reduces the options to six.[3]

The determination of the *isotropic* (**Annex 4.5**) or anisotropic (**Annex 4.6**) *character* of the mineral provides clues that can be decisive, especially since Picot and Johan (1977, 1982) showed the diagnostic value of *colors of anisotropism* (or polarisation tints) listed in the second of these tables (**Annex 4.6**). This diagnostic value of color requires extreme care in dealing with photomicrographs, either capturing or reproducing the digital images, to ensure the fidelity of the colors obtained (virtual or printed) to those visible under the microscope. For this reason, all images have been processed and defined in the *sRGB color space* and to ensure color fidelity must be viewed on a monitor with color calibration.

Box Annex 5.1 Recommendations for the Beginner It is advisable to write down one's observations **orderly** on a *worksheet* for each mineral. This sheet or card is a basic personal document that allows you to remember, follow up, correct and finally memorize the descriptions, as well as to control and evaluate your own progress. It is recommended to select ore samples from available collections in the following sequence:

1. Start with easy samples (comprising one or, at best, several ores selected in the Mineral Indices or in **Annex 1.2** as Learning Priority 1: e.g. pyrite, magnetite, sphalerite, galena, silver or native gold), observe them with a dry objective (in air) and with single polarizer (no analyzer: ***dsp***):
   a. Compare and order the species found by their brightness and color, estimating relative values of their reflectances (to start with, take as a reference quartz gangue as R$\approx$ 5% and pyrite as R $\approx$ 55%);
   b. Do the same for a sample containing magnetite (R$\approx$ 20%);
   c. Do the same for a sample containing galena (R$\approx$ 43%) and/or sphalerite (R$\approx$ 17%);
   d. Do the same for a sample containing either native silver (reference as R$\approx$ 90%) or native gold (reference as R$\approx$ 70%);

[2] Actually, notable differences have been found in the various *PolH* tables. Nevertheless, the data of Ramdohr (1980) and those of Schneiderhöhn (1952) are generally consistent with each other, and also with the author's observations, for which reason they have been given preference. The greatest divergences were found in those published by Uytenbogaardt and Burke (1971), and these were discarded when they proved irreconcilable.

[3] Of all ores under study (common ores).

e. Do the same on all of them for their polishing hardnesses (*PolH*); compare, particularly, PolH of pyrite, magnetite, sphalerite and galena; describe cleavage, where this is visible (*triangular pits* of galena).

2. Observe the same samples with the analyzer crossed at 90° (***d + p***), and distinguish isotropic from anisotropic species (if any); compare and describe the **IR**, internal reflections of sphalerite and gangue; describe the effects of anisotropism, if anisotropic species are present
3. Observe the same samples in immersion (**osp**), and note the changes in their appearance (brightness, color); compare the effect of immersion on gangue, magnetite and sphalerite
4. Do the same with the analyzer crossed at 90° (***o + p***); compare the effect of immersion on gangue and sphalerite IR
5. Compare your own observations with the descriptions of each mineral given in the text or elsewhere. Resolve any discrepancies with new observations or by consulting an expert
6. Move now to samples—still *Priority* 1—of anisotropic minerals (such as hematite, covellite, chalcopyrite, arsenopyrite, marcasite, nickelite, pyrrhotite, molybdenite, graphite) and observe them following steps 1 to 5 for them all (not forgetting any of the four basic settings: ***dsp***, ***d + p***, **osp**, ***o + p***; and using the most suitable objective in each case). Enrich the observations until all outlined properties have been completed (Boxes 1.1 and 1.2, in vol. 1, Chapter 1). For example, describe and compare the bireflectance of hematite and molybdenite; the pleochroism of covellite, pyrrhotite and graphite; the twinning of chalcopyrite; the anisotropism effects (colors) or polarization tints of pyrrhotite, marcasite, molybdenite or graphite
7. Continue with the same observation methodology until you are familiar with all Priority 1 and then Priority 2 ores (cf. Mineral Indices).

Box Annex 5.2 Most Frequent Minerals in Common Ores

Common ores: basic selection (priority 1)		
Class	Mineral	Symbol
Elements	Native Gold	Au
	Native Silver	Ag
	Native Platinum	Pt
Sulfides/Arsenides	Acanthite	ac
	Arsenopyrite	asp
	Bornite	bn
	Chalcopyrite	ccp
	Chalcocite	cc
	Cinnabar	cin
	Covellite	cv
	Digenite	dg
	Sphalerite	sp
	Stibnite	sbt
	Galena	gn
	Linneite	ln
	Marcasite	mc
	Molybdenite	mol
	Nickeline	nc
	Pentlandite	pn
	Pyrite	py
	Pyrrhotite	po
Sulfosalts	Tetrahedrite (*s.l.* or fahlore)	td
Oxides, hydroxides, tungstates, etc.	Cassiterite	cst
	Chromite	chr
	Cuprite	cup
	Goethite (and Limonite)	gth
	Hematite	ht
	Ilmenite	il
	Magnetite	mt
	Pyrolusite	prl
	Romanechite (Psilomelane)	rom (psi)
	Rutile	rt
	Uraninite	urn
	Wolframite	wf
Gangue	Quartz	q
	Carbonates	Carb
	Coal and Organic matter	COM

Box Annex 5.3 Classification of Ores by Reflectance Ranges[(*)] (R %, white light) in typical mineral associations[(**)]

In Ultramafic rocks and Ophiolithic Complexes	
1	**Gold** (72), **PGE** (65)
2	**Arsenopyrite** (52), **Pyrite** (51), Pentlandite (49), Nickeline (50′5–45), **Chalcopyrite** (45)
3	**Pyrrhotite** (39)
4	Martite (29), **Hematite** (29), Rutile (23.5)
5	**Ilmenite** (19.5), **Sphalerite** (19), **Magnetite** (17), **Chromite** (12.5), **Spinel** (8)
6	**Gangue** (6)
Pegmatitic[(***)]—Pneumatolytic	
1	**Gold** (72), Bismuth (67)
Sulfide Phases:	
2	Arsenopyrite (52), **Pyrite** (51), **Chalcopyrite** (45), **Molybdenite** (42–19), **Galena** (43), Stannite (28), **Pyrrhotite** (39)
Oxidized Phases:	
3	Rutile (23.5)
4	**Wolframite** (17), **Cassiterite** (13), **Scheelite** (11)
Silicates:	
5	Beryl, Tourmaline (6), Mica (5), Quartz (5)
Hydrothermal Pb–Zn-Ag	
1	**Silver** (85), ***Electrum*** (80)
2	**Arsenopyrite** (52), **Pyrite** (51)
3	**Chalcopyrite** (45), **Galena** (43)
4	Bournonite (36), Acanthite (32), **Tetrahedrite** (31) and other sulfosalts
5	**Sphalerite** (19)
6	Gangue (6)
Volcanogenic—sedimentary massive sulfides (VMS)	
1	**Silver** (85), ***Electrum*** (80), **Gold** (72)
2	**Arsenopyrite** (52), **Pyrite** (51), White Arsenides (51)
3	**Chalcopyrite** (45), **Galena** (43), **Pyrrhotite** (39)
4	**Hematite** (29), Rutile (23.5), **Magnetite** (17)
5	**Goethite**(18–15), **Cassiterite** (13), Jarosite (7)
In metamorphic terrains: orogenic gold deposits	
1	***Electrum*** (80), **Gold** (72)
2	**Arsenopyrite (52), Pyrite** (51)
3	Chalcopyrite (45)
4	Galena (43), Stibnite (47–30), Molybdenite (42–19), Pyrrhotite (39)
5	Hematite (29), Rutile (23.5)
6	Sphalerite (19), Graphite (16–6)z
7	Gangue (6)

[(*)]Average reflectance values (R %), calculated in white light and given in parenthesis after the mineral name, group the ores into categories according to their R range ; these give a first indication of spectral differences. Important minerals in each type are highlighted in bold: these are not always the most common

[(**)]According to the typology of mineral deposits presented in **Annexes 2.1** and **2.2**

[(***)]Complex pegmatites (REM or Rare-Earth Minerals, gems, semi-precious stones, etc.) require individual treatment

MIX
Papier aus verantwortungsvollen Quellen
Paper from responsible sources
FSC® C105338

If you have any concerns about our products,
you can contact us on
ProductSafety@springernature.com

In case Publisher is established outside the EU,
the EU authorized representative is:
Springer Nature Customer Service Center GmbH
Europaplatz 3, 69115 Heidelberg, Germany

Printed by Libri Plureos GmbH
in Hamburg, Germany